Fruit Tree Physiology and Molecular Biology

Fruit Tree Physiology and Molecular Biology

Guest Editors

Dajiang Wang
Yuan Gao

MDPI

Basel • Beijing • Wuhan • Barcelona • Belgrade • Novi Sad • Cluj • Manchester

Guest Editors

Dajiang Wang
Research Institute of Pomology
Chinese Academy of Agricultural Sciences (CAAS)
Xingcheng
China

Yuan Gao
Research Institute of Pomology
Chinese Academy of Agricultural Sciences (CAAS)
Xingcheng
China

Editorial Office
MDPI AG
Grosspeteranlage 5
4052 Basel, Switzerland

This is a reprint of the Special Issue, published open access by the journal *Horticulturae* (ISSN 2311-7524), freely accessible at: https://www.mdpi.com/journal/horticulturae/special_issues/PZPCT0JC46.

For citation purposes, cite each article independently as indicated on the article page online and as indicated below:

Lastname, A.A.; Lastname, B.B. Article Title. *Journal Name* **Year**, *Volume Number*, Page Range.

ISBN 978-3-7258-6280-1 (Hbk)
ISBN 978-3-7258-6281-8 (PDF)
https://doi.org/10.3390/books978-3-7258-6281-8

Contents

About the Editors

Dajiang Wang

Dajiang Wang, from the Research Institute of Pomology, Chinese Academy of Agricultural Sciences, specializes in *Malus* germplasm resources and rootstocks, with research interests including the following: the identification and evaluation of functional substances (e.g., polyphenols, aroma compounds) in *Malus* resources; the mining of regulatory genes associated with the apple aroma; the identification and screening of stress-tolerant elite *Malus* germplasms; the mining of stress-resistance genes as well as the development of molecular markers; the breeding and application of novel stress-tolerant and dwarfing apple rootstock; the exploration of rootstock effects on scions and related molecular mechanisms; and the functional characterization of non-coding RNAs in apples. He has established a prominent academic track record through sustained efforts in *Malus* germplasm innovation, functional gene research, and industrial technology promotion. He has presided over ten key projects at the national, provincial, and municipal levels, including projects funded by the National Key Research and Development Plan and the Liaoning Provincial Science and Technology Program. Additionally, he has actively participated in more than ten national- and provincial-level projects. In terms of academic publications, he has published over 80 papers in prestigious journals, among which he authored 40 as the first author or corresponding author. He has also made remarkable achievements in variety innovation, successfully registering 25 new apple varieties. Furthermore, Dr. Wang has obtained the copyright for four pieces of software, participated in formulating one national agricultural industry standard, compiled three monographs, co-authored ten academic works, and obtained five invention patents. His outstanding academic contributions have been widely recognized: he has won one Third-Class Award of Liaoning Provincial Natural Science Academic Achievement as the first completer.

Yuan Gao

Yuan Gao, from the Research Institute of Pomology, Chinese Academy of Agricultural Sciences, primarily engages in research on *Malus* germplasm resources, focusing on collection and conservation, phenotypic and genotypic identification/evaluation, and germplasm innovation and exploitation. Specifically, she employs molecular approaches such as fluorescent SSR, SNP, and chloroplast haplotype analysis to identify *Malus* germplasm resources and construct their molecular IDs, providing technical support for apple variety property rights protection and market supervision. At the phenotypic and genetic levels, she investigates the genetic diversity, genetic relationships, and origin/evolutionary mechanisms of *Malus* germplasms, offering scientific references and theoretical foundations for the collection, conservation, systematic classification, and effective utilization of China's *Malus* germplasm resources. Additionally, she conducts phenotypic–genotypic association studies to map and mine genes controlling key traits, clarifying their genetic mechanisms and laying a foundation for the predictive selection of innovative germplasm. She has presided over seven national- and provincial-/municipal-level projects, including the National Natural Science Foundation of China (NSFC) and the National Key Research and Development Plan, and participated as a core executor in 22 provincial-/municipal-level projects. She has published 81 academic papers (25 as the first author), compiled four monographs and co-authored four others, bred 23 new apple varieties (eight as the first breeder), formulated four agricultural industry standards, obtained one national invention patent and two international invention patents as the first inventor, secured the copyright for 11 pieces of software (seven as the first completer), and participated in six provincial-/municipal-level scientific and technological achievement awards.

Preface

Fruit tree physiology and molecular biology, a vital pillar of agricultural science, has emerged as a frontier driving the sustainable development of the global fruit industry amid escalating climate change—characterized by frequent extreme events like high temperatures, waterlogging, and saline–alkali stress—and evolving consumer demands for high-quality, diversified products. This Reprint focuses on four core research areas—extreme environment adaptation, efficient resource utilization, precise quality improvement, and variety innovation—encompassing in-depth studies on major fruit crops such as apples, pears, grapes, longans, cherries, raspberries, passion fruits, and citrus. By integrating state-of-the-art multi-omics technologies (including genomics for pan-genome construction, transcriptomics for stress response profiling, and metabolomics for quality component analysis), gene editing tools (e.g., CRISPR/Cas9 and single-base editing), and advanced cultivation techniques (such as aerated irrigation and tree shape optimization), the collected works delve into the molecular mechanisms underlying stress resistance (e.g., cold, salt, and disease tolerance), nutrient regulation (nitrogen, phosphorus, boron, and manganese utilization), quality formation (aroma, color, and texture development), and germplasm innovation (dwarfing and continuous flowering traits). The overarching goal is to provide systematic theoretical support and practical technical paths for addressing industry challenges, such as low resource use efficiency and insufficient stress resilience, thereby facilitating industrial upgrading.

The motivation for compiling this Reprint stems from two critical needs: first, the need to address pressing industry challenges, including nutritional imbalance, disease outbreaks (e.g., pear ring rot), and the gap between biotechnological advances and field applications; and second, the need to consolidate cutting-edge research progress that has laid the groundwork for innovation, such as the identification of key genes like MdTCP11 (apple dwarfing) and CYP76F14 (grape aroma), the development of high-precision variety identification models (e.g., YOLOv10-based leaf classification), and the construction of gap-free haplotype genomes for pears. Despite these breakthroughs, significant gaps remain: linking molecular mechanisms (e.g., gene expression and methylation) to physiological phenotypes (e.g., fruit ripening and stress-induced organ plasticity), applying multi-omics joint analysis to dissect complex traits (e.g., synergistic nutrient deficiencies), and translating lab-scale biotechnologies into large-scale breeding practices. This Special Issue Reprint aims to bridge these gaps by showcasing original research that combines basic theory with industrial applications, fostering academic exchange and technological dissemination.

The target audience of this Reprint is broad and interdisciplinary, including researchers and technicians specializing in fruit tree physiology, molecular biology, and genetic breeding, who will gain insights into novel mechanisms and techniques; agronomists engaged in orchard management and technical promotion, who can apply the findings to optimize field practices (e.g., aerated irrigation duration and disease-resistant variety selection); and students and educators in agricultural science disciplines, who will benefit from the comprehensive overview of current research trends. We anticipate that the diverse research findings presented herein will inspire cross-disciplinary collaborations (e.g., between molecular biologists and agronomists), promote the translation of scientific discoveries into agricultural practices (e.g., molecular marker-assisted breeding for high nitrogen use efficiency), and contribute to the development of a climate-resilient, resource-efficient, and high-quality fruit industry. It is our sincere hope that this Special Issue will serve as a valuable platform for sharing knowledge, exchanging ideas, and advancing the frontiers of fruit tree science to meet global food security and sustainability goals.

Dajiang Wang and Yuan Gao
Guest Editors

horticulturae

Editorial

Advances in Fruit Tree Physiology and Molecular Biology

Dajiang Wang * and Yuan Gao *

Research Institute of Pomology, Chinese Academy of Agricultural Sciences (CAAS), Key Laboratory of Horticulture Crops Germplasm Resources Utilization, Ministry of Agriculture, Xingcheng 125100, China
* Correspondence: wangdajiang@caas.cn (D.W.); gaoyuan02@caas.cn (Y.G.)

1. Introduction

As an important branch of agricultural science, fruit tree physiology and molecular biology have in recent years been driven by the intensification of global climate change and the upgrading of consumer demands towards high-quality and diversified products. The research focus has gradually shifted to four core issues: adaptation to extreme environments, efficient resource utilization, precise quality improvement, and variety innovation. By leveraging multi-omics technologies (genomics, transcriptomics, metabolomics, etc.) and gene editing methods (CRISPR/Cas9, single-base editing, etc.), breakthroughs have been made in areas such as stress response mechanisms, nutrient regulatory networks, quality formation pathways, and molecular breeding applications, providing key theoretical support and technical paths for improving the quality and efficiency of the fruit tree industry and promoting stress-resistant breeding.

In the research on adaptation to extreme environments, Asharf et al. [1] systematically expounded how fruit trees precisely regulate the plasticity adaptation of lateral organs (such as advendent roots, aerated tissues, and stem elongation) through hormone interaction networks during hypoxia and reoxygenation processes, in response to abiotic stresses like high temperature, waterlogging, and saline–alkali conditions. This provided a new theoretical framework and strategic direction for the waterlogging resistance breeding and cultivation management of fruit trees. By integrating single-cell omics to locate key response cell types, core regulatory factors such as ERF-VIIs and PIN proteins were identified, providing precise targets for the breeding of stress-resistant varieties.

The field of efficient resource utilization focuses on nutrients such as nitrogen, phosphorus, boron, and manganese. Wang et al. [2] identified a transcription factor MhNAC1 that was highly correlated with the nitrogen absorption efficiency of apples. Interfering with the expression of *MhNAC1* would promote nitrogen absorption and citric acid secretion in the root system and enhance the tolerance of apple rootstocks to low-nitrogen stress. Further research demonstrated that under low-nitrogen stress, *MhNAC1* directly bound to and inhibited the activity of the nitrate transporter *MhNRT2.4* and the citric acid transporter *MhMATE* promoter. Xiao et al. [3] cloned a gene related to nitrogen assimilation in peaches. This gene actively responded to low-nitrogen signals, promoted lateral root development and nitrogen absorption and utilization in peaches, and effectively enhanced the nitrogen use efficiency of transgenic plants, providing genetic resources for high nitrogen use efficiency breeding in peach trees. It can solve the current situation of excessive use of nitrogen fertilizer due to low nitrogen utilization efficiency in peach tree production.

Quality improvement research focuses on key traits such as flavor, color, and storability. Liang et al. found [4] that the α subunit of peach SNF kinase 1 (PpSnRK1α) could respond to nitrogen regulation, and its transcriptional level was strongly induced by

 https://doi.org/10.3390/horticulturae11121455

nitrogen. Through the functional analysis of overexpressed/silenced peach fruit materials, it was found that PpSnRK1α inhibited the biosynthesis of γ-decanolactone. Further analysis confirmed that PpSnRK1α could directly interact with PpNAC6/PpNAC36 and phosphorylate PpNAC36, thereby regulating the biosynthesis of γ-decanolactone. This study proposed the regulatory pathway of PpSnRK1α–PpNAC6/PpNAC36–peach aroma biosynthesis, revealing the mystery of nitrogen signaling regulating peach aroma biosynthesis. It provided a key theoretical basis for improving the flavor quality of peach fruits through molecular breeding and promoting precise nitrogen fertilizer management. Cao et al. [5] found that PpNAC1 could directly bind to the promoter of *PpDML1* and activate its expression, thereby affecting the DNA methylation levels of its own and downstream ripening-related genes. Transcription factor PpNAC1 and DNA demethylase PpDML1 synergistically regulated the ripening of peach fruits and the formation of flavor quality. Cao et al. [6] found that the interaction among PbHB.G7.2, PbHB.G1, and PbHB.G2.1 affected the expression of the ethylene synthesis gene *PbACS1b* and the ethylene yield in pear fruits, thereby regulating the ripening of pear fruits and revealing the molecular mechanism by which HB family proteins interact to regulate ethylene synthesis during the ripening process of pear fruits.

In the field of genomic research, Sun et al. [7] used the representative varieties of Asian and Western pears, 'Dangshan Su Pear' and 'Hongba Pear', as test materials to construct a high-quality, gap-free haplotype typing genome. By systematically integrating high-quality genomes, multi-omics data, and population and quantitative genetic analyses, they were the first to systematically explore the genetic differences between Asian pears and Western pears at the haplotype level, revealing that long-term differentiation and independent domestication were important reasons for the significant differences in quality and adaptability between Asian and Western pears. The structural variations and key functional genes closely related to important agronomic traits such as self-incompatibility, fruit quality, and yield of pears were precisely located using methods such as SV-GWAS and SV-gene expression association, providing rich genetic information resources for the mining of key genes and the development of molecular markers for superior traits of pears. It also provided a theoretical foundation and technical support for accelerating the molecular design breeding of pears and the aggregation of superior traits. Based on 30 high-quality genomes of *Malus* plants, Li et al. [8] systematically analyzed the evolution of the genus *Malus* over tens of millions of years and constructed the first graphical pan-genome of the genus *Malus*, releasing a panoramic view of genetic diversity of the genus *Malus*. Li et al. [9] successfully identified the key genes causing dwarfing in apples, deciphered the genetic code of dwarfing in apple rootstocks, and lifted the veil of mystery surrounding apple dwarfing. This laid a solid foundation for initiating molecular design breeding of dwarfing rootstocks in the field of woody economic fruits such as apples, accelerating the realization of the "green revolution". Cai et al. [10] found that the short-branch trait existed in multiple branches of the 'Fuji' lineage, indicating that short-branch varieties may have different origins. The main short-branch varieties had a 167 bp heterozygous sequence deletion in the upstream promoter region of the *MdTCP11* gene, which overlapped with the MITE transpose element and was modified by methylation. The methylation level of short-branch varieties in this region was relatively low, while the expression of *MdTCP11* significantly increased. The deletion of MITE sequences could enhance the gene expression activity. Overexpression of *MdTCP11* led to a decrease in apple plant height, indicating that MdTCP11 played an important role in the internode development of short-branch varieties.

Although biotechnology has not been applied in practice in the innovation and breeding of fruit tree varieties, its research on other crops provides a reference for its application in fruit trees. Scientists have targeted the modification of Rubisco enzyme through gene editing to enhance the efficiency of photosynthetic carbon assimilation, laying the foundation for the breeding of high-quality varieties [11]. Innovations in molecular breeding techniques, such as the combination of CRISPR/Cas9-targeted editing and whole-genome selection, have significantly shortened the breeding cycle, accelerated the industrial application of stress-resistant and high-quality varieties [12], and promoted the upgrading of the fruit tree industry towards "strong climate resilience, high resource efficiency, and excellent product quality".

2. Overview of Published Articles

Huang et al. [13] compared gene expression differences between mature leaves of continuous-flowering and once-flowering longan by transcriptome sequencing. The high levels of transcription factor (TF) expression and the high number of gibberellic acid (GA)-signaling-pathway-specific genes expressed at high levels in continuous-flowering longan were shown. Floral-induction-gene expression levels in continuous-flowering longan, such as levels of GA-signaling-related and FT genes, were always high. Wang et al. [14] detected the components and contents of polyphenols in the peel and pulp of the red flesh *Malus* plant 'Hongxun 2' (*Malus neidzwetzkyana* (Dieck) Langenf.) and green flesh *Malus* plant 'Xinye 13-11' (*Malus sieversii* (Led.) Roem.) during the development period, and the dynamic changes and differences in the polyphenols between the two kinds of fruit were discussed. The accumulation of major polyphenol components in the peel and flesh of 'Hongxun 2' and 'Xinye 13-11' apples was significant in the period before and after 65 days after flowering, and the contents of procyanidin B1 and procyanidin C1 were the highest in this period. The results supported the viewpoint that *Malus neidzwetzkyana* (Dieck) Langenf was a separate species to *Malus sieversii* (Led.) Roem. Wang et al. [15] collected tissues from pear trees infected with ring rot from orchards located in Liaoning, Hebei, Shandong, and other regions throughout China. A total of 21 strains belonging to the *Botryosphaeria* spp. were utilized to evaluate the resistance levels of various pear varieties against ring rot disease. The resistance was assessed by inoculating different isolates onto distinct pear varieties; it established the criteria for evaluating resistance while minimizing identification errors stemming from the variable responses exhibited by certain varieties towards individual strains, and provided a theoretical foundation for effective prevention and treatment strategies against pear ring rot. Zhao et al. [16] used 'Zaosu pear' as a test material, and five different tree shapes—Y, trunk, single-arm, double-arm, and spindle shapes—were studied regarding three aspects, branching and leafing parameters, light radiation parameters, and root distribution. The single-arm and double-arm shapes had the highest fruit weight per fruit values. The trees mainly had short fruiting branches when they entered the fruiting stage, and the double-arm shape had the highest number of short fruiting branches, which was significantly higher than those of the other tree shapes. It provided a theoretical basis for the selection of pear tree shapes in arid areas. Liang et al. [17] investigated the effects of varying durations of aerated irrigation, administered at a consistent frequency, on the growth of greenhouse grape seedlings and the structure of the rhizosphere soil microbial community. It was found that 20 min aeration duration could effectively promote grape seedling growth, enhance the diversity of rhizosphere soil microbial communities, increase beneficial aerobic microorganisms, and reduce harmful ones. The study provided a theoretical basis for optimizing aerated irrigation practices in facility grape cultivation. Mou et al. [18] carried out research on 156 crisp pear fruit germplasm samples to analyze the diversity level of texture traits. It showed that puncture parameters were significantly different between crisp pear cultivars, and the analysis of the influencing factors of flesh texture showed that fruit maturity and shelf life had significant effects on flesh quality. It provided an important reference for the standardization, evaluation, and utilization of crisp pear variety resources. Li et al. [19] collected leaf images of 33 pear varieties against natural backgrounds to introduce a novel approach to classifying pear varieties. YOLOv10 based on the PyTorch 1.11.0 framework was applied to train the leaf dataset, and constructed a pear leaf identification and classification model. The results showed that the model precision could reach 99.71%, and that it provided a reference for the conservation, utilization, and classification research of pear resources, as well as for the identification of other crops. Shi et al. [20] utilized PMAT to assemble the mitogenomes of six *Rubus* species according to long-read HiFi reads, and annotated them through homologous alignment. A large number of gene transfers were detected between organellar genomes of six *Rubus* species. It clarified the mitogenome characteristics of *Rubus* and provided valuable insights into the evolution of the genus. Song et al. [21] revealed cultivar-specific linalool accumulation patterns among the three aroma-distinct cultivars. It revealed that cytochrome P450 CYP76F14 mediated the conversion of its substrate linalool in table grape berries, especially of Muscat-type grapes, and proposed the CYP76F14 polymorphic variants as molecular markers for aroma-type breeding. Zavala et al. [22] addressed the relevance of endocarp lignification and *PavSPLs* expression for the transition to ripening. *PavSPL2* and *PavSPL9* were targeted by mtr-miR156a and gma-miR156f. Both *PavSPLs* and miRNAs were expressed in fruits and seeds at the yellow stage, an advanced point in the transition to ripening in sweet cherry. Shi et al. [23] comprehensively characterized the *CDPK* gene family in the apple cultivar 'Hanfu' at the genome-wide level, and 38 *MdCDPKs* were identified. *MdCDPK24* played a positive role in apple bitter rot resistance through *MdCDPK24*-overexpressing calli. Chen et al. [24] analyzed the physiological, transcriptomic, and metabolomics of two yellow passion fruit cultivars, including heat-tolerant and heat-sensitive cultivars. It showed that transcription factors (e.g., MYB, HSF, WRKY) played pivotal roles in heat adaptation, and knockdown of *PeGDA* and *PeXDH* markedly altered the heat tolerance of heat-tolerant cultivar. It offered new insights into the molecular mechanisms of heat tolerance and potential targets for breeding climate-resilient passion fruit varieties. Li et al. [25] found that Mn deficiency exacerbated B-deficiency-induced corky split vein not only by intensifying photosynthetic dysfunction and carbohydrate accumulation but also by promoting lignin biosynthesis in citrus. The findings highlighted the synergistic nature of B and Mn deficiencies in impairing leaf function and structure, providing new insights into the physiological and molecular mechanisms underlying corky split vein development.

3. Conclusions and Prospects

Studies on fruit tree physiology and molecular biology focus on the core mechanisms of fruit tree growth and development, stress adaptation, and quality formation, providing important theoretical support for improving the quality and efficiency of the industry. In this Special Issue, researchers have focused on aspects such as the identification and evaluation of germplasm resources, the determination of resistance to biological and abiotic stresses, the formation of fruit quality, soil nutrition, and cultivation techniques. The physiology and molecular biology of fruit trees such as apples, pears, grapes, longans, cherries, raspberries, passion fruits, and citrus fruits were studied using techniques such as model construction, physiological index determination, transcriptome and metabolome, and genome assembly. Despite having achieved many breakthrough results, there are still many problems that need to be urgently solved. Future research needs to further strengthen the verification of the correlation between molecular mechanisms and physiological phenotypes, and deepen the application of multi-omics joint analysis in the analysis of complex traits. In terms of breeding direction, efforts should be focused on target traits such as continuous flowering, stress resistance, and quality, and the transformation and application of molecular marker-assisted breeding and gene editing technologies should be accelerated. At the same time, it is necessary to enhance the coupled research on cultivation techniques and environmental adaptability, and improve the intelligent identification and precise management technology system. In addition, in response to industrial pain points such as nutritional imbalance and disease prevention and control, research on multi-factor interaction mechanisms is carried out to provide more comprehensive theoretical and technical support for the high-quality and sustainable development of the fruit tree industry.

Conflicts of Interest: The authors declare no conflicts of interest.

References

1. Ashraf, M.A.; Ateeq, M.; Zhu, K.J.; Asim, M.; Mohibullah, S.; Riza, T.; Huang, X.; Pan, H.Q.; Li, G.H.; Shabala, S.; et al. Phytohormone network orchestrating lateral organ adaptations to hypoxia and reoxygenation in fruit crops. *Plant Cell Environ.* **2025**, 1–16. [CrossRef] [PubMed]
2. Wang, X.N.; Zhou, Y.; Chai, X.F.; Foster, T.M.; Deng, C.H.; Wu, T.; Zhang, X.Z.; Han, Z.H.; Wang, Y. miR164-*MhNAC1* regulates apple root nitrogen uptake under low nitrogen stress. *New Phytol.* **2024**, *242*, 1218–1237. [CrossRef]
3. Xiao, Y.S.; Yu, C.W.; Peng, F.J.; Cai, H.L.; Guo, J.; Gao, H.F. Peach Tree MADS BOX Transcription Factor PpAGL24 gene and Its applications. CN120099022A, 6 June 2025.
4. Liang, J.H.; Zheng, X.; Wu, X.L.; Wang, Z.; Li, Z.X.; Xiao, Y.S.; Guo, J.; Chen, Q.J.; Luo, J.J.; Gao, H.F.; et al. PpSnRK1α-PpNAC6/PpNAC36 module mediates nitrogen-regulated biosynthesis of γ-decalactone in peach fruit. *Hortic. Res.* **2025**, *12*, uhaf256. [CrossRef]
5. Cao, X.M.; Li, X.Z.; Su, Y.K.; Zhang, C.; Wei, C.Y.; Chen, K.S.; Grierson, D.; Zhang, B. Transcription factor PpNAC1 and DNA demethylase PpDML1 synergistically regulate peach fruit ripening. *Plant Physiol.* **2023**, *194*, 2049–2068. [CrossRef]
6. Cao, S.H.; Guo, Z.H.; Liu, H.; Wang, G.M.; Qi, K.J.; Wang, Z.W.; Tian, R.P.; Sha, S.F.; Zhang, S.L.; Gu, C. Interaction among homeodomain transcription factors mediates ethylene biosynthesis during pear fruit ripening. *Hortic. Res.* **2024**, *11*, uhae086. [CrossRef]
7. Sun, M.S.; Cao, B.B.; Li, K.; Li, J.M.; Liu, J.; Xue, C.; Xu, S.Z.; Li, Y.J.; Li, Q.Y.; Qu, M.N.; et al. Haplotype-resolved, gap-free genome assemblies provide insights into the divergence between Asian and European pears. *Nat. Genet.* **2025**, *57*, 2040–2051. [CrossRef]
8. Li, W.; Chu, C.; Zhang, T.K.; Sun, H.C.; Wang, S.Y.; Liu, Z.Y.; Wang, Z.J.; Li, H.; Li, Y.Q.; Zhang, X.T.; et al. Pan-genome analysis reveals the evolution and diversity of *Malus*. *Nat. Genet.* **2025**, *57*, 1274–1286. [CrossRef]
9. Li, W.; Chu, C.; Li, H.; Zhang, H.T.; Sun, H.C.; Wang, S.Y.; Wang, Z.J.; Li, Y.Q.; Foster, T.M.; López-Girona, E.; et al. Near-gapless and haplotype-resolved apple genomes provide insights into the genetic basis of rootstock-induced dwarfing. *Nat. Genet.* **2024**, *56*, 505–516. [CrossRef]
10. Cai, Y.D.; Gao, X.H.; Mao, J.P.; Liu, Y.; Tong, L.; Chen, X.L.; Liu, Y.D.; Kou, W.Y.; Chang, C.J.; Foster, T.; et al. Genome sequencing of 'Fuji' apple clonal varieties reveals genetic mechanism of the spur-type morphology. *Nat. Commun.* **2024**, *15*, 10082. [CrossRef] [PubMed]
11. Martin-Avila, E.; Lim, Y.L.; Birch, R.; Dirk, L.M.A.; Buck, S.; Rhodes, T.; Sharwood, R.E.; Kapralow, M.V.; Whitney, S.M. Modifying plant photosynthesis and growth via simultaneous chloroplast transformation of Rubisco large and small subunits. *Plant Cell* **2020**, *32*, 2898–2916. [CrossRef] [PubMed]
12. Wang, X.H.; Tu, M.X.; Wang, Y.; Yin, W.C.; Zhang, Y.; Wu, H.S.; Gu, Y.C.; Xi, Z.M.; Wang, X.P. Whole-genome sequencing reveals rare off-target mutations in CRISPR/Cas9-edited grapevine. *Hortic. Res.* **2021**, *8*, 114.
13. Huang, S.L.; Lv, X.M.; Wei, J.B.; Han, D.M.; Li, J.G.; Guo, D.L. Comparative transcriptome analysis of mature leaves of *Dimocarpus longan* cv. 'Sijimi' provides insight into its continuous-flowering trait. *Horticulturae* **2024**, *10*, 974. [CrossRef]
14. Wang, D.J.; Wang, G.Y.; Lu, X.; Liu, Z.; Sun, S.M.; Guo, H.X.; Tian, W.; Li, Z.C.; Wang, L.; Li, L.W.; et al. Dynamic changes in polyphenols in fruit development of red flesh apple 'Hongxun 2'. *Horticulturae* **2024**, *10*, 1125. [CrossRef]
15. Wang, C.; Zhang, Y.J.; Ou, C.Q.; Wang, F.; Ma, L.; Wang, C.H.; Jiang, S.L. Isolation and identification of pear ring rot fungus and resistance evaluation of different pear varieties. *Horticulturae* **2024**, *10*, 1152. [CrossRef]
16. Zhao, M.X.; Ma, M.Z.; Wang, W.; Cao, S.F.; Cao, G.; Li, H.X. Effects of different tree shapes on canopy characteristics of Zaosu pear (*Pyrus* ssp. Va.) in the Northwest arid zone. *Horticulturae* **2024**, *10*, 1254. [CrossRef]
17. Liang, Y.C.; Wang, W.; Xi, J.S.; Zhao, F.Y.; Zhou, Y.F.; Zhang, H.H.; Yu, K.; Zheng, Z.; Zhao, F.Y. Impact of aerated irrigation duration on the growth of greenhouse grape seedlings and rhizosphere soil microorganisms. *Horticulturae* **2024**, *10*, 1351. [CrossRef]

18. Mou, Y.L.; Dong, X.G.; Zhang, Y.; Tian, L.M.; Huo, H.L.; Qi, D.; Xu, J.Y.; Liu, C.; Li, N.M.; Yin, C.; et al. Identification and evaluation of flesh texture of crisp pear fruit based on penetration test using texture analyzer. *Horticulturae* **2025**, *11*, 359. [CrossRef]
19. Li, N.M.; Wu, Y.Q.; Jiang, Z.Y.; Mou, Y.L.; Ji, X.H.; Huo, H.L.; Dong, X.G. Efficient identification and classification of pear varieties based on leaf appearance with YOLOv10 model. *Horticulturae* **2025**, *11*, 489. [CrossRef]
20. Shi, Y.J.; Chen, Z.; Jiang, J.Y.; Li, Q.F.; Zeng, W. De novo assembly and comparative analysis of the mitochondrial genomes for six *Rubus* species. *Horticulturae* **2025**, *11*, 559. [CrossRef]
21. Song, Z.Z.; Zhang, J.J.; Shi, M.; Li, D.; Liu, X.H. Cytochrome P450 CYP76F14 mediates the conversion of its substrate linalool in table grape berries. *Horticulturae* **2025**, *11*, 651. [CrossRef]
22. Zavala, M.; Menares, M.; Acevedo, O.; Melo, M.; Nunez, C.; Arancibia, C.; Pedreschi, R.; Donoso, J.M.; Meisel, L.A.; Maldonado, J.E.; et al. *PavSPL* Expression dynamics in fruits and seeds and in relation to endocarp lignification status during the transition from development to ripening in sweet cherry. *Horticulturae* **2025**, *11*, 706. [CrossRef]
23. Shi, J.J.; Ma, Y.X.; Wang, D.J.; Wang, F. MdCDPK24 encoding calcium-dependent protein kinase enhances apple resistance to Colletotrichum gloeosporioides. *Horticulturae* **2025**, *11*, 942. [CrossRef]
24. Chen, L.H.; Dong, J.; Fan, B.L.; Huang, Y.C.; Yang, L.; Cai, W.G.; Chen, L.L. Integrated transcriptomics and metabolomics analyses provide insights into heat resistance in passion fruit (*P. edulis* f. *flavicarpa*). *Horticulturae* **2025**, *11*, 1037. [CrossRef]
25. Li, Y.H.; Fu, Y.P.; Gan, Z.L.; Wei, Q.J.; Yang, M.; Yao, F.X.; Zhou, G.F. Manganese deficiency exacerbates boron deficiency-induced corky split vein in citrus by disrupting photosynthetic physiology and enhancing lignin metabolism. *Horticulturae* **2025**, *11*, 1172. [CrossRef]

horticulturae

MDPI

Article

Comparative Transcriptome Analysis of Mature Leaves of *Dimocarpus longan* cv. 'Sijimi' Provides Insight into Its Continuous-Flowering Trait

Shilian Huang, Xinmin Lv, Junbin Wei, Dongmei Han, Jianguang Li and Dongliang Guo *

Institute of Fruit Tree Research, Guangdong Academy of Agricultural Sciences, Key Laboratory of South Subtropical Fruit Biology and Genetic Resource Utilization, Ministry of Agriculture and Rural Affairs, Guangdong Provincial Key Laboratory of Science and Technology Research on Fruit Trees, Guangzhou 510640, China; shiil_huang@163.com (S.H.); lvxinmin@gdaas.cn (X.L.); weijunbin@gdaas.cn (J.W.); handongmei@gdaas.cn (D.H.); lijianguang@gdaas.cn (J.L.)

* Correspondence: guodongliang@gdaas.cn; Tel.: +86-20-38765541

Abstract: Longan (*Dimocarpus longan* Lour.) is an important tropical and subtropical fruit, and most of its cultivars bloom once a year (once-flowering, OF). *Dimocarpus longan* cv. 'Sijimi' (SJ) is a tropical ecotype variety that blooms several times a year (continuous-flowering, CF) without the need for low-temperature induction. Several studies have focused on the mechanism of continuous flowering in SJ longan; however, none used leaves as research material. As leaves are a key organ in sensing floral-induction signals, we compared gene-expression differences between mature leaves of CF (SJ) and OF (*D. longan* cv. 'Shixia' (SX) and *D. longan* cv. 'Chuliang' (CL)) longan by transcriptome sequencing. An average of 47,982,637, 43,833,340 and 54,441,291 clean reads were obtained for SJ, SX and CL respectively, and a total of 6745 differentially expressed genes (DEGs) were detected. Following Metabolic pathways, Plant-pathogen interaction and Biosynthesis of secondary metabolites, most of the other genes were assigned to the KEGG classifications of MAPK signaling pathway-plant, Plant hormone signal transduction, Amino sugar and nucleotide sugar metabolism and Starch and sucrose metabolism. WGCNA analysis clustered genes into 27 modules, among which bisque4 and darkorange2 module genes specifically were expressed at low and high levels in SJ, respectively. Different gene-expression patterns were detected between CF and OF longan in bisque4 and darkorange2 modules, especially the high levels of transcription factor (TF) expression and the large number of gibberellic acid (GA)-signaling-pathway-specific genes expressed at high levels in CF longan (SJ). Floral-induction-gene expression levels in CF longan, such as levels of GA-signaling-related and *FT* genes, were always high. In CF longan, after vegetative-growth accumulation, flowers could be directly induced, thereby eliminating the need for low-temperature induction.

Keywords: continuous flowering; *Dimocarpus longan*; *FT*; MAPK; plant hormone signal transduction; sugar metabolism

1. Introduction

Dimocarpus longan Lour. (Sapindaceae) is an important tropical and subtropical fruit native to China and Southeast Asia [1], and it is cultivated in more than 20 countries [2]. In addition to being used for fresh food, longan can be processed into dry longan, which is used as traditional Chinese medicine to treat leucorrhea, kidney disorders, allergies, and cardiovascular diseases [3,4]. The main longan cultivated varieties such as 'Shixia' (SX) and 'Chuliang' (CL), bloom once a year (once-flowering, OF), induced by low temperature and drought under natural conditions [5]. However, *Dimocarpus longan* cv. 'Sijimi' (SJ) is a tropical ecotype variety that can complete flower-bud differentiation without low-temperature induction and bloom several times a year (continuous-flowering, CF) [6] and is ideal for studying the longan flower formation mechanism.

Research on the mechanism of continuous flowering has focused on horticultural flowers, mainly roses and orchids. Molecular labeling in a large number of varieties has targeted the *RoKSN* gene, a *AtTFL1* homologue, which in the CF variety has a reverse-transcription rotor insertion in the second intron, resulting in *RoKSN* transcription blocking and the loss of its flowering-inhibition function [7,8]. RoKSN performs the flowering-inhibition function by competing with RoFT for RoFD [9]. Transcriptome analysis comparing CF and OF varieties revealed that gibberellin and the gibberellin-responsive gene *CO-like 2* may be involved in regulation of continuous flowering [10]. GA suppressed flowering by inducing *RoKSN* expression in spring [11]. In orchids, transcriptome analysis revealed genetic integrators, MADS-box genes, miRNAs and transcription factors involved in regulating continuous flowering [12,13]. The unique FT splicing variant with intron retention in *Liriodendron chinense slb1* mutants may be responsible for its continuous flowering [14].

Several studies have also reported the mechanism of continuous flowering in *Dimocarpus longan* cv. 'Sijimi'. Differentially expressed genes (DEGs) between 'SJ' (CF) and 'SX' or 'Lidongben' (OF) were detected by transcriptome analysis, and most identified flowering-related DEGs involved in photoperiod and circadian-clock pathways, such as *CONSTANS-like* (*COL*), *two-component response regulator-like* (*APRRs*), *GIGANTEA* (*GI*), *EARLY FLOWERING* (*ELF*), and *F-BOX 1* (*FKF1*). Of these, *ELF4* may be the key gene [5,6]. There were no differences in *TFL* homologous gene sequences between CF and OF, indicating that the mechanism of continuous flowering in longan may be different from that in other woody plants [5]. High-throughput microRNA sequencing indicated that novel-miR137, novel-miR76, novel-miR101, novel-miR37, and csi-miR3954 may play vital regulatory roles in the regulation of continuous flowering in longan [15].

Photoperiod is an important parameter controlling plant flowering and links with the circadian clock [16]. Leaves are key organs that sense floral-inductive signals, such as a change in light regime or mobile florigen production [17]. After receiving the light signal, leaves can accumulate flower-related substances, such as FLOWER LOCUS T (FT) protein [17]. Previous studies on continuous flowering in longan mainly used terminal buds as study material. In this study, leaves were used, and transcriptome analysis was conducted to compare CF and OF gene-expression differences. Our study aims to provide more evidence to reveal the regulatory mechanism of continuous flowering in *Dimocarpus longan* cv. 'Sijimi'.

2. Materials and Methods

2.1. Plant Materials

Twelve-year-old *Dimocarpus longan* cv. 'Sijimi' (SJ, continuous-flowering), *D. longan* cv. 'Shixia' and *D. longan* cv. 'Chuliang' (SX and CL, single-flowering) were selected. They were cultivated at the Guangdong Longan Germplasm Resource Nursery (in Institute of Fruit Tree Research, Guangdong Academy of Agricultural Sciences), Guangzhou City, Guangdong Province, China and grown under similar planting conditions. The varieties were identified by the builder of the nursery, Jianguang Li, who is also the leader of our lab. Mature leaves (from last autumn shoot of the previous year) of SX, CL and non-flowering SJ branches (with diameter around 1 cm) were collected on 20 September 2023. Three replicates of each variety were collected, and each replicate contained mature leaves in the same condition from three trees. After they had been picked, the leaves were placed immediately in liquid nitrogen and stored at −80 °C until RNA-Seq and quantitative reverse-transcription polymerase chain reactions (qRT-PCR).

2.2. RNA Extraction and RNA-Seq

Total RNA was isolated using Trizol Reagent (Invitrogen, Carlsbad, CA, USA). RNA concentration and purity were determined using NanoDrop 2000C (Thermo Scientific, Waltham, MA, USA), and RNA integrity was assessed by using an Agilent 2100 Bioanalyzer (Agilent Technologies, Santa Clara, CA, USA) (Figure S1). After enrichment and purification, RNA was fragmented into short fragments by fragmentation buffer. The first cDNA strand was synthesized with random hexamers and MMuLV Reverse Transcriptase (RNase H), and the second was synthesized using DNA polymerase I and RNase H. Double-stranded cDNA was purified; segment sizes were selected by AMPure XP beads; and the sample was finally enriched by PCR to obtain the final cDNA library. Each library was loaded onto the Illumina Hiseq2000 platform for high-throughput sequencing [18]. The raw sequence data reported in this paper have been deposited in the Genome Sequence Archive [19] in National Genomics Data Center [20], China National Center for Bioinformation/Beijing Institute of Genomics, Chinese Academy of Sciences (GSA: CRA015692) and are publicly accessible at https://ngdc.cncb.ac.cn/gsa (accessed on 10 April 2024).

2.3. De Novo Assembly and Functional Annotation

Clean reads were obtained by checking the sequencing error rate and GC content distribution and filtering the raw sequencing data. Then, clean reads were aligned to the longan genome (http://gigadb.org/dataset/view/id/100276 (accessed on 10 November 2023)) [21] using the hisat 2 program [22]. Fragments per kilobase of transcript per million fragments mapped (FPKM) value, an indicator for transcript or gene-expression level, was calculated by featureCounts [23]. Genes with a $|\log2\text{fold change}| \geq 1$ and false-discovery rate (FDR) < 0.05 were identified as DEGs using DESeq2 [24]. To elucidate the functions of the genes, DEGs were annotated through Kyoto Encyclopedia of Genes and Genomes (KEGG; https://www.genome.jp/kegg (accessed on 12 November 2023)) [25], Gene Ontology (GO) [26], Clusters of Orthologous Groups of Proteins (KOG; https://www.ncbi.nlm.nih.gov/COG/ (accessed on 12 November 2023)) [27] and Non-redundant protein sequences (Nr; https://ftp.ncbi.nlm.nih.gov/blast/db/FASTA/ (accessed on 12 November 2023)) database.

2.4. WGCNA

Coexpression of genes and metabolites was estimated using the weighted coexpression network analysis (WGCNA) (v1.47) package in R [28]. After genes were filtered, FPKM or relative abundance values were imported into WGCNA to construct coexpression modules. The threshold of merge cut height was set to 0.25, and the minimum module size was set to 50. Gene modules with significant correlations with traits were enriched to KEGG pathway terms based on the hypergeometric test (adjusted p-value < 0.05). For genes involved in important pathways, Cytoscape software (v3.3.0) [29] was used to visualize their coexpression patterns.

2.5. Quantitative Real-Time PCR (qRT-PCR) Validation

Total RNA was extracted using TRIzol reagent (Invitrogen Life Technologies). One µg of RNA was used for cDNA synthesis using a PrimeScriptTM RT reagent Kit with gDNA Eraser (Perfect Real Time) (Takara, Shiga, Japan). qRT-PCR was conducted in a total volume of 10 µL containing 30 ng cDNA, 200 nM each of gene-specific primers and 5 µL SYBR Premix Ex Taq reagent (Takara, Shiga, Japan) using a Mini Option Real-Time PCR System (Bio-Rad, Hercules, CA, USA) according to the manufacturer's instructions. The *DlActin* gene was used as internal control [30]. All qRT-PCR experiments were repeated at least three times, and relative expression was calculated based on $2^{-\triangle\triangle Ct}$. The primers used for qRT-PCR and their sequences are listed in Table S1.

2.6. Statistical Analysis

All data were collected from at least three replicates and analyzed using a completely randomized design in the SPSS 25.0 program (IBM, Armonk, NY, USA). Differences among means from different time-point were evaluated by Duncan's test at the 0.05 probability level. Histograms were constructed using Microsoft Office Excel 2016 (Microsoft, Los Angeles, CA, USA).

3. Results

3.1. Illumina Sequencing, Reads Assembly and Annotation

To identify differentially expressed genes in SJ, SX and CL mature leaves, transcriptome sequencing was conducted using the Illumina sequencing platform. After the raw data had been trimmed, an average of 47,982,637, 43,833,340 and 54,441,291 clean reads were obtained for SJ, SX and CL, respectively. Furthermore, the average numbers of reads mapped onto the reference genome were 44,354,326 (92.45%), 40,157,857 (91.62%) and 48,710,060 (91.32%) for SJ, SX and CL, respectively. The Q30 percentages, on average, for SJ, SX and CL were 93.32, 93.40 and 93.13%, respectively. Additionally, the average GC contents were 44.40, 44.55 and 44.12% for the SJ, SX and CL transcriptome, respectively (Table 1).

Table 1. Summary of Illumina transcriptome sequencing.

Sample	Raw Reads	Clean Reads	Reads Mapped	Error Rate (%)	Q20 (%)	Q30 (%)	GC Content (%)
CL1	56,508,126	54,531,106	50,215,847 (92.09%)	0.03	97.69	93.24	43.94
CL2	58,805,112	56,520,954	51,230,072 (90.64%)	0.03	97.68	93.18	44.24
CL3	53,907,766	52,271,814	47,684,261 (91.22%)	0.03	97.58	92.96	44.19
SJ1	51,469,290	50,029,874	46,120,702 (92.19%)	0.03	97.65	93.16	44.29
SJ2	48,643,342	47,113,166	43,536,518 (92.41%)	0.03	97.79	93.5	44.34
SJ3	48,346,330	46,804,872	43,405,759 (92.74%)	0.03	97.7	93.3	44.57
SX1	46,602,468	44,987,712	41,640,487 (92.56%)	0.03	97.76	93.43	45.18
SX2	44,219,408	42,643,896	39,537,158 (92.71%)	0.03	97.79	93.49	44.25
SX3	45,368,976	43,868,412	39,295,927 (89.58%)	0.03	97.7	93.27	44.21

The unigenes were annotated against five public databases. The number of annotated unigenes in the databases Non-redundant protein sequences (Nr), Gene Ontology (GO), EuKaryotic Orthologous Groups (KOG), Kyoto Encyclopedia of Genes and Genomes (KEGG) and Swiss-Prot were 35,537, 29,620, 34,690, 28,096 and 26,665, respectively. The Nr annotation result indicated that *Citrus sinensis* (21.09%), *C. clementina* (13.84%) and *C. unshiu* (7.21%) showed the highest homology with *Dimocarpus longan* (Figure 1c). Furthermore, 34,690 unigenes were clustered into 25 KOG classifications. The top 3 classifications were General function prediction only (7426), Signal transduction mechanisms (2198) and Posttranslational modification, protein turnover, chaperones (1878) (Figure 1a). A total of 2039 transcription factors were identified, including MYB (128, 6.28%), NAC (120, 5.89%), AP2/ERF-ERF (116, 5.69%), bHLH (107, 5.25%), C2H2 (100, 4.9%), MADS-M-type (91, 4.46%), B3 (86, 4.22%), MYB-related (70, 3.43%), GRAS (60, 2.94%) and WRKY (60, 2.94%) (Figure 1b).

3.2. Functional Annotation of Differentially Expressed Genes (DEGs)

A total of 6745 DEGs were detected. The groups CL_vs_SJ, SX_vs_CL and SX_vs_SJ had 4198 (2192 down-regulated and 2006 up-regulated), 3749 (1842 down-regulated and 1907 up-regulated) and 3924 (2026 down-regulated and 1898 up-regulated) DEGs, respectively (Figure 2a,b). To determine the gene-expression patterns of different genes, the FPKMs of the genes were first centralized and standardized, and then hierarchical cluster analysis and Kmeans cluster analysis were performed. The cluster heat map for each differential group was drawn (Figure 2d) and showed that six subclasses emerged from the clustering and that the SJ gene-expression patterns in subclasses 1 and 3 differed between SX and CL (Figure 2c). Genes were selected from these for further validation.

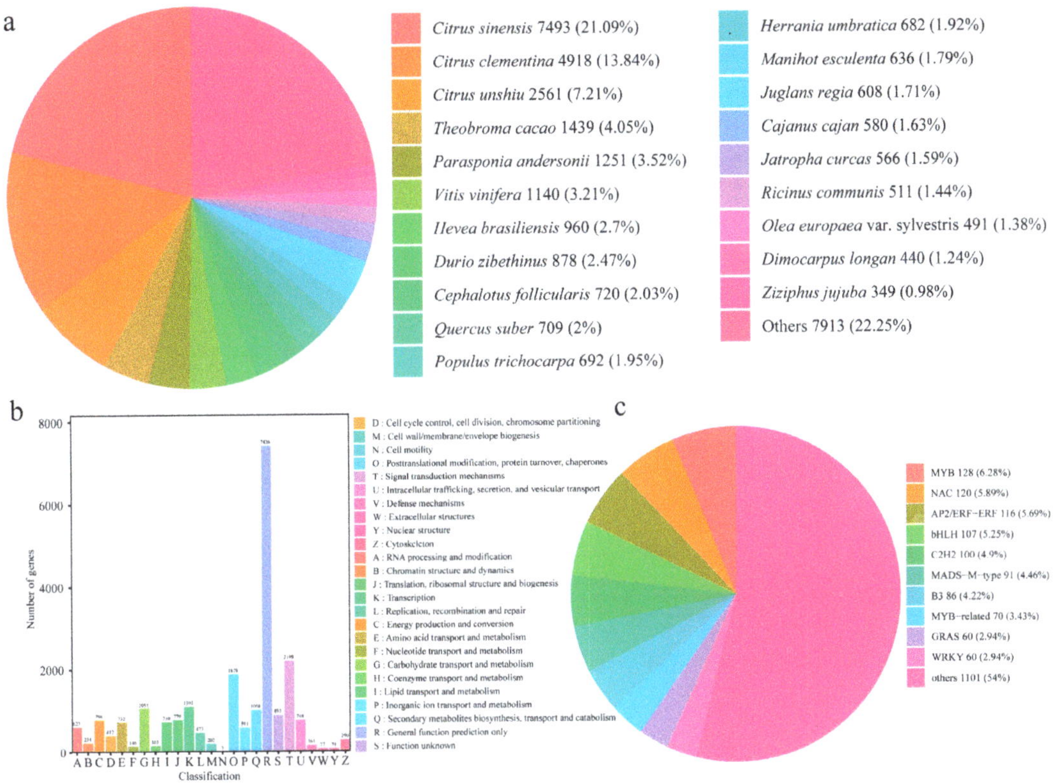

Figure 1. Annotation of unigenes. (**a**) KOG annotation of unigenes; (**b**) Transcription factor (TF) annotation of unigenes; (**c**) Nr annotation of unigenes.

GO analysis indicated that the DEGs of the groups CL_vs_SJ, SX_vs_CL and SX_vs_SJ were categorized into 50, 50 and 51 GO terms, respectively, consisting of three domains: biological process, cellular component and molecular function (Figure S2). KOG analysis clustered the DEGs into KOG classifications, and the most common classifications were 'General function prediction only', 'Signal transduction mechanisms' and 'Posttranslational modification, protein turnover, chaperones' (Figure S3).

In the KEGG pathway analysis, DEGs of the groups CL_vs_SJ, SX_vs_CL and SX_vs_SJ were classified into 61, 57 and 50 classifications, respectively. After Metabolic pathways, Plant-pathogen interaction and Biosynthesis of secondary metabolites, the next-most-common classifications were MAPK signaling pathway-plant, Plant hormone signal transduction, Amino sugar and nucleotide sugar metabolism and Starch and sucrose metabolism (Figure S4). KEGG enrichment was measured by Rich factor, Q-value and the number of genes enriched in each pathway. DEGs classified in the pathways Plant-pathogen interaction, MAPK signaling pathway-plant and Amino sugar and nucleotide sugar metabolism showed significant enrichment in all three groups (Figure S4). The enrichment of DEGs classified as Photosynthesis-antenna proteins and Anthocyanin biosynthesis in the groups CL_vs_SJ and SX_vs_SJ was more significant than in the group SX_vs_CL (Figure S4), which may reflect the difference between CF (SJ) and OF (SX, CL).

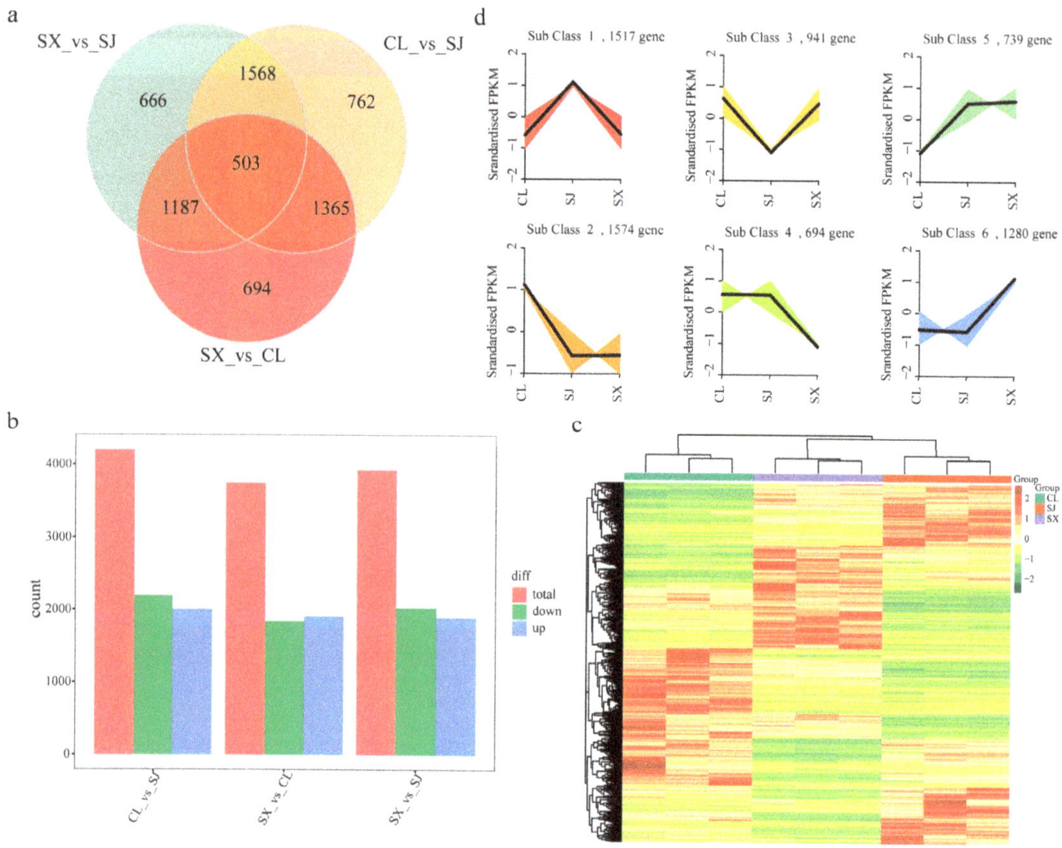

Figure 2. Analysis of differentially expressed genes (DEGs). (**a**) DEGs Venn diagram; (**b**) Number of DEGs; (**c**) Kmeans clustering of DEGs; (**d**) DEGs heat map. SX, CL and SJ represent *Dimocarpus longan* cv. 'Shixia', *D. longan* cv. 'Chuliang' and *D. longan* cv. 'Sijimi' respectively.

3.3. Key Modules Obtained by WGCNA Analysis Related to Flowering Habits

After genes with FPKM< 0.1 were filtered out in all samples, WGCNA analysis clustered genes into 27 modules. Among them, the grey module included genes that had not been assigned to other modules, which had little significance and did not require special attention. The 26 modules (all except for grey) were divided into 4 groups based on module eigenvalues (Figure 3a). The heat map of sample and module correlations indicated that the bisque4 module was positively correlated with all CJ and SX samples and negatively correlated with all SJ samples (Figure 3b,c), while thedarkorange2 module was negatively correlated with all CJ and SX samples and positively correlated with all SJ samples (Figure 3b,d). A total of 2634 and 2225 genes were included in the

bisque4 and darkorange2 modules, respectively. Of these, 935 and 727 genes were annotated to the KEGG database, respectively. After Metabolic pathways, Biosynthesis of secondary metabolites and Plant-pathogen interaction, the classifications of MAPK signaling pathway-plant, Plant hormone signal transduction, and Starch and sucrose metabolism had more genes (Figure 4a,b). This result was similar to KEGG annotation of DEGs. The MAPK signaling pathway-plant and Peroxisome pathways were significantly enriched in the bisque4 module, while Photosynthesis-antenna proteins, sulfur metabolism and Ribosome pathways were significantly enriched in the darkorange2 module (Figure 4c,d).

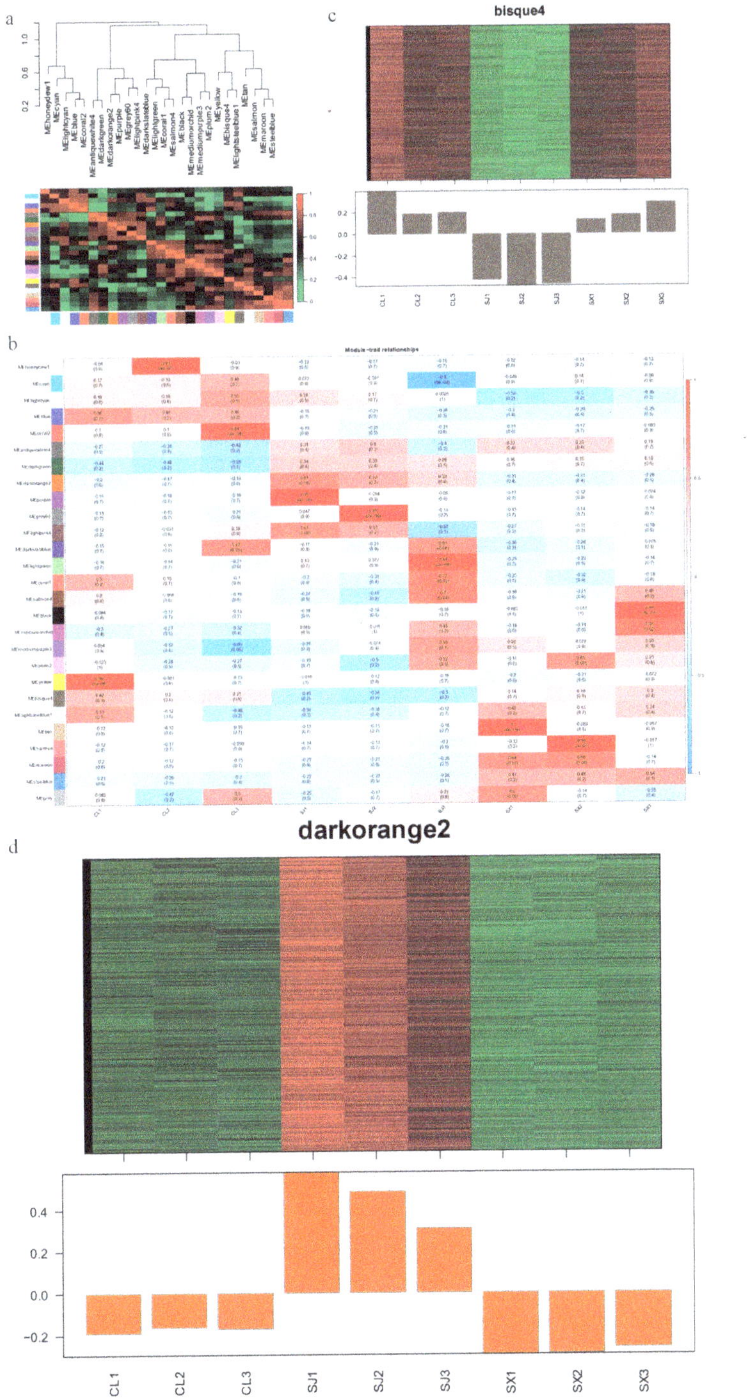

Figure 3. WGCNA analysis. (**a**) Heat map of module correlations (the smaller the value on the upper vertical axis, the higher the similarity between the two modules, and the darker the color (redder) in

the lower squares, the stronger the correlation); (**b**) heat map of sample and module correlations (the horizontal axis represents the sample, and the vertical axis represents the module. The number in each cell represents the correlation between the module and the sample. The closer the value is to 1, the stronger the positive correlation between the module and the sample; the closer it is to −1, the stronger the negative correlation between the module and the sample. The number in parentheses represents P value significance, and the smaller the value, the stronger the significance.); gene-expression patterns of the bisque4 (**c**) and darkorange2 (**d**) modules; the upper part shows the clustering heatmap of genes within the module, with red indicating high expression and green indicating low expression, and the lower part shows the expression patterns of module feature values in different samples. SX, CL and SJ represent *Dimocarpus longan* cv. 'Shixia', *D. longan* cv. 'Chuliang' and *D. longan* cv. 'Sijimi', respectively.

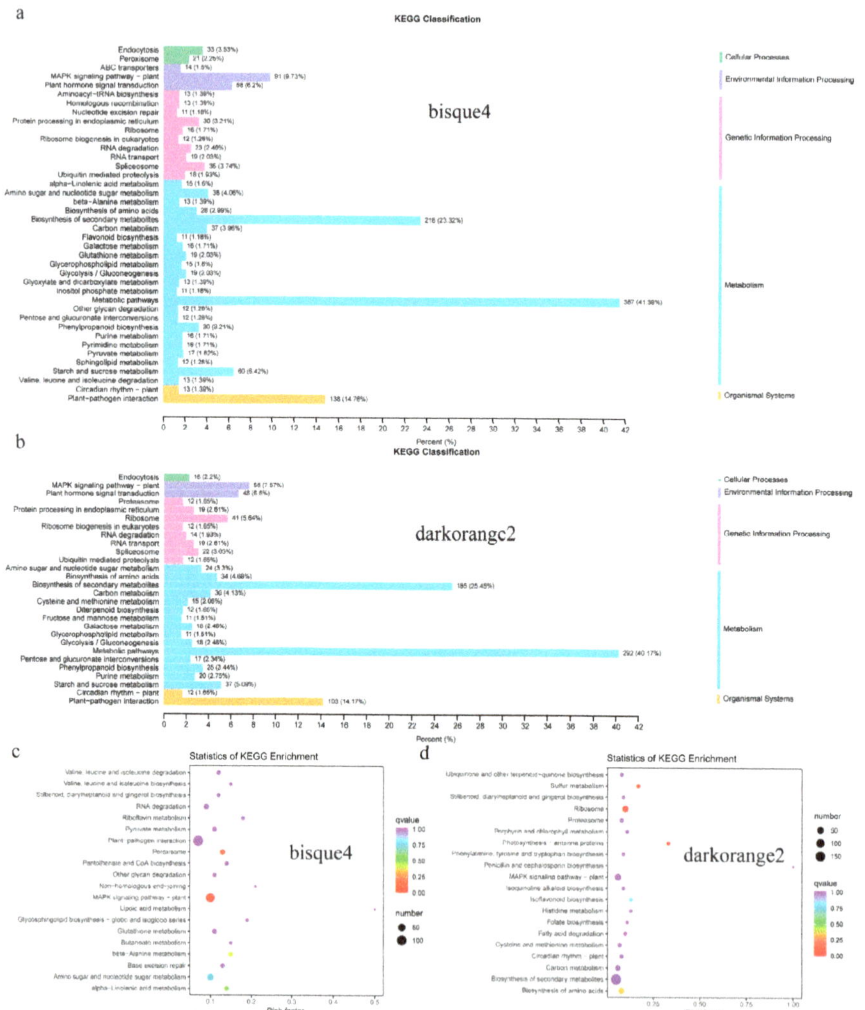

Figure 4. KEGG annotation of the bisque4 and darkorange2 modules. KEGG classification of the bisque4 (**a**) and darkorange2 (**b**) modules; KEGG enrichment of the bisque4 (**c**) and darkorange2 (**d**) modules.

Genes annotated to Photosynthesis, Circadian rhythm—plant, Starch and sucrose metabolism, MAPK signaling pathway—plant and Plant hormone signal transduction pathways in the bisque4 and darkorange2 modules were compared. There were 5, 44 and 89 genes expressed at low levels in SJ (bisque4 module) and annotated to Photosynthesis, Starch and sucrose metabolism and MAPK signaling pathway—plant pathways, respectively (Table S2). Totals of 17, 27 and 53 genes highly expressed in SJ (darkorange2 module) were annotated to Photosynthesis, Starch and sucrose metabolism and MAPK signaling pathway—plant pathways, respectively (Table S3). There were 10 and 11 genes expressed at low and high levels, respectively, in SJ that were annotated to a Circadian rhythm—plant pathway (Tables S2 and S3). Moreover, 3 *FT* genes were detected in the darkorange2 module (Table S3). There were 34 genes annotated to a Plant hormone signal transduction pathway expressed at low levels in SJ, including 2 abscisic acid (ABA), 9 auxin, 13 brassinosteroid (BR) and 5 gibberellin (GA)-related genes (Table S2). There were 31 genes belonging to Plant hormone signal transduction pathways that were highly expressed in SJ, including 1 abscisic acid (ABA), 7 auxin, 8 brassinosteroid (BR) and 13 gibberellin (GA)-related genes (Table S3).

3.4. The qRT-PCR Validation of Selected DEGs

Combining the results of DEGs annotation and WGCNA analysis, 10 genes related to glycometabolism, plant-hormone signal transduction, the MAPK signaling pathway and *Flowering locus T* (*FT*) were further validated by qRT-PCR (Table S1). The qRT-PCR expression levels of most of the DEGs were consistent with transcriptomic expression levels. Genes that were specifically expressed at high or low levels in SJ were verified (Figure 5a–l). Most of the MAPK signaling pathway genes were found in the bisque4 module, which had low SJ FPKM values in the transcriptome. The qRT-PCR results indicated that the expression level was low not only in SJ, but also in SX and CL (Figure 5o–t).

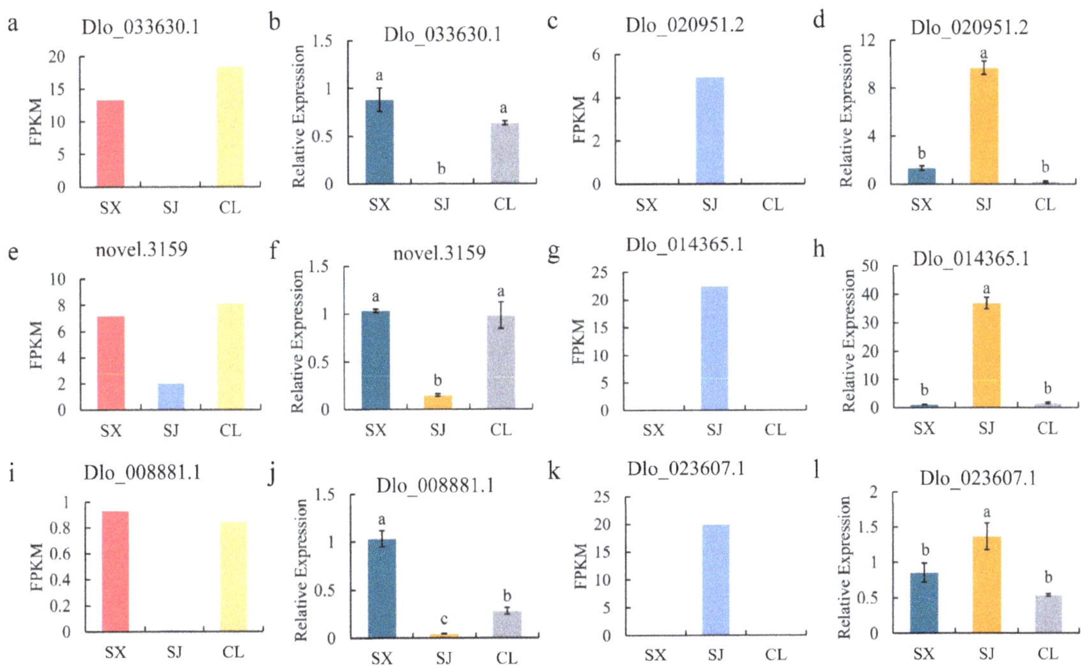

Figure 5. *Cont.*

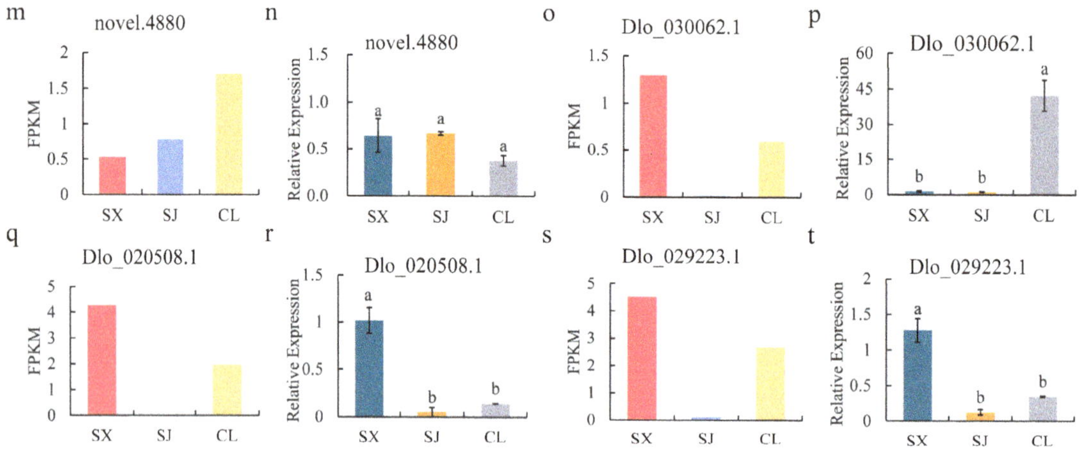

Figure 5. Comparison of gene expression in CF(SJ) and OF (SX, CL). FPKM values of ten genes: (Dlo_033630.1 (**a**), Dlo_020951.2 (**c**), novel.3159 (**e**), Dlo_014365.1 (**g**), Dlo_008881.1 (**i**), Dlo_023607.1 (**k**), novel.4880 (**m**), Dlo_030062.1 (**o**), Dlo_020508.1 (**q**), Dlo_029223.1 (**s**)) in SX, SJ and CL and relative expression level of ten genes (Dlo_033630.1 (**b**), Dlo_020951.2 (**d**), novel.3159 (**f**), Dlo_014365.1 (**h**), Dlo_008881.1 (**j**), Dlo_023607.1 (**l**), novel.4880 (**n**), Dlo_030062.1 (**p**), Dlo_020508.1 (**r**), Dlo_029223.1 (**t**)) in SX, SJ and CL were compared. SX, CL and SJ represent *Dimocarpus longan* cv. 'Shixia', *D. longan* cv. 'Chuliang' and *D. longan* cv. 'Sijimi', respectively. Data are presented as the mean ± standard error, and different lowercase letters above the bars indicate significant differences (Duncan's test, $p < 0.05$).

4. Discussion

We analyzed DEGs of mature longan leaves from CF (SJ) and OF (SX, CL) by transcriptome sequencing. CF and OF DEGs were mainly clustered into Photosynthesis-antenna proteins, MAPK signaling pathway-plant, plant hormone signal transduction and Starch and sucrose metabolism pathways, suggesting different patterns of nutrient accumulation in leaves, leading in turn to different flowering patterns.

In leaves, the photoperiodic pathway mainly transmits signals through GI and CO [31]. Under the regulation light signals, CO protein accumulates stably, inducing the expression of their direct downstream genes *FT* and *TWINSISTER OF FT* (*TSF*) in leaves, as well as FT transport to the terminal buds [32]. PHYA and PHYB are involved in the regulation of CO protein stability and *FT* expression [16]. FT integrates signals that regulate flowering from almost all regulatory pathways and then delivers them downstream to *SUPPRESSOR OF OVEREXPRESSION OF CO 1* (*SOC1*) and *APETALA1* (*AP1*) [33]. There are 3 *CO* (Dlo_010015.1, Dlo_029217.1, Dlo_029296.1) genes, 3 *FT* (Dlo_000296.1, Dlo_012579.1, Dlo_014365.1) genes and 1 *PHYA* (novel.246) gene specifically highly expressed in CF longan (SJ) (Figure 3d, Table S3). These results indicate that flowering-related genes were continuously expressed in SJ leaves.

Sugars are important for plant growth and development, both as energy metabolites and as signaling molecules [34]. Sugars regulate flowering, senescence, embryo formation, seed germination, pollen germination, root development, tillering and stress tolerance through a complex network of pathways [34]. It was experimentally confirmed in the early 20th century that a higher C/N ratio benefits flower-bud differentiation, suggesting that sugar accumulation in plants, including fruit trees, is the material basis for bud differentiation [35]. Many DEGs of the CL_vs_SJ, SX_vs_CL and SX_vs_SJ groups and genes in the bisque4 and darkorange2 modules were annotated to Starch and sucrose metabolism (Figure 3c,d, Tables S2 and S3). The number of genes annotated to Starch and sucrose metabolism pathways that were specifically highly expressed in OF longan (SX, CL) was greater than the number that were highly expressed in CF longan (SJ) (Tables S2 and S3). This suggests that in OF longan, which only has nutritional growth in September, sugar metabolism was more active and more genes were involved, while in CF longan, substances required for flower formation accumulated during vegetative growth.

Mitogen-activated protein kinase (MAPK) is a highly conserved biological signal-transduction module that plays a crucial role in plant growth, development, and stress resistance [36]. This signaling pathway can amplify and transmit environmental signals sensed by plants, thereby activating downstream enzymes and transcription factors, ultimately activating target gene expression and responding to external signals [37]. Numerous DEGs of the CL_vs_SJ, SX_vs_CL and SX_vs_SJ groups and genes in the bisque4 and darkorange2 modules were annotated to MAPK signaling pathway—plant pathways (Figure 3c,d, Tables S2 and S3), demonstrating that MAPK may participate in light signal transduction, sugar metabolism and signal transduction during flower formation. Besides, MAPK can also transmit hormonal signals [38].

Plant endogenous hormones participate in the whole plant life cycle [39], while hormonal signaling plays a crucial role in flower formation [40]. Regulation of hormonal signaling is often achieved by pooling different hormone signals and then altering expression levels of key genes involved in flowering [41]. GA is the most well-studied hormone that regulates flowering and is considered the most important. GA promotes flowering by inducing *SOC1*, *FT*, *TSF*, *SPL* and *LFY* expression [42]. The number of GA-related genes that were specifically highly expressed (13) in SJ was significantly higher than the number expressed at low levels (5) (Tables S2 and S3), suggesting an important role for a GA signaling pathway in CF longan flowering induction. Some IAA- and BR-related genes were also specifically expressed at low or high levels in SJ (Tables S2 and S3). IAA can promote flowering by regulating GA content and promoting DELLA degradation [43,44]. BR signaling integrates with environmental cues to fine-tune flowering time through *FT* and the other flowering pathways [45].

Floral induction is regulated by different regulatory pathways, which contain various genes. While the expression of genes is determined by transcription factors (TFs), the importance of TFs in floral induction is evident [46]. Several TFs, such as MYB, NAC, AP2/ERF and WRKY, have been detected in the transcriptome and have been confirmed to be involved in regulation of flowering. Overexpression of *TaMYB72* from *Triticum aestivum* in rice could up-regulate the florigen genes *Hd3a* and *RFT1* and ultimately promote rice flowering under long-day conditions [47]. The AP2/ERF transcription factor TOE4b from *Glycine max* could directly bound to the promoters and gene bodies of the key flowering integration factor genes *FT2a* and *FT5a* to inhibit their transcription and delay flowering in lines displaying overexpression [48]. WRKY TFs are implicated in phytohormone pathways, such as ethylene, auxin, and abscisic acid pathways, through which they modulate flowering time [49].

5. Conclusions

Through transcriptome analysis, we found that the DEGs between mature leaves of CF and OF longan were mainly clustered into Photosynthesis - antenna proteins, MAPK signaling pathway - plant, plant hormone signal transduction and Starch and sucrose metabolism pathways. Especially significant were the high *TF* expression and the large number of GA signaling pathway genes that were highly expressed in CF longan (SJ). We speculate that OF longan initiates expression of genes related to vegetative growth, such as photosynthesis, auxin, brassinosteroid, and MAPK pathway genes, promoting plant vegetative growth. After low-temperature induction, expression of flowering-related genes is initiated. But for CF longan, the expression of genes related to vegetative and reproductive growth are simultaneously induced. Expression levels of floral-induction genes, such as GA-signaling-related and *FT* genes, are always high, and after accumulation of vegetative growth, flowering can be directly induced.

Supplementary Materials: The following supporting information can be downloaded at: https://www.mdpi.com/article/10.3390/horticulturae10090974/s1, Figure S1: RNA electropherogram of nine samples; Figure S2: GO annotation of DEGs; Figure S3: KOG annotation of DEGs; Figure S4: KEGG annotation of DEGs; Table S1: Primers used in qRT-PCR; Table S2: KEGG annotation of genes in bisque4 module; Table S3: KEGG annotation of genes in darkorange2 module.

Author Contributions: S.H.: data curation, funding acquisition, methodology, investigation, writing-original draft, writing-review & editing. X.L.: data curation, methodology, writing-original draft. J.W.: methodology, formal analysis, writing-original draft. D.H.: data curation, validation, writing-original draft. J.L.: supervision, project administration, writing-review & editing. D.G.: conceptualization, funding acquisition, project administration, writing-review & editing. All authors have read and agreed to the published version of the manuscript.

Funding: This research was funded by the Innovation Fund project of Guangdong Academy of Agricultural Sciences (202212).

Data Availability Statement: The raw sequence data for this study can be found in the Genome Sequence Archive in National Genomics Data Center (https://ngdc.cncb.ac.cn/gsa (accessed on 10 April 2024); GSA: CRA015692). The original contributions presented in the study are included in the article/Supplementary Material.

Conflicts of Interest: The authors declare no conflicts of interest.

References

1. Lai, Z.; Chen, C.; Zeng, L.; Chen, Z. Somatic embryogenesis in longan [*Dimocarpus longan* Lour.]. In *Somatic Embryogenesis in Woody Plants*; Jain, S.M., Gupta, P., Newton, R., Eds.; Springer: Dordrecht, The Netherlands, 2000; Volume 67, pp. 415–431.
2. Lithanatudom, S.K.; Chaowasku, T.; Nantarat, N.; Jaroenkit, T.; Smith, D.R.; Lithanatudom, P. A first phylogeny of the genus *Dimocarpus* and suggestions for revision of some taxa based on molecular and morphological evidence. *Sci. Rep.* **2017**, *7*, 6716. [CrossRef] [PubMed]
3. Huang, S.; Han, D.; Wang, J.; Guo, D.; Li, J. Floral induction of longan (*Dimocarpus longan*) by potassium chlorate: Application, mechanism, and future perspectives. *Front. Plant Sci.* **2021**, *12*, 670587. [CrossRef] [PubMed]
4. Yue, X.; Chen, Z.; Zhang, J.; Huang, C.; Zhao, S.; Li, X.; Qu, Y.; Zhang, C. Extraction, purification, structural features and biological activities of longan fruit pulp (Longyan) polysaccharides: A review. *Front. Nutr.* **2022**, *9*, 914679. [CrossRef] [PubMed]
5. Jue, D.; Sang, X.; Liu, L.; Shu, B.; Wang, Y.; Liu, C.; Wang, Y.; Xie, J.; Shi, S. Comprehensive analysis of the longan transcriptome reveals distinct regulatory programs during the floral transition. *BMC Genom.* **2019**, *20*, 126. [CrossRef] [PubMed]
6. Jia, T.; Wei, D.; Meng, S.; Allan, A.C.; Zeng, L. Identification of regulatory genes implicated in continuous flowering of longan (*Dimocarpus longan* L.). *PLoS ONE* **2014**, *9*, e114568. [CrossRef]
7. Iwata, H.; Gaston, A.; Remay, A.; Thouroude, T.; Jeauffre, J.; Kawamura, K.; Oyant, L.H.S.; Araki, T.; Denoyes, B.; Foucher, F. The *TFL1* homologue *KSN* is a regulator of continuous flowering in rose and strawberry. *Plant J.* **2012**, *69*, 116–125. [CrossRef]
8. Bai, M.; Liu, J.; Fan, C.; Chen, Y.; Chen, H.; Lu, J.; Sun, J.; Ning, G.; Wang, C. *KSN* heterozygosity is associated with continuous flowering of *Rosa rugosa* Purple branch. *Hortic. Res.* **2021**, *8*, 26. [CrossRef]
9. Randoux, M.; Davière, J.M.; Jeauffre, J.; Thouroude, T.; Pierre, S.; Toualbia, Y.; Perrotte, J.; Reynoird, J.P.; Jammes, M.J.; Oyant, L.H.S.; et al. RoKSN, a floral repressor, forms protein complexes with RoFD and RoFT to regulate vegetative and reproductive development in rose. *New Phytol.* **2014**, *202*, 161–173. [CrossRef]
10. Yi, X.; Gao, H.; Yang, Y.; Yang, S.; Luo, L.; Yu, C.; Wang, J.; Cheng, T.; Zhang, Q.; Pan, H. Differentially expressed genes related to flowering transition between once- and continuous-flowering Roses. *Biomolecules* **2021**, *12*, 58. [CrossRef]
11. Randoux, M.; Jeauffre, J.; Thouroude, T.; Vasseur, F.; Hamama, L.; Juchaux, M.; Sakr, S.; Foucher, F. Gibberellins regulate the transcription of the continuous flowering regulator, *RoKSN*, a rose *TFL1* homologue. *J. Exp. Bot.* **2012**, *63*, 6543–6554. [CrossRef]
12. Ahmad, S.; Lu, C.; Gao, J.; Ren, R.; Wei, Y.; Wu, J.; Jin, J.; Zheng, C.; Zhu, G.; Yang, F. Genetic insights into the regulatory pathways for continuous flowering in a unique orchid *Arundina graminifolia*. *BMC Plant Biol.* **2021**, *21*, 587. [CrossRef] [PubMed]
13. Ahmad, S.; Peng, D.; Zhou, Y.; Zhao, K. The genetic and hormonal inducers of continuous flowering in Orchids: An emerging view. *Cells* **2022**, *11*, 657. [CrossRef] [PubMed]
14. Sheng, Y.; Hao, Z.; Peng, Y.; Liu, S.; Hu, L.; Shen, Y.; Shi, J.; Chen, J. Morphological, phenological, and transcriptional analyses provide insight into the diverse flowering traits of a mutant of the relic woody plant *Liriodendron chinense*. *Hortic. Res.* **2021**, *8*, 174. [CrossRef] [PubMed]
15. Waheed, S.; Liang, F.; Zhang, M.; He, D.; Zeng, L. High-throughput sequencing reveals novel microRNAs involved in the continuous flowering trait of longan (*Dimocarpus longan* Lour.). *Int. J. Mol. Sci.* **2022**, *23*, 15565. [CrossRef] [PubMed]
16. Song, Y.H.; Shim, J.S.; Kinmonth-Schultz, H.A.; Imaizumi, T. Photoperiodic flowering: Time measurement mechanisms in leaves. *Annu. Rev. Plant Biol.* **2015**, *66*, 441–464. [CrossRef]

17. Del Prete, S.; Molitor, A.; Charif, D.; Bessoltane, N.; Soubigou-Taconnat, L.; Guichard, C.; Brunaud, V.; Granier, F.; Fransz, P.; Gaudin, V. Extensive nuclear reprogramming and endoreduplication in mature leaf during floral induction. *BMC Plant Biol.* **2019**, *19*, 135. [CrossRef]
18. Huang, S.; Lv, X.; Han, Y.; Han, D.; Wei, J.; Li, J.; Guo, D. Disrupted sugar transport and continued sugar consumption lead to sugar decline in ripe 'Shixia' longan fruit. *LWT* **2024**, *191*, 115620. [CrossRef]
19. Chen, T.; Chen, X.; Zhang, S.; Zhu, J.; Tang, B.; Wang, A.; Dong, L.; Zhang, Z.; Yu, C.; Sun, Y.; et al. The Genome Sequence Archive family: Toward explosive data growth and diverse data types. *Genom. Proteom. Bioinform.* **2021**, *19*, 578–583. [CrossRef]
20. CNCB-NGDC Members and Partners. Database resources of the National Genomics Data Center, China National Center for Bioinformation in 2022. *Nucleic Acids Res.* **2022**, *50*, D27–D38. [CrossRef]
21. Lin, Y.; Min, J.; Lai, R.; Wu, Z.; Chen, Y.; Yu, L.; Cheng, C.; Jin, Y.; Tian, Q.; Liu, Q.; et al. Genome-wide sequencing of longan (*Dimocarpus longan* lour.) provides insights into molecular basis of its polyphenol-rich characteristics. *GigaScience* **2017**, *6*, 1–14. [CrossRef]
22. Kim, D.; Langmead, B.; Salzberg, S.L. HISAT: A fast spliced aligner with low memory requirements. *Nat. Methods* **2015**, *12*, 357–360. [CrossRef]
23. Liao, Y.; Smyth, G.K.; Shi, W. FeatureCounts: An efficient general purpose program for assigning sequence reads to genomic features. *Bioinformatics* **2014**, *30*, 923–930. [CrossRef] [PubMed]
24. Love, M.I.; Huber, W.; Anders, S. Moderated estimation of fold change and dispersion for rna-seq data with DEseq2. *Genome Biol.* **2014**, *15*, 550. [CrossRef]
25. Kanehisa, M.; Araki, M.; Goto, S.; Hattori, M.; Hirakawa, M.; Itoh, M.; Katayama, T.; Kawashima, S.; Okuda, S.; Tokimatsu, T.; et al. KEGG for linking genomes to life and the environment. *Nucleic Acids Res.* **2008**, *36*, D480–D484. [CrossRef]
26. Ashburner, M.; Ball, C.A.; Blake, J.A.; Botstein, D.; Butler, H.; Cherry, J.M.; Davis, A.P.; Dolinski, K.; Dwight, S.S.; Eppig, J.T.; et al. Gene ontology: Tool for the unification of biology. *Nat. Genet.* **2000**, *25*, 25–29. [CrossRef] [PubMed]
27. Koonin, E.V.; Fedorova, N.D.; Jackson, J.D.; Jacobs, A.R.; Krylov, D.M.; Makarova, K.S.; Mazumder, R.; Mekhedov, S.L.; Nikolskaya, A.N.; Rao, B.S.; et al. A comprehensive evolutionary classification of proteins encoded in complete eukaryotic genomes. *Genome Biol.* **2004**, *5*, R7. [CrossRef]
28. Langfelder, P.; Horvath, S. WGCNA: An R package for weighted correlation network analysis. *BMC Bioinform.* **2008**, *9*, 559. [CrossRef]
29. Shannon, P.; Markiel, A.; Ozier, O.; Baliga, N.S.; Wang, J.T.; Ramage, D.; Amin, N.; Schwikowski, B.; Ideker, T. Cytoscape: A software environment for integrated models of biomolecular interaction networks. *Genome Res.* **2003**, *13*, 2498–2504. [CrossRef] [PubMed]
30. Luo, T.; Shuai, L.; Lai, T.; Liao, L.; Li, J.; Duan, Z.; Xue, X.; Han, D.; Wu, Z. Upregulated glycolysis, TCA, fermentation and energy metabolism promoted the sugar receding in 'Shixia' longan (*Dimocarpus longan* Lour.) pulp. *Sci. Hortic.* **2021**, *281*, 109998. [CrossRef]
31. Mishra, P.; Panigrahi, K.C. GIGANTEA-an emerging story. *Front. Plant Sci.* **2015**, *6*, 8. [CrossRef]
32. Shim, J.S.; Kubota, A.; Imaizumi, T. Circadian clock and photoperiodic flowering in Arabidopsis: CONSTANS is a hub for signal integration. *Plant Physiol.* **2017**, *173*, 5–15. [CrossRef]
33. Takagi, H.; Hempton, A.K.; Imaizumi, T. Photoperiodic flowering in *Arabidopsis*: Multilayered regulatory mechanisms of *CONSTANS* and the florigen *FLOWERING LOCUS T*. *Plant Commun.* **2023**, *4*, 100552. [CrossRef] [PubMed]
34. Chen, Q.; Zhang, J.; Li, G. Dynamic epigenetic modifications in plant sugar signal transduction. *Trends Plant Sci.* **2022**, *27*, 379–390. [CrossRef] [PubMed]
35. Corbesier, L.; Bernier, G.; Périlleux, C. C:N ratio increases in the phloem sap during floral transition of the long-day plants Sinapis alba and *Arabidopsis thaliana*. *Plant Cell Physiol.* **2002**, *43*, 684–688. [CrossRef] [PubMed]
36. Xu, J.; Zhang, S. Mitogen-activated protein kinase cascades in signaling plant growth and development. *Trends Plant Sci.* **2015**, *20*, 56–64. [CrossRef]
37. Zhang, M.; Su, J.; Zhang, Y.; Xu, J.; Zhang, S. Conveying endogenous and exogenous signals: MAPK cascades in plant growth and defense. *Curr. Opin. Plant Biol.* **2018**, *45*, 1–10. [CrossRef]
38. Sun, T.; Zhang, Y. MAP kinase cascades in plant development and immune signaling. *EMBO Rep.* **2022**, *23*, e53817. [CrossRef]
39. Santner, A.; Estelle, M. Recent advances and emerging trends in plant hormone signalling. *Nature* **2009**, *459*, 1071–1078. [CrossRef]
40. Wolters, H.; Jürgens, G. Survival of the flexible: Hormonal growth control and adaptation in plant development. *Nat. Rev. Genet.* **2009**, *10*, 305–317. [CrossRef] [PubMed]
41. Conti, L. Hormonal control of the floral transition: Can one catch them all? *Dev. Biol.* **2017**, *430*, 288–301. [CrossRef]
42. Bao, S.; Hua, C.; Shen, L.; Yu, H. New insights into gibberellin signaling in regulating flowering in Arabidopsis. *J. Integr. Plant Biol.* **2020**, *62*, 118–131. [CrossRef] [PubMed]
43. Fu, X.; Harberd, N.P. Auxin promotes Arabidopsis root growth by modulating gibberellin response. *Nature* **2003**, *421*, 740–743. [CrossRef]
44. Frigerio, M.; Alabadí, D.; Pérez-Gómez, J.; García-Cárcel, L.; Phillips, A.L.; Hedden, P.; Blázquez, M.A. Transcriptional regulation of gibberellin metabolism genes by auxin signaling in Arabidopsis. *Plant Physiol.* **2006**, *142*, 553–563. [CrossRef]
45. Li, Z.; He, Y. Roles of brassinosteroids in plant reproduction. *Int. J. Mol. Sci.* **2020**, *21*, 872. [CrossRef] [PubMed]

46. Koyama, T. Regulatory mechanisms of transcription factors in plant morphology and function. *Int. J. Mol. Sci.* **2023**, *24*, 7039. [CrossRef] [PubMed]
47. Zhang, L.; Liu, G.; Jia, J.; Zhao, G.; Xia, C.; Zhang, L.; Li, F.; Zhang, Q.; Dong, C.; Gao, S.; et al. The wheat MYB-related transcription factor TaMYB72 promotes flowering in rice. *J. Integr. Plant Biol.* **2016**, *58*, 701–704. [CrossRef]
48. Li, H.; Du, H.; Huang, Z.; He, M.; Kong, L.; Fang, C.; Chen, L.; Yang, H.; Zhang, Y.; Liu, B.; et al. The AP2/ERF transcription factor TOE4b regulates photoperiodic flowering and grain yield per plant in soybean. *Plant Biotechnol. J.* **2023**, *21*, 1682–1694. [CrossRef]
49. Song, H.; Duan, Z.; Zhang, J. WRKY transcription factors modulate flowering time and response to environmental changes. *Plant Physiol. Biochem.* **2024**, *210*, 108630. [CrossRef]

horticulturae

MDPI

Article

Dynamic Changes in Polyphenols in Fruit Development of Red Flesh Apple 'Hongxun 2'

Dajiang Wang [1,†], Guangyi Wang [1,†], Xiang Lu [1,2], Zhao Liu [1,2], Simiao Sun [1], Hanxin Guo [1], Wen Tian [1,2], Zichen Li [1], Lin Wang [1], Lianwen Li [1], Yuan Gao [1,*] and Kun Wang [1,*]

1 Research Institute of Pomology, Chinese Academy of Agricultural Sciences (CAAS), Key Laboratory of Horticulture Crops Germplasm Resources Utilization, Ministry of Agriculture, Xingcheng 125100, China; wangdajiang@caas.cn (D.W.); wanggy0315@163.com (G.W.); xianglu1997@163.com (X.L.); lz__0427@163.com (Z.L.); sunsimiao@caas.cn (S.S.); 82101235244@caas.cn (H.G.); tianwen8025@163.com (W.T.); lishencheno@163.com (Z.L.); juziwanglin@163.com (L.W.); lilianwen@caas.cn (L.L.)

2 Xinjiang Production and Construction Corps Key Laboratory of Special Fruits and Vegetables Cultivation Physiology and Germplasm Resources Utilization, Agricultural College, Shihezi University, Shihezi 832003, China

* Correspondence: gaoyuan02@caas.cn (Y.G.); wangkun@caas.cn (K.W.); Tel.: +86-429-3598369 (Y.G.); +86-429-3598120 (K.W.)

† These authors contributed equally to this work.

Abstract: In this study, fruits of the red flesh *Malus* plant 'Hongxun 2' (*Malus neidzwetzkyana* (Dieck) Langenf.) and green flesh *Malus* plant 'Xinye 13-11' (*Malus sieversii* (Led.) Roem.) were used as experimental materials. Both of them came from Xinjiang, China, and *Malus neidzwetzkyana* (Dieck) Langenf. is believed to be a variant of *Malus sieversii* (Led.) Roem. The components and contents of polyphenols in the peel and pulp of the two kinds of fruit during the development period were detected, and the dynamic changes and differences in the polyphenols between the two kinds of fruit were discussed. The results showed that the total polyphenol content of 'Xinye 13-11' was higher in the peel and pulp than that of 'Hongxun 2', and the content of peel was higher than that of pulp in the two kinds of fruit. An analysis of five types of polyphenols showed that anthocyanins were only contained in the peel and pulp of 'Hongxun-2', and the peel had a higher content than the pulp. Cyanidin 3-O-galactoside was the main anthocyanin component. Four other types of substances, except hydroxycinnamics, were higher in 'Hongxun-2' than 'Xinye 13-11', while the contents of other substances in 'Xinye 13-11' were higher than those of 'Hongxun 2'. The accumulation of major polyphenol components in the peel and flesh of 'Hongxun 2' and 'Xinye 13-11' apples was significant in the period before and after 65 days after flowering, and the contents of procyanidin B1 and procyanidin C1 were the highest in this period. In addition to the difference in anthocyanin content between 'Hongxun 2' and 'Xinye 13-11', the chlorogenic acid content in the peel and pulp of 'Hongxun 2' was significantly higher than that of 'Xinye 13-11', and the contents of other components were lower than those of 'Xinye 13-11'. Moreover, based on the components and contents of polyphenol components, this paper supports the viewpoint that *Malus neidzwetzkyana* (Dieck) Langenf is a separate species to *Malus sieversii* (Led.) Roem.

Keywords: apple; red flesh; polyphenol; dynamic changes

1. Introduction

Polyphenols are an indispensable component of the human diet and are widely found in fruits, vegetables, nuts, and other crops. Apples are the third-largest dietary source of polyphenols after tea and onions [1,2]. Polyphenols are closely related to their potential health value in apples, and substances such as procyanidins and catechins have a strong antioxidant capacity and can resist low-density lipoprotein oxidation [3]. Chlorogenic acid has the ability to scavenge alkyl peroxide free radicals and plays an important role in anti-tumor capabilities [4,5]. Flavonols can inhibit platelet aggregation, calcium activation, and tyrosine protein phosphorylation and effectively resist the occurrence of cardiovascular diseases [6]. Anthocyanins belong to a class of polyphenols stored in plant vacuoles with strong antioxidant, anti-cancer, and free radical scavenging capabilities, in addition to enhancing resistance, restoring vision, and preventing cardiovascular diseases [7].

Malus neidzwetzkyana (Dieck) Langenf. has become a research hotspot due to its high anthocyanin content and strong anti-oxidation, anti-cancer, and free radical scavenging potential. In addition, the flowers, leaves, fruits, and branches of *Malus neidzwetzkyana* (Dieck) Langenf. show different degrees of red, and it is considered to be an ornamental tree. The attractive color is popular with consumers [7–9]. In recent years, Shandong Agricultural University has selected and bred a series of new red flesh varieties with high flavonoids as its main line [10]. Qingdao Agricultural University bred a red flesh variety 'Daihong' [11]. The Research Institute of Pomology of Chinese Academy of Agricultural Sciences bred a red flesh variety 'Hongyun' [12]. Red flesh apple varieties such as 'JPP35', 'Weirouge', 'Baya Marise', and 'Redlove' have also been bred abroad [13,14].

The breeding parents of domestic red flesh varieties mostly come from *Malus neidzwetzkyana* (Dieck) Langenf. Two red flesh superior strains, "Hongxun 1" and "Hongxun 2", with excellent traits, were obtained through the local selection of red flesh apples. In the fruit ripening stage, red flesh apples mainly contain anthocyanins and flavanols, while non-red flesh apples mainly contain flavonols and flavanols. The contents of polyphenolic substances and the antioxidant capacity of red flesh apples are higher than those of non-red flesh apples. In the process of fruit development, except anthocyanin, the polyphenolic substances of red flesh and non-red flesh apples showed a downward trend in the early stage of development, and the change trends of different red flesh types were different. The anthocyanin content of 'Hongxun 1' was low in the early stage, then continued to increase, and significantly decreased before maturity. The initial content of "Xiahongrou" (*Malus neidzwetzkyana* (Dieck) Langenf.) was very high, then it continued to decline, and increased again at maturity [15]. It was found that there were differences in the anthocyanin accumulation patterns among different red flesh types. 'Maypole' always had a large amount of anthocyanin accumulation in the flesh during the whole development period of the fruit, while 'JPP35' only had anthocyanin accumulation at the mature stage [16]. A physical and chemical analysis of "Hongxun 1" cider found that the "Hongxun 1" produced in Tacheng had a rich flavor and red color after aging, the color gradually deepened during the aging process, and had better fermentation and aging habits [17].

There have been many reports on the polyphenol content and dynamic changes in red flesh apples, and many data have shown that *Malus neidzwetzkyana* (Dieck) Langenf. is a variant of the *Malus sieversii* (Led.) Roem. However, there are few studies on the comparison of the polyphenol contents and dynamics between red flesh apples and *Malus sieversii* (Led.) Roem. Polyphenols are a basis for apple classification, and whether red flesh apples are varieties of *Malus sieversii* (Led.) Roem can also be verified by polyphenol studies. In this study, the components and contents of polyphenols in the peel and pulp of 'Hongxun 2' (*Malus neidzwetzkyana* (Dieck) Langenf.) and 'Xinye 13-11' (*Malus sieversii* (Led.) Roem.) during the fruit growth period were determined and analyzed. The differences in the polyphenol components and contents between them were analyzed. The periods of 'Hongxun 2' suitable for extracting polyphenol components for processing were determined. The relationship between *Malus neidzwetzkyana* (Dieck) Langenf. and *Malus sieversii* (Led.) Roem. was studied.

2. Materials and Methods

2.1. Experimental Material

'Hongxun 2' and 'Xinye 13-11' with good growth were selected as experimental materials. The first samples were collected about 50 days after flowering (21 June 2023), and then samples were collected at 65, 80, 95, and 110 days after flowering. A total of five samples were collected from the outer crown of the tree with a uniform size. There were 30–50 ripe fruits free of pests, diseases, and mechanical damage in each collection, 10–15 per replicate, and 3 biological replicates were set. The samples were transported to the laboratory at room temperature, then the fruit cores were removed, the peel was sliced, and the flesh was cut into pieces. The peel and flesh were frozen in liquid nitrogen and stored in the refrigerator at −80 °C to be tested.

2.2. Methods

The extraction method for polyphenols referred to the method of Nie et al. and Wang et al. [18,19], which was slightly modified. The apple peel and pulp were frozen and ground into a powder, weighed at 5.0 g (fresh weight, FW), placed in a 50 mL centrifuge tube (BD Falcon®, Corning, New York, NY, USA) with 25 mL of 80% ethanol, shaken well, and placed away from light for 12 h, in an ultrasonic apparatus (SB 25-12 DTD, Ningbo, China). After 20 min of vibration, they were centrifuged for 5 min at 10,000 r/min (CF 16 RX, Hitachi, Japan), the supernatant was absorbed by a pipette carefully, and then discarded. The residue was mixed with 20 mL of 80% ethanol for repeated extraction, the supernatant was combined twice, and the volume of 80% ethanol was fixed to 50 mL. The ethanol was removed by the evaporation of 10 mL of the extract on a 40 °C rotary evaporator (R-215, Buchi, Switzerland). An OasisHLB solid-phase extraction column (Oasis®HLB, Waters, MA, USA) was activated with 10 mL of methanol and 10 mL of purified water. The solid-phase extraction column was washed with 5 mL of deionized water twice and the waste liquid was discarded. With 5 mL of methanol, the solid-phase extraction column was washed twice, and the filtrate was collected. The filtrate collected by the rotary evaporator was evaporated at 40 °C to nearly dry, the methanol volume was fixed to 5 mL, and the filtrate was filtered through a 0.22 μm nylon (Nylon66) organic-phase filter membrane (Jinteng, Tianjin, China) into a brown vial to be measured.

Polyphenols were detected using a UPLC-XeVo/TQ ultra-high-performance liquid chromatograph (UPLC-XEVO/TQ) with a PDA eλ detector (Waters, MA, USA), and the LC-10 ATvp high-performance liquid chromatograph (Shimadzu, Shimane, Japan). The columns were ACQUIITY UPLC® HSS T3 1.8 μm and XSelect® HSS T3 5 μm from the Waters Corporation.

The conditions for the UPLC-Xevo/TQ ultra-high-performance liquid chromatograph and ultra-performance liquid chromatography (UPLC) were as follows: the flow rate was 0.3 mL/min, the sample size was 2.0 μL, the column temperature was 40 °C, the wavelength scanning range was 200–600 nm, the quantitative detection wavelengths were 280 nm (flavanol), 320 nm (hydroxycinnamic acid), 360 nm (flavonol), the mobile phase A was 0.5% formic acid solution, and the mobile phase B was acetonitrile. Gradient dewashing, liquid B, 0% (0 min) → 10% (1 min) → 20% (10 min) → 25% (16 min) → 40% (18 min) → 100% (19 min), 20 min back to the initial state, balance for 5 min.

Dihydrochalcone polyphenols should be detected separately under UPLC conditions. The UPLC conditions were as follows: a flow rate of 0.3 mL·min^{-1}, a sample size of 2.0 μL, a column temperature of 40 °C, a quantitative detection wavelength of 280 nm, mobile phase A as 0.5% formic acid solution, and mobile phase B as acetonitrile. Gradient dewashing, B solution, 0% (0 min) → 8% (2 min) → 15% (10 min) → 23% (20 min) → 100% (20.5 min) → 100% (21.5 min), 22 min return to the initial state, balance for 5 min.

The conditions for liquid chromatography–mass spectrometry (LC-MS) were as follows: in the electrospray ionization (ESI) and multiple reaction monitoring (MRM) modes, the ion source temperature was 150 °C, the desolvent temperature was 450 °C, and the desolvent gas flow rate was 650 L/h. The flow rate of cone hole gas was 50 L/h, and the flow rate of collision gas (high purity argon) was 0.14 mL/min.

The LC-10 ATvp High-Performance Liquid Chromatography (High-Performance Liquid Chromatography, HPLC) conditions were as follows: the flow rate was 0.7 mL·min^{-1}, the sample size was 1 μL, the column temperature was 40 °C, and the wavelength was quantitatively detected at 520 nm (anthocyanins). The mobile phase A was 5% formic acid, and B was a 1:1 mixture of formic acid and acetonitrile. Gradient dewashing, solution B: 5% (0 min) →10% (10 min) →20% (30 min) →30% (40 min) →90% (40.5 min) →90% (44.5 min) →5% (50 min), return to the initial state, balance for 20 min.

2.3. Quantitative and Qualitative Measurements of Polyphenol Components

The polyphenolic compositions without standards were identified using LC-MS and quantified using UPLC. Polyphenol standards of catechin, chlorogenic acid, epicatechin, rutin, quercetin 3-O-xyloside, cyanidin 3-O-galactoside, cyanidin 3-O-glucoside, cyanidin 3-O-arabinoside, peonidin 3-O-galactoside, cyanidin 3-O-xyloside, and phlorizin were purchased from Sigma-Aldrich (St. Louis, MO, USA); procyanidin B1, procyanidin B2, procyanidin C1, quercetin 3-O-galactoside, quercetin 3-O-glucoside, quercetin 3-O-arabinoside, and quercetin3-O-rhamnoside were purchased from ChromaDex (Irvine, CA, USA). 4-*p*-coumarylquinic acid and 5-*p*-coumarylquinic acid were quantified by the chlorogenic acid standard. 3-hydroxyphloretin-xylglucoside, 3-hydroxyphloretin-glucoside, phloridin-hexose-hexose, and phloretin xyloglucoside were quantified by phloridzin.

2.4. Statistical Analysis

Microsoft Office Excel 2021 was used for data analysis, and SPSS 27 was used for single-factor ANOVA tests to calculate significance, mean values, and standard deviations. Origin 2021 was used to plot the dynamic changes in the total polyphenol and polyphenol component contents. The total polyphenol content was calculated as the sum of the contents of each polyphenol component, and the contents of five types of polyphenols were calculated as the sum of the contents of each type of polyphenol component.

3. Results

3.1. Dynamic Changes in Total Polyphenol During Fruit Development

The total polyphenol contents of 'Hongxun No. 2' and 'Xinye 13-11' were different during development. The content of total polyphenols in 'Xinye 13-11' was higher than that in 'Hongxun 2' in all periods, in both the peel and pulp. From 50 days to 110 days after flowering, with the gradual development and ripening of fruits, the overall content showed a decreasing trend, with the reduction becoming getting smaller and smaller. In all periods, the total polyphenol contents of 'Hongxun 2' and 'Xinye 13-11' were higher in the peel than in the pulp.

During the development stage from 50 to 110 days after flowering, the total polyphenol content decreased by 66.20% and 63.15% in the peel and pulp of 'Hongxun 2', while the total polyphenol content decreased by 47.74% and 59.78% in the peel and pulp of 'Xinye 13-11' (Figure 1).

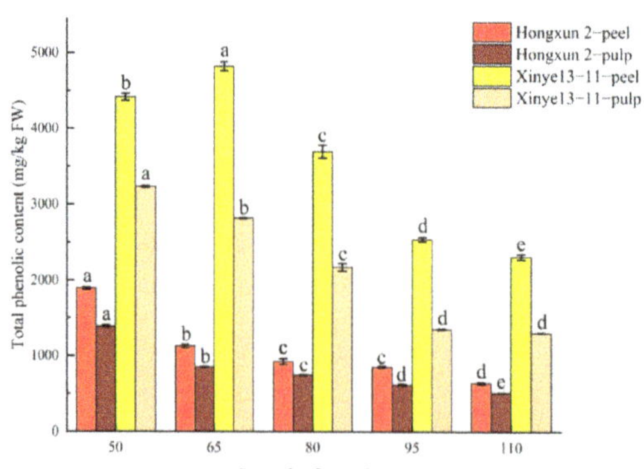

Figure 1. Total polyphenolic contents during fruit development of 'Hongxun 2' and 'Xinye 13-11'. Note: Different letters indicate differences at $p < 0.05$ levels.

3.2. Dynamic Changes in Five Types of Polyphenols During Fruit Development

The flavanol contents of 'Xinye 13-11' were significantly higher than those of 'Hongxun 2' in the peel and pulp, and the flavanol content reached 3203.45 mg/kg in the peel of 'Xinye 13-11' at 65 days after flowering, and then gradually decreased. The flavanol contents were 2762.99 mg/kg and 394.00 mg/kg in the pulp of 'Xinye 13-11' and the peel of 'Hongxun 2' at 50 days after flowering, respectively, and gradually decreased after flowering. The flavanol content of the pulp was 62.80 mg/kg in 'Hongxun 2' at 50 days after flowering, which decreased from 50 days to 65 days after flowering and then began to rise again, reaching 61.31 mg/kg at 95 days after flowering, and then decreased again (Figure 2A).

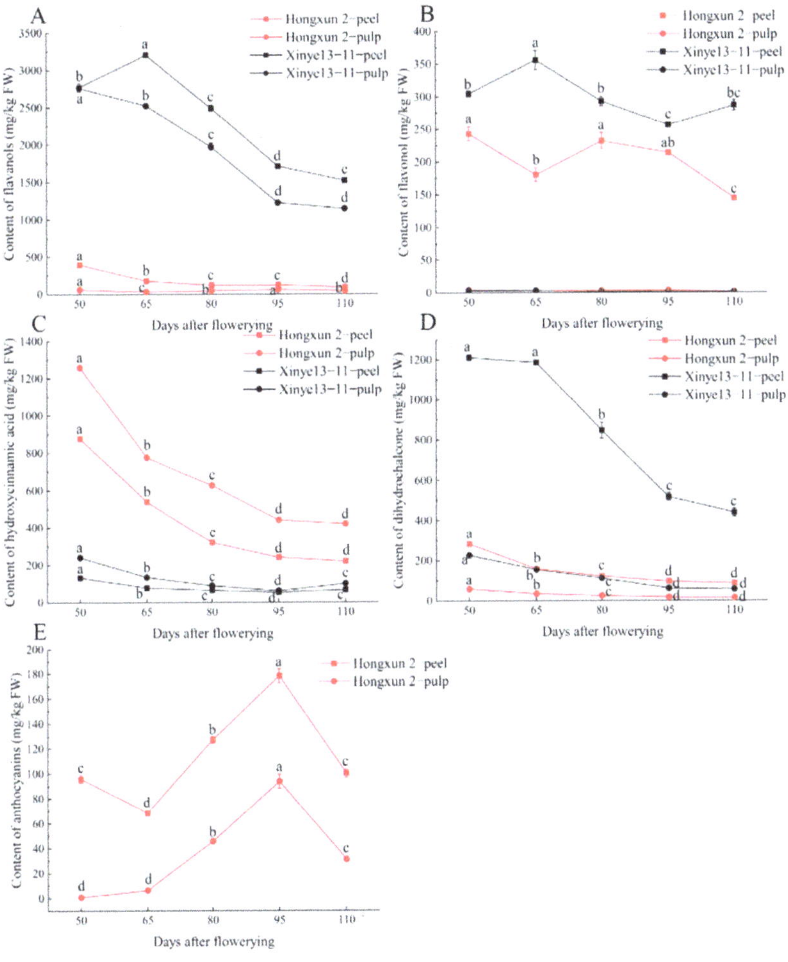

Figure 2. Changes in the contents of five types of polyphenolic substances at different developmental stages in 'Hongxun 2' and 'Xinye 13-11'. (**A**) Changes in the contents of flavanol at different developmental stages in 'Hongxun 2' and 'Xinye 13-11'. (**B**) Changes in the contents of flavonol at different developmental stages in 'Hongxun 2' and 'Xinye 13-11'. (**C**) Changes in the contents of hydroxycinnamic acid at different developmental stages in 'Hongxun 2' and 'Xinye 13-11'. (**D**) Changes in the contents of dihydrochalcone at different developmental stages in 'Hongxun 2' and 'Xinye 13-11'. (**E**) Changes in the contents of anthocyanins at different developmental stages in 'Hongxun 2' and 'Xinye 13-11'. Note: Different letters indicate differences at $p < 0.05$ levels.

The flavonol content was lower in the flesh of 'Hongxun 2' and 'Xinye 13-11', but higher in the peel, and its variation trends were different. The flavonol content was 243.24 mg/kg in the peel of 'Hongxun 2' at day 50 after flowering, decreased at day 65, then increased, and decreased again at 80 days after flowering. The flavonol content was 355.82 mg/kg in the peel of 'Xinye 13-11' at 75 days after flowering, the content increased from 50 days to 65 days, then gradually decreased, and gradually increased at 95 days after flowering (Figure 2B).

The content of hydroxycinnamic acid in 'Hongxun 2' was significantly higher than that of 'Xinye 13-11' in the peel and pulp. The contents of hydroxycinnamic acid in 'Hongxun-2' and 'Xinye 13-11' were 877.36 mg/kg, 1258.80 mg/kg, 132.55 mg/kg, and 239.96 mg/kg in the peel and pulp, respectively, 50 days after flowering. Except for in the pulp of 'Xinye 13-11', the content of 'Xinye 13-11' decreased gradually with development, and the content of 'Xinye 13-11' increased significantly from 95 days to 110 days after flowering (Figure 2C).

The dihydrochalcone content of 'Hongxun 2' was higher than that of 'Xinye 13-11' in the pulp, and it showed a downward trend, tending to be stable after 95 days of flowering. The contents of dihydrochalcone were 283.13 mg/kg, 58.74 mg/kg, 1210.43 mg/kg, and 226.13 mg/kg in the peel and pulp of 'Hongxun 2' and 'Xinye 13-11' at 50 days after flowering, respectively (Figure 2D).

The dynamic change trend of the anthocyanin content of 'Hongxun 2' was similar in the peel and pulp, and the highest contents were 178.51 mg/kg and 93.93 mg/kg at 95 days after flowering for the peel and pulp, respectively. However, the anthocyanin content decreased, to a certain extent, in the peel of 'Hongxun 2' at 65 days after flowering, then gradually increased, and reached its highest at 95 days after flowering (Figure 2E).

3.3. The Difference in Polyphenol Components and Contents During Fruit Development

There were significant differences in the contents of polyphenol components in 'Hongxun 2' in the peel and pulp during the development of fruits (Tables 1 and 2). In the peel, the content of polyphenols decreased significantly from 50 to 65 days after flowering. The contents of 29 polyphenol components, including procyanidin B1, catechin, procyanidin B2, epicatechin, etc., were significantly decreased, while the contents of kaempferol 3-O-galactoside, and 5-*p*-coumaryl quinic acid showed no significant difference, and paeoniflorin 3-o-galactoside was not detected during this period. The contents of flavonols such as 3-hydroxyphloretin-xyloglucose, rutin, and quercetin 3-O-galactoside and anthocyanins such as cyanidin 3-O-galactoside and cyanidin 7-O-arabinoside increased significantly at this stage from 65 to 80 days after flowering. The other components were significantly reduced. From 80 to 95 days after flowering, the contents of most components tended to be stable without significant differences. The contents of rutin, quercetin 3-O-glucoside, and anthocyanin components, except cyanidin 7-O-arabinoside, increased significantly at this stage. The contents of quercetin 3-O-xyloside, quercetin 3-O-rhamnoside, chlorogenic acid, 3-hydroxyphloretin-xyloglucose, and phloretin-xyloglucose were significantly reduced at this stage. Kaempferol 3-O-rhamnoside and 4-*p*-coumarylquinic acid were not detected from 95 to 110 days after flowering. The contents of catechin, procyanidin B2, procyanidin C1, kaempferol 3-O-galactoside, 3-hydroxyphloretin-xyloglucose, phloridin-hexose-hexose, 3-hydroxyphloretin-glucose, phloretin-xyloglucose, and phloretin showed no significant difference in the maturation stage, while the contents of other components were significantly reduced (Table 1).

Table 1. Polyphenol components and content in the peel of 'Hongxun 2' during fruit development.

Polyphenol Components	Contents of Polyphenol Components (mg·kg^{-1} FW)				
	50 d	**65 d**	**80 d**	**95 d**	**110 d**
PROB1	17.93 ± 0.39 a	7 ± 0.51 b	4.66 ± 0.07 c	4.97 ± 0.14 c	3.18 ± 0.3 d
CATE	15.97 ± 0.51 a	3.63 ± 0.08 b	2.01 ± 0.1 c	2.02 ± 0.03 c	1.35 ± 0.07 c
PROB2	121.98 ± 4.34 a	65.47 ± 3.58 b	49.81 ± 3.31 c	52.72 ± 0.6 c	41.25 ± 0.83 c
EPI	153.08 ± 2.6 a	60.11 ± 1.93 b	35.03 ± 1.91 c	33.42 ± 0.44 c	23.42 ± 0.79 d
PROC1	85.03 ± 8.16 a	42.21 ± 6.17 b	30.8 ± 1.28 bc	28.67 ± 0.58 bc	20.86 ± 0.46 c
RUTIN	0.35 ± 0.01 bc	0.2 ± 0.02 c	0.43 ± 0.03 b	0.46 ± 0.04 a	0.32 ± 0.01 bc
QUEGA	79.46 ± 5.29 ab	64.7 ± 3.75 c	100.33 ± 7.19 a	98.97 ± 3.04 a	67.92 ± 2.82 b
QUEGL	7.05 ± 0.23 c	5.6 ± 0.2 d	9.47 ± 0.4 b	12.31 ± 0.17 a	8.71 ± 0.13 b
QUEXY	34.41 ± 1.26 a	24.43 ± 0.42 c	29.12 ± 1.58 b	25.2 ± 0.07 c	15.98 ± 0.18 d
QUEPY	3.27 ± 0.13 ab	2.56 ± 0.13 c	3.6 ± 0.19 a	3.78 ± 0.05 a	2.94 ± 0.1 bc
QUEFU	84.06 ± 3.17 a	59.64 ± 6.39 b	59.77 ± 2.41 b	49.21 ± 0.62 b	30.03 ± 0.37 c
QUERH	33.25 ± 0.56 a	22.55 ± 0.16 c	28.32 ± 1.11 b	23.24 ± 0.44 c	18.08 ± 0.24 d
KAEGA	0.26 ± 0.02 a	0.19 ± 0.03 a	0.29 ± 0.03 a	0.28 ± 0.02 a	0.2 ± 0.03 a
KAEGL	nd	nd	nd	nd	nd
KAEFU	0.9 ± 0.01 a	0.43 ± 0.01 c	0.5 ± 0.02 b	0.29 ± 0.01 d	0.19 ± 0.02 e
KAERH	0.23 ± 0.01 a	0.13 ± 0.01 b	0.17 ± 0.03 ab	0.13 ± 0.02 b	nd
CHLAC	850.86 ± 11.61 a	525.93 ± 8.02 b	312.65 ± 12.34 c	235.76 ± 1.96 d	215.14 ± 4.88 d
4COUAC	19.87 ± 0.49 a	6.81 ± 0.4 b	3.06 ± 0.17 c	2.24 ± 0.28 c	nd
5COUAC	6.63 ± 0.21 a	7.09 ± 0.11 a	4.52 ± 0.28 b	2.82 ± 0.16 b	4.02 ± 0.17 c
HYDXY	5.61 ± 0.3 a	4.55 ± 0.08 b	5.3 ± 0.25 ab	3.48 ± 0.07 c	3.49 ± 0.37 c
PHLHE	1.07 ± 0.06 a	0.77 ± 0.01 b	0.59 ± 0.06 bc	0.44 ± 0.06 c	0.5 ± 0.03 c
HYDGL	12.45 ± 0.32 a	6.19 ± 0.38 b	5.21 ± 0.47 bc	4.12 ± 0.18 cd	3.01 ± 0.21 d
PHLXY	110.21 ± 2.09 a	74.68 ± 3.27 b	65.48 ± 0.33 c	52.29 ± 1.61 d	52.87 ± 2.41 d
PHLPE	2.33 ± 0.19 a	1.77 ± 0.17 b	1.47 ± 0.14 b	1.2 ± 0.07 b	1.3 ± 0.07 b
PHLPE1	5.81 ± 0.41 a	4.21 ± 0.31 b	3.4 ± 0.51 b	2.99 ± 0.26 b	2.88 ± 0.18 b
PHLZI	145.66 ± 6.44 a	66.64 ± 1.57 b	40.63 ± 4.22 c	31.18 ± 1.71 cd	22.12 ± 2.37 d
CYAGA	72.45 ± 2.24 d	53.89 ± 1.65 e	106.6 ± 2.36 b	155.08 ± 4.83 a	87.23 ± 2.72 c
CYAGL	0.38 ± 0.02 c	0.28 ± 0.02 c	0.75 ± 0.05 b	1.11 ± 0.1 a	0.61 ± 0.01 b
CYA3AR	1.41 ± 0.05 c	0.93 ± 0.03 d	1.99 ± 0.01 b	3.2 ± 0.12 a	1.88 ± 0.11 b
PEOGA	nd	nd	0.17 ± 0.01 c	0.37 ± 0.02 a	0.29 ± 0.03 b
CYA7AR	10.51 ± 0.32 a	5.9 ± 0.16 b	6.72 ± 0.18 b	6.02 ± 0.21 b	3.73 ± 0.19 c
CYAXY	11.34 ± 0.4 b	7.46 ± 0.2 c	10.97 ± 0.24 b	12.73 ± 0.52 a	6.57 ± 0.18 c

Note: PROB1: procyanidin B1; CATE: catechin; PROB2: procyanidin B2; EPI, epicatechin; PROC1: procyanidin C1; RUTIN: rutin; QUEGA: quercetin 3-O-galactoside; QUEGL: quercetin 3-O-glucoside; QUEXY: quercetin 3-O-xyloside; QUEPY: quercetin 3-O-arabinopyranoside; QUEFU: quercetin 3-O-arabinofuranoside; QUERH: Quercetin 3-O-rhamnoside; KAEGA: kaempferol 3-O-galactoside; KAEGL: kaempferol 3-O-glucoside; KAEFU: kaempferol 3-O-arabinofuranoside; KAERH: kaempferol 3-O-rhamnoside; CHLAC: chlorogenic acid; 4COUAC: 4-*p*-coumaryl quinic acid; 5COUAC: 5-*p*-coumaryl quinic acid; HYDGL: 3-hydroxyphloridin-glucose; HYDXY: 3-hydroxyphloridin-xyloglucose; PHLHE: phloridin-hexose-hexose; PHLXY: phloridin-xyloglucose; PHLPE: phloridin-pentose-hexose; PHLPE1: phloridin-pentose-hexose 1; PHLZI, phloridin; CYAGA: cyanidin 3-O-galactoside; CYAGL: cyanidin 3-O-glucoside; CYA3AR: cyanidin 3-O-arabinoside; PEOGA: Peonidin 3-O-galactoside; CYA7AR: cyanidin 7-O-arabinoside; CYAXY: cyanidin 3-O-xyloside; and nd: No detected. Different letters in the same line indicate differences at the $p < 0.05$ level, as shown in Tables 2–4.

Table 2. Polyphenol components and content in the pulp of 'Hongxun 2' during fruit development.

Polyphenol Components	Contents of Polyphenol Components (mg·kg^{-1} FW)				
	50 d	**65 d**	**80 d**	**95 d**	**110 d**
PROB1	5.71 ± 0.15 a	3.05 ± 0.14 b	2.39 ± 0.2 b	2.86 ± 0.15 b	2.63 ± 0.16 b
CATE	1.66 ± 0.04 a	1.13 ± 0.06 b	1.17 ± 0.01 b	1.26 ± 0.08 b	1.14 ± 0.04 b
PROB2	25.13 ± 1.24 a	10.73 ± 0.36 c	15.95 ± 1.27 b	25.42 ± 0.33 a	19.65 ± 1.28 b
EPI	14.16 ± 0.48 b	7.96 ± 0.24 c	9.58 ± 0.23 c	16.45 ± 0.66 a	9.25 ± 0.42 c
PROC1	16.14 ± 1.29 a	7.57 ± 0.09 b	16.62 ± 1.16 a	15.31 ± 0.45 a	10.41 ± 0.66 b
RUTIN	nd	nd	nd	nd	nd
QUEGA	1.44 ± 0.1 a	0.67 ± 0.12 c	1.01 ± 0.08 b	1.37 ± 0 a	nd
QUEGL	nd	nd	nd	nd	nd

Table 2. *Cont.*

Polyphenol Components	Contents of Polyphenol Components (mg·kg^{-1} FW)				
	50 d	65 d	80 d	95 d	110 d
QUEXY	nd	nd	nd	nd	nd
QUEPY	nd	nd	nd	nd	nd
QUEFU	nd	nd	nd	nd	nd
QUERH	2.44 ± 0.36 a	1.68 ± 0.17 ab	1.89 ± 0.32 ab	1.76 ± 0.16 ab	1.1 ± 0.12 b
KAEGA	nd	nd	nd	nd	nd
KAEGL	nd	nd	nd	nd	nd
KAEFU	nd	nd	nd	nd	nd
KAERH	nd	nd	nd	nd	nd
CHLAC	1222.6 ± 11.37 a	758.14 ± 6.47 b	609.74 ± 4.96 c	431.51 ± 7.41 d	412.55 ± 0.61 d
4COUAC	22 ± 0.37 a	8.41 ± 0.46 b	5.88 ± 0.05 c	3.89 ± 0.12 d	2.12 ± 0.14 e
5COUAC	14.19 ± 0.45 a	9.91 ± 0.43 c	11.19 ± 0.32 b	5.6 ± 0.07 d	5.78 ± 0.15 d
HYDXY	2.44 ± 0.2 a	2.03 ± 0.17 a	1.38 ± 0.08 b	1.2 ± 0.06 b	1.21 ± 0.06 b
PHLHE	1.24 ± 0.09 a	1.15 ± 0.11 a	1.26 ± 0.05 a	1.14 ± 0.06 a	1.2 ± 0.08 a
HYDGL	2.44 ± 0.1 b	3.62 ± 0.13 a	2.03 ± 0.2 b	1.1 ± 0.09 c	0.95 ± 0.02 c
PHLXY	19.35 ± 0.59 a	10.97 ± 0.32 b	7.82 ± 0.21 c	5.67 ± 0.22 d	5.43 ± 0.44 d
PHLPE	2.51 ± 0.03 a	1.69 ± 0.13 b	1.66 ± 0.06 b	nd	nd
PHLPE1	3.69 ± 0.51 a	2.21 ± 0.03 b	2.17 ± 0.21 b	1.86 ± 0.2 b	nd
PHLZI	27.07 ± 0.93 a	13.79 ± 0.77 b	8.48 ± 1.05 c	6.56 ± 0.27 c	5.88 ± 0.38 c
CYAGA	0.59 ± 0.05 d	4.76 ± 0.32 d	38.07 ± 1.38 b	81.12 ± 4.87 a	26.65 ± 1.06 c
CYAGL	nd	nd	0.31 ± 0.03 b	0.62 ± 0.03 a	0.21 ± 0.01 c
CYA3AR	nd	nd	0.62 ± 0.02 b	1.48 ± 0.11 a	0.34 ± 0.02 c
PEOGA	nd	nd	nd	nd	nd
CYA7AR	0.16 ± 0.02 d	0.66 ± 0.06 cd	2.13 ± 0.13 b	2.83 ± 0.23 a	1.07 ± 0.11 c
CYAXY	0.18 ± 0.02 d	1.07 ± 0.14 d	4.62 ± 0.13 b	7.87 ± 0.45 a	2.79 ± 0.16 c

nd: No detected. Different letters in the same line indicate differences at the $p < 0.05$ level.

Table 3. Polyphenol components and contents in the peel of 'Xinye 13-11' during fruit development.

Polyphenol Components	Contents of Polyphenol Components (mg·kg^{-1} FW)				
	50 d	65 d	80 d	95 d	110 d
PROB1	116.3 ± 6.62 c	214.64 ± 1 a	187.94 ± 7.03 b	131.94 ± 6.25 c	118.99 ± 4.12 c
CATE	386.16 ± 7.5 a	321.57 ± 1.4 b	187.47 ± 3.59 c	81.97 ± 2.05 d	61.03 ± 1.52 e
PROB2	556.65 ± 18.8 c	811.89 ± 3.62 a	729.51 ± 9.97 b	600.97 ± 21.9 c	544.93 ± 6.87 c
EPI	928.6 ± 7.37 a	879.27 ± 7.84 b	692.33 ± 15.22 c	452.82 ± 14.47 d	382.88 ± 7.05 e
PROC1	782.53 ± 27.12 b	976.09 ± 34.77 a	695.34 ± 6.67 c	444.54 ± 6.93 d	410.21 ± 13.94 d
RUTIN	5.25 ± 0.31 a	4.37 ± 0.56 a	4.08 ± 0.15 a	4.4 ± 0.3 a	4.97 ± 0.48 a
QUEGA	73.62 ± 1.28 c	90.49 ± 2.85 b	82.49 ± 3.3 bc	84.92 ± 0.6 b	113.7 ± 2.31 a
QUEGL	37.88 ± 1.01 c	49.5 ± 1.91 b	44.5 ± 1.69 b	48.3 ± 0.07 b	56.07 ± 1.11 a
QUEXY	26.08 ± 0.99 a	28.91 ± 1.73 a	21.97 ± 0.07 b	16.97 ± 0.17 c	16.28 ± 0.65 c
QUEPY	2.22 ± 0.06 ab	2.64 ± 0.22 ab	2.32 ± 0.06 ab	1.97 ± 0.07 c	2.88 ± 0.27 a
QUEFU	91.91 ± 1.84 a	100.66 ± 3.18 a	75.63 ± 2.44 b	54.14 ± 0.2 c	52.2 ± 2.94 c
QUERH	64.49 ± 1.56 b	76.78 ± 4.07 a	59.84 ± 0.22 b	44.33 ± 1.04 c	38.78 ± 1.22 c
KAEGA	0.25 ± 0.03 ab	0.24 ± 0.01 ab	0.22 ± 0.01 ab	0.19 ± 0.01 b	0.28 ± 0.02 a
KAEGL	0.82 ± 0.02 a	0.64 ± 0.03 b	0.55 ± 0.01 c	0.5 ± 0.02 c	0.53 ± 0.03 c
KAEFU	1.17 ± 0.02 a	0.95 ± 0.03 b	0.66 ± 0.01 c	0.4 ± 0.02 d	0.33 ± 0.02 d
KAERH	0.49 ± 0.02 a	0.49 ± 0.02 a	0.34 ± 0.01 b	0.24 ± 0.02 c	0.21 ± 0.02 c
CHLAC	121.16 ± 0.88 a	73.93 ± 0.66 b	62.03 ± 0.44 d	51.92 ± 1.44 e	65.92 ± 1.21 c
4COUAC	11.38 ± 0.48 a	3.76 ± 0.09 b	2.21 ± 0.21 c	1.52 ± 0.2 c	nd
5COUAC	nd	nd	nd	nd	nd
HYDXY	9.58 ± 0.57 a	9.02 ± 0.57 a	9.72 ± 1.08 a	7.89 ± 0.08 a	4.94 ± 0.39 b
PHLHE	2.85 ± 0.12 a	2.93 ± 0.14 a	2.51 ± 0.12 a	1.78 ± 0.1 b	1.63 ± 0.09 b
HYDGL	43.54 ± 1.34 a	45.7 ± 0.78 a	36.69 ± 2.06 b	28.61 ± 1.03 c	18.47 ± 0.75 d
PHLXY	252.11 ± 4.1 b	304.63 ± 10.25 a	228.52 ± 7.76 b	162.25 ± 4.33 c	146.17 ± 6.67 c
PHLPE	6.54 ± 0.95 a	6.97 ± 0.12 a	4.56 ± 0.4 b	3.25 ± 0.2 b	3.29 ± 0.36 b
PHLPE1	11.36 ± 0.95 ab	12.52 ± 0.58 a	9.97 ± 0.6 b	5.64 ± 0.31 c	5.53 ± 0.26 c

Table 3. *Cont.*

Polyphenol Components	Contents of Polyphenol Components (mg·kg^{-1} FW)				
	50 d	65 d	80 d	95 d	110 d
PHLZI	884.45 ± 11.54 a	803.03 ± 9.03 b	554.24 ± 28.07 c	306.62 ± 15.66 d	258.35 ± 12.73 d
CYAGA	nd	nd	nd	nd	nd
CYAGL	nd	nd	nd	nd	nd
CYA3AR	nd	nd	nd	nd	nd
PEOGA	nd	nd	nd	nd	nd
CYA7AR	nd	nd	nd	nd	nd
CYAXY	nd	nd	nd	nd	nd

nd: No detected. Different letters in the same line indicate differences at the $p < 0.05$ level.

Table 4. Polyphenol components and contents in the pulp of 'Xinye 13-11' during fruit development.

Polyphenol Components	Contents of Polyphenol Components (mg·kg^{-1} FW)				
	50 d	65 d	80 d	95 d	110 d
PROB1	99.72 ± 2.14 c	142.66 ± 6.05 a	122.93 ± 5.74 b	76.69 ± 2.16 d	74.72 ± 0.22 d
CATE	241.76 ± 12.1 a	181.07 ± 4.63 b	103.81 ± 4.66 c	45.79 ± 1.02 d	38.18 ± 0.21 d
PROB2	792.75 ± 9.71 a	736.77 ± 10.11 b	659.1 ± 23.05 c	452.5 ± 9.8 d	447.7 ± 2.28 d
EPI	870.56 ± 6.71 a	778.33 ± 7.6 b	580.98 ± 11.43 c	357.51 ± 1.46 d	316.03 ± 1.36 e
PROC1	758.21 ± 2.05 a	686.98 ± 4.77 b	505.35 ± 11.95 c	289.76 ± 0.44 d	267.28 ± 1.12 d
RUTIN	nd	nd	nd	nd	nd
QUEGA	nd	nd	nd	nd	nd
QUEGL	nd	nd	nd	nd	nd
QUEXY	nd	nd	nd	nd	nd
QUEPY	nd	nd	nd	nd	nd
QUEFU	1.37 ± 0.12 a	0.98 ± 0.13 b	nd	nd	nd
QUERH	1.64 ± 0.2 a	1.66 ± 0.28 a	1.41 ± 0.17 ab	0.81 ± 0.1 b	0.79 ± 0.06 b
KAEGA	nd	nd	nd	nd	nd
KAEGL	nd	nd	nd	nd	nd
KAEFU	nd	nd	nd	nd	nd
KAERH	nd	nd	nd	nd	nd
CHLAC	223.7 ± 0.51 a	129.53 ± 1.77 b	87.16 ± 2.24 d	59.32 ± 0.66 e	98.11 ± 1.63 c
4COUAC	16.26 ± 0.1 a	3.99 ± 0.1 b	2.16 ± 0.12 c	1.96 ± 0.06 c	0.8 ± 0.19 d
5COUAC	nd	nd	nd	nd	nd
HYDXY	nd	nd	6.15 ± 0.21 a	5.81 ± 0.27 a	4.29 ± 0.14 b
PHLHE	8.88 ± 0.28 a	7.5 ± 0.21 b	4.81 ± 0.25 c	3.1 ± 0.06 d	2.9 ± 0.09 d
HYDGL	4.86 ± 0.11 b	6.73 ± 0.08 a	4.67 ± 0.12 b	2.38 ± 0.08 c	1.94 ± 0.24 c
PHLXY	49.86 ± 0.41 a	42.96 ± 0.33 b	31.86 ± 0.91 c	18.64 ± 0.29 d	16.71 ± 0.53 d
PHLPE	2.36 ± 0.52 a	2.51 ± 0.04 a	2.32 ± 0.33 a	nd	nd
PHLPE1	16.15 ± 0.94 a	12.79 ± 0.43 b	9.13 ± 0.65 c	4.65 ± 0.39 d	4.42 ± 0.26 d
PHLZI	144.03 ± 3.38 a	81.95 ± 1.35 b	52.34 ± 1.44 c	26.6 ± 0.73 d	26.16 ± 0.2 d
CYAGA	nd	nd	nd	nd	nd
CYAGL	nd	nd	nd	nd	nd
CYA3AR	nd	nd	nd	nd	nd
PEOGA	nd	nd	nd	nd	nd
CYA7AR	nd	nd	nd	nd	nd
CYAXY	nd	nd	nd	nd	nd

nd: No detected. Different letters in the same line indicate differences at the $p < 0.05$ level.

In the pulp of 'Hongxun 2', rutin, quercetin 3-O-glucoside, quercetin 3-O-xyloside, quercetin 3-O-arabinopyranoside, quercetin 3-O-arabinofuranoside, kaempferol 3-O-galactoside, kaempferol 3-O-glucoside, kaempferol 3-O-arabinoside, kaempferol 3- O-glucoside, kaempferol 3-O-arabinoside, kaempferol 3-O-rharnoside, and paeoniflorin 3-O-galactoside were not detected (Table 2). From 50 days to 65 days after flowering, the contents of 3-hydroxyphloretin-glucose were significantly increased, while the contents of 3-hydroxyphloretin-xyloglucose, phloridin-hexose-hexose, quercetin 3-O-rhamnoside, and anthocyanin were not significantly different, and the contents of other polyphenol components were significantly decreased. Cyanidin 3-O-xyloside and entaurin 7-O-arabinoside were not detected at this stage. From 65 to 80 days after flowering, the contents of procyanidin B2, procyanidin C1, quercetin 3-O-galactoside, 5-*p*-coumaryl quinic acid, and anthocyanin polyphenols were significantly increased. The contents of chlorogenic acid, 4-*p*-coumaryl quinic acid, 3-hydroxyphloretin-xyloglucose, 3-hydroxyphloretin-glucose, phloretin-xyloglucose, and phloretin were significantly decreased, while the contents of other polyphenol components saw no significant difference. From 80 to 95 days after flowering, the contents of 3-hydroxyphloretin-glucose, phloretin-xyloglucose, and hydroxy-cinnamic acid polyphenols decreased significantly, while the contents of procyanidin B2, epicatechin, quercetin 3-O-galactoside, and anthocyanin increased significantly, and the contents of other polyphenols saw no significant difference. Phloretin-pentose-hexose was not detected at 95 days after flowering. The contents of procyanidin B2, epicatechin, procyanidin C1, 4-*p*-coumaryl quinic acid, phloretin-pentose hexose 1, and anthocyanin polyphenols were significantly decreased from 95 days to 110 days after flowering, while the contents of other polyphenol components were not significantly different. Quercetin 3-O-galactoside was not detected at 110 days after flowering (Table 2).

'Xinye 13-11' is a green pulp apple, so anthocyanins were not detected at any time, and 5-*p*-coumaryl quinic acid was not detected either (Tables 3 and 4). In the peel of 'Xinye 13-11', the contents of procyanidin B1, procyanidin B2, procyanidin C1, quercetin 3-O-galactoside, quercetin 3-O-glucoside, quercetin 3-O-rhamnoside, and phloretin-xyloglucose were significantly increased from 50 days to 65 days after flowering. The contents of catechin, epicatechin, kaempferol 3-O-glucoside, kaempferol 3-o-arabinofuranoside, chlorogenic acid, 4-*p*-coumaryl quinic acid, and phloridzin were significantly decreased, while the other polyphenol components saw no significant difference. From 65 to 80 days after flowering, the contents of quercetin 3-O-galactoside, quercetin 3-O-glucoside, quercetin 3-O-arabopyranoside, kaempferol 3-O-galactoside, 3-hydroxyphloretin-xyloglucose, and phloretin-hexosaccharide saw no significant difference, while the contents of other polyphenol components were significantly decreased. From 80 to 95 days after flowering, the contents of quercetin 3-O-galactoside, quercetin 3-O-glucoside, kaempferol 3-O-galactoside, kaempferol 3-O-glucoside, 4-*p*-coumaryl quinic acid, 3-hydroxyphloretin-xyloglucose, and phloretin-pentose hexose saw no significant difference, while the contents of other polyphenol components were significantly decreased. From 95 days to 110 days after flowering, the quercetin 3-O-galactoside, quercetin 3-O-glucoside, quercetin 3-O-arabopyranoside, kaempferol 3-O-galactoside, and chlorogenic acid contents increased significantly, while the catechin, epicatechin, 3-hydroxyphloretin-xyloglucose, and phloretin-xyloglucose contents decreased significantly. There were no significant differences in other polyphenol components, and 4-*p*-coumaryl quinic acid was not detected at 110 days after flowering (Table 3).

In the pulp of 'Xinye 13-11', rutin, quercetin 3-O-galactoside, quercetin 3-O-glucoside, quercetin 3-O-xyloside, quercetin 3-O-xyloside, quercetin 3-O-arabopyranoside, kaempferol 3-O-galactoside, kaempferol 3-O-glucoside, kaempferol 3-o-arabinoside, kaempferol 3-O-rhamnoside, and 5-*p*-coumarylquinic acid were not detected (Table 4). From 50 days to 65 days after flowering, the contents of procyanidin B1 and 3-hydroxyphloretin-glucose were significantly increased, while the contents of quercetin 3-O-rhamnoside and phloretin-pentose hexose were not significantly different. The contents of other polyphenol components were significantly decreased, and 3-hydroxyphloretin-xyloglucose was not detected. There were no significant differences in the contents of quercetin 3-O-rhamnoside and phloretin-pentose hexose from 65 days to 80 days after flowering, 3-hydroxyphloretin-xyloglucose was detected at 80 days after flowering, and the contents of other polyphenol components were significantly reduced. From 80 to 95 days after flowering, the quercetin 3-O-rhamnoside, 4-*p*-coumaryl quinic acid, and 3-hydroxyphloretin-xyloglucose contents were not significantly different, phloretin-pentose hexose was not detected at 95 days after flowering, and the contents of other polyphenol components were significantly decreased. From 95 days to 110 days after flowering, the contents of all polyphenol components tended to be stable, only the chlorogenic acid content increased significantly, and the contents of epicatechin, 4-*p*-coumaryl quinic acid, and 3-hydroxyphloretin-xyloglucose decreased significantly, while the contents of other polyphenol components showed no significant differences (Table 4).

3.4. Dynamic Changes in Main Polyphenol Components During Fruit Development

From 50 days to 110 days after flowering, the fruits developed and matured gradually, and the main polyphenol components and contents of the peel and pulp saw significant changes (Figure 3). The content of procyanidin B2 reached 121.98 mg/kg in the peel of 'Hongxun 2' at 50 days after flowering, then gradually decreased, and slightly increased at 95 days after flowering, but this difference was not significant. In peel, the content of procyanidin B2 decreased gradually at first, reached its highest value of 25.42 mg/kg at 95 days after flowering, and then decreased again. The content of procyanidin B2 in the peel of 'Xinye 13-11' was the highest at 65 days after flowering, reaching 811.89 mg/kg, and then gradually decreased, but in the pulp, its content was 792.75 mg/kg at 50 days after flowering, and then gradually decreased (Figure 3A).

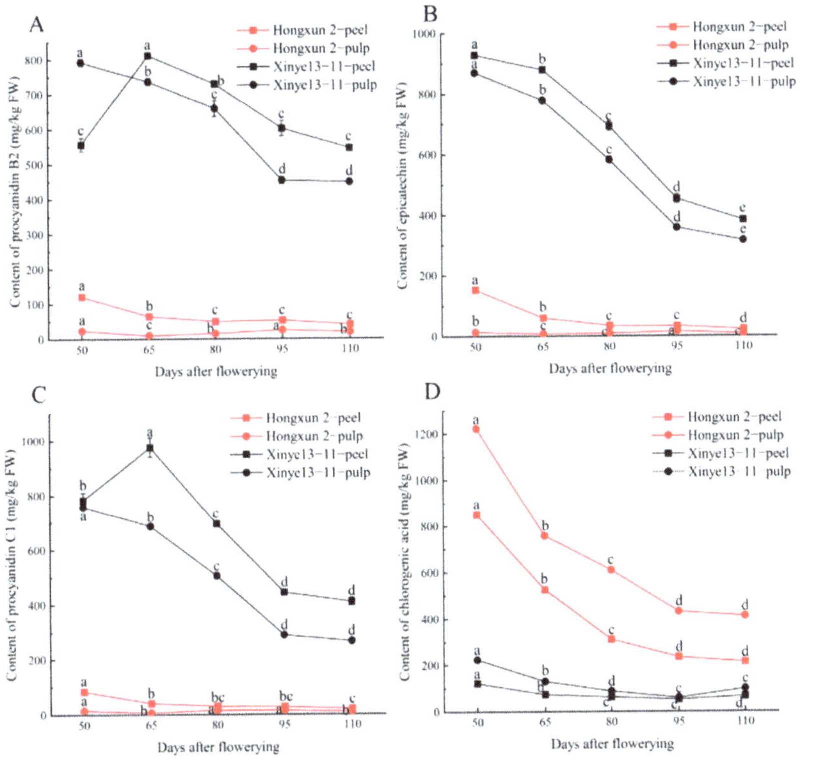

Figure 3. Dynamic changes in the contents of major polyphenol components during fruit development. (**A**) Dynamic changes in the contents of procyanidin B2 during fruit development. (**B**) Dynamic changes in the contents of epicatechin during fruit development. (**C**) Dynamic changes in the contents of procyanidin C1 during fruit development. (**D**) Dynamic changes in the contents of chlorogenic acid during fruit development. Note: Different letters indicate differences at $p < 0.05$ levels.

The dynamic changes in epicatechin content in the peel and pulp of 'Hongxun2' and 'Xinye 13-11' were similar to those of catechin, with values of 153.08 mg/kg, 928.60 mg/kg, and 870.56 mg/kg at 50 days after flowering, respectively, when it then gradually decreased.

The content in the pulp of 'Hongxun 2' was the highest at 95 days after flowering at 16.45 mg/kg (Figure 3B).

The contents of procyanidin C1 in the peel of 'Hongxun 2' and the pulp of 'Xinye 13-11' were 85.03 mg/kg and 758.21 mg/kg, respectively, and then gradually decreased at 50 days after flowering. The highest procyanidin C1 content was 16.62 mg/kg in the pulp of 'Hongxun 2' at 95 days after flowering, while the highest content was 976.09 mg/kg in the peel of 'Xinye 13-11' at 65 days after flowering (Figure 3C).

The contents of chlorogenic acid in the peel and pulp of 'Hongxun 2' and 'Xinye 13-11' were 850.86 mg/kg, 1222.60 mg/kg, 121.16 mg/kg, and 223.76 mg/kg at 50 days after flowering, respectively. Among them, the chlorogenic acid content in the peel and pulp of 'Hongxun 2' and 'Xinye 13-11' gradually decreased and then tended to be stable, but the content of chlorogenic acid in the pulp of 'Xinye 13-11' saw a great increase at 110 days after flowering (Figure 3D).

4. Discussion

4.1. Dynamic Changes in Total Phenols and Five Types of Polyphenols in Different Developmental Stages of Fruit

These results show that the contents of polyphenols in the peel and pulp of apple gradually decreased with the ripening of the fruit, and the contents of flavanols, flavonols, and chlorogenic acid showed a downward trend during the growth and development of the apple fruits. The contents of flavanols, flavonols, and chlorogenic acid decreased rapidly in the early stage of apple development, then this decline rate gradually slowed down, and finally tended to be stable or slightly decreased [20,21]. In the young fruit stage, the proportions of flavanols and flavonols were small, and chlorogenic acid was the main polyphenol substance in the pulp. During fruit development and expansion, the proportion of chlorogenic acid gradually decreased, while the proportions of flavanols and flavonols gradually increased. The contents of flavanols and procyanidins in the ripe apple fruits were more than that of chlorogenic acid, becoming the main polyphenols [22]. In this study, the total polyphenol contents of 'Hongxun 2' and 'Xinye 13-11' gradually decreased with fruit development, and the magnitude of this decline diminished and tended to be stable from 95 to 110 days after flowering, which is consistent with previous results [23]. However, the total polyphenol contents of the 'Xinye 13-11' pericardium increased at 65 days after flowering and then gradually decreased, showing some differences from previous studies. Kevers et al. also conducted a related study on the effect of harvesting time on the total polyphenol content of apples. In total, 14 kinds of ripe apple fruits were collected three times, each with an interval of two weeks, and the total polyphenol contents of the apples were detected. The results showed that there were no significant differences in the total polyphenol contents of the three harvests [24].

The five types of polyphenol components in the apples gradually decreased with the development of fruits. Renard et al. studied two wine apple varieties and found that the total phenol contents of the fruits decreased from 35 to 100 days after flowering, and the contents of flavanols, dihydrochalcone, and flavonols all decreased rapidly, while that of hydroxyl cinnamic acid decreased slowly [25]. In this study, the flavonol content fluctuated with the development period, and its overall trend decreased, while the other four polyphenol contents gradually decreased. In general, flavanol contents decreased rapidly during the rapid fruit expansion phase [26]. However, in this study, the flavanol content in the peel of 'Xinye 13-11' increased significantly from 50 to 65 days after flowering, which was an important reason for the increase in the total polyphenol content. Studies have shown that flavanols are synthesized in large quantities in the early stage of fruit development and then increase slowly during the development process, with the expansion of fruit development leading to a reduction in flavanol content. Increases in the total polyphenol content may be due to the fact that the fresh weight proportion of fruit expansion is less than that of flavanol content [27].

Previous studies on the dynamic changes in anthocyanin content in red flesh apples were of two types, the single-peak type and the double-peak type [28]. However, Sun et al. found that red flesh apples had two content peaks in early fruit development (30 d after flowering) and near maturity (105 d to 120 d after flowering) [11]. Wang et al. also found that the contents of red flesh apples gradually increased during fruit development and reached a peak value 150 days after flowering [29]. In this study, the peel of 'Hongxun 2' showed a decreasing trend from 50 to 65 days after flowering, indicating that there was a peak content in the peel before 50 days after flowering, which was similar to the results of Wang et al. [30]. The contents of anthocyanins in the peel and pulp were the highest at 95 days after flowering, which was consistent with research results on the change trend of polyphenol contents in 'Hongmantang' during development [31]. Studies have shown that red flesh apples have a higher antioxidant activity than non-red flesh apples, and anthocyanins play a major role in this [9]. Wang et al. studied four red flesh apple varieties and two white flesh varieties and found that the red flesh apples had a higher antioxidant activity. The key period for anthocyanin extraction and utilization was 95 days after flowering [32].

4.2. Dynamic Changes and Differences in Polyphenol Components in Fruits

Catechins, epicatechin, chlorogenic acid, and phloridzin have strong antibacterial abilities and can effectively inhibit food spoilage [33,34]. In this study, the contents of catechin, epicatechin, chlorogenic acid, and phlophorin in 'Hongxun 2' and 'Xinye 13-11' gradually decreased with fruit development from 50 days after flowering. Similar results were obtained in a study of "*Malus rockii* Schneid." and "Tsugaru" [35]. The extraction and utilization of these components should be carried out in the early stage of fruit development. The contents of procyanidin B1, procyanidin B2, and procyanidin C1 all decreased after 65 days of flowering, but the contents of procyanidin B1 and procyanidin C1 decreased significantly more than that of procyanidin B2. Flavonols gradually accumulate during fruit development, and other flavanols are transformed into each other, mainly into procyanidin B2. The content of each component decreases with an increase in single fruit weight, and the content of procyanidin B2 decreases the least and is the highest in mature fruits [25]. It can be inferred that procyanidin B2 can be obtained from mature fruits when a specific flavanol component is required for production and processing.

The main polyphenols of flavanols are procyanidin C1 and procyanidin B2, and the main polyphenol of hydroxycinnamic acid is chlorogenic acid [34,36]. The differences in polyphenol contents between 'Hongxun 2' and 'Xinye 13-11' were mainly reflected in these three components. The contents of chlorogenic acid in 'Hongxun 2' and flavonols in 'Xinye 13-11' were higher. Flavonols and chlorogenic acid are the main substances contributing to the astringency of fruit [37,38]. However, other studies have shown that the contents of flavonols and chlorogenic acid are positively correlated with the degree of fruit browning, and the contributions of flavonols and chlorogenic acid to the degree of browning are affected by the level of polyphenol content and variety due to differences in polyphenol oxidase substrates [27,39,40]. Chlorogenic acid is the main cause of enzymatic browning. Although procyanidins are positively correlated with fruit browning, they have a high antioxidant activity. Fresh-cut apple preservation was prolonged by applying the procyanidins from *Aronia melanocarpa* (Michx.) Elliott [41]. Therefore, in the breeding process of processed varieties, resources with a low chlorogenic acid content and a high procyanidin content can be selected as parents to reduce the degree of fruit browning and achieve a high antioxidant activity level. However, in the breeding process of fresh food varieties, the edible property of fruits should be considered, so the flavonol content should not be too high, which will lead to severe fruit astringency and affect the fruit flavor.

4.3. Dynamic Variation in Flavonol Components in Fruits

The overall flavonol content in fruit development was decreased [34]. In this study, quercetin 3-O-galactoside, quercetin 3-O-arabinofuranoside, and quercetin 3-O-rhanoside in the peel of 'Hongxun 2' and 'Xinye 13-11' were the main flavonol components. Their contents fluctuated during fruit ripening and showed a downward trend as a whole, which was consistent with previous studies [42,43]. The content of flavanols in the peel of 'Honghun 2' decreased first from 50 to 80 days after flowering and then increased gradually from 80 to 110 days after flowering, while the content of 'Xinye 13-11' increased first from 50 to 95 days and then decreased from 95 to 110 days after flowering, showing an opposite trend to that of 'Honghun 2'. The contents of flavonols and anthocyanins in quercetin glycosides are both affected by light. Light can affect the substrate synthesis and use of quercetin glycosides and anthocyanins by controlling enzymes. Generally, fruits with sufficient light have higher contents of quercetin glycosides and anthocyanins. Meanwhile, cyanidin 3-O-galactoside, the main substance of quercetin glycosides and anthocyanins, has the same biosynthetic pathway as other quercetin substances [44–48]. From 60 to 75 days after flowering is the key period for the rapid expansion of fruit and light absorption, and usually, the quercetin glycoside and anthocyanin contents increase [27,43]. However, the flavonol content of Honghun 2 decreased during this period. It is speculated that 'Hongxun 2' is rich in anthocyanins, which begin to accumulate gradually 65 days after flowering, while the enzymes used to synthesize quercetin glycoside simultaneously synthesize anthocyanins, resulting in a decrease in quercetin glycoside synthesis and a decrease in flavanol content. The dynamic changes between the components of flavonols of 'Hongxun 2' and 'Xinye 13-11' during the development process were different, and whether they are applicable to other red flesh apples and non-red flesh apples needs to be further studied and verified.

4.4. *Relationship Between Malus neidzwetzkyana (Dieck) Langenf and Malus sieversii (Led.) Roem*

Based on fruit morphological diversity, early studies generally believed that *Malus neidzwetzkyana* (Dieck) Langenf was a variant of *Malus sieversii* (Led.) Roem. This view was adopted in books such as "*Malolog*", "*Annals of Chinese Fruit Trees (Apple Scroll)*", and "*Research on Germplasm Resources of Apple Genus*" [49–51]. Through many years of field investigation, field records, and molecular biological technology research, Yan et al. concluded that the genetic distance between *Malus sieversii* (Led.) Roem. and *Malus neidzwetzkyana* (Dieck) Langenf was far, and red flesh apples should be regarded as an independent species [52]. The composition and contents of polyphenols such as flavonoids can be used as the basis for the classification of apple plants [53]. In this study, it was found that, during the fruit development of 'Hongxun 2' and 'Xinye 13-11', the change trends of the five polyphenols were different, except for the same trend observed for hydroxylcinnamic acid. Therefore, this paper supports the viewpoint that *Malus neidzwetzkyana* (Dieck) Langenf is a separate species to *Malus sieversii* (Led.) Roem.

5. Conclusions

There were significant differences in the components and contents of polyphenols between red flesh apples and green flesh apples, and the total polyphenol contents in the peel and pulp of 'Xinye 13-11' were significantly higher than those of 'Hongxun 2'. Anthocyanins were the main substances displaying differences in the components of polyphenols in the peel and pulp of red flesh apples and green flesh apples, and they are also the reason for the red color of red flesh apples. In addition, the hydroxycinnamic acid of 'Hongxun 2' was higher than that of 'Xinye 13-11' during fruit development, especially chlorogenic acid. Red flesh apples can be used as raw materials for chlorogenic acid extraction, and the extraction period was better at 50 days after flowering. Based on the classification of polyphenols, the differences in the components and contents of polyphenols support the existence of *Malus neidzwetzkyana* (Dieck) Langenf as an independent species of *Malus*.

Author Contributions: Conceptualization, D.W., G.W. and K.W.; investigation, X.L. and Z.L. (Zhao Liu); resources, S.S., H.G., L.W., W.T., Z.L. (Zichen Li) and L.L.; writing—original draft preparation,

D.W. and G.W.; writing—review and editing, D.W.; supervision, Y.G. and K.W. All authors have read and agreed to the published version of the manuscript.

Funding: This research was partly funded by the National Key Research and Development Plan [grant number 2023YFD1200104]; and by the Agricultural Science and Technology Innovation Program (CAAS-ASTIP-2021-RIP-02).

Data Availability Statement: The original contributions presented in the study are included in the article, further inquiries can be directed to the corresponding authors.

Conflicts of Interest: The authors declare no conflicts of interest.

References

1. Heron, M.G.L.; Holman, P.C.H.; Katan, M.B.; Kromhout, D. Intake of potentially anticarcenogenic flavonoids and their determinants in adults in the Netherlands. *Nutr. Cancer* **1993**, *20*, 21–29.
2. Lin, Q.L.; Shi, Z.P. Relationship between the structure of natural antioxidants such as flavonoids and phenolic acids and their antioxidant power. *Food Sci.* **2001**, *22*, 85–91.
3. Porto, P.A.L.D.S.; Laranjinha, J.A.N.; Freitas, V.A.P.D. Antioxidant protection of low density lipoprotein by procyanidins: Structure/activity relationships. *Biochem. Pharmacol.* **2003**, *66*, 947–954. [CrossRef] [PubMed]
4. Sawa, T.; Nakao, M.; Akaike, T.; Ono, K.; Maeda, H. Alkylperoxyl radical scavenging activity of various flavonoids and other phenolic compounds: Implications for the anti-tumor promoter effect of vegetables. *J. Agric. Food Chem.* **1999**, *47*, 397–402. [CrossRef]
5. Kasai, H.; Fukada, S.; Yamaizumi, Z.; Sugie, S.; Mori, H. Action of chlorogenic acid in vegetables and fruits as an inhibitor of 8-hydrox-ydeoxyguanosine formation in vitro and in a rat carcinogenesis model. *Food Chem. Toxical.* **2000**, *38*, 467–471. [CrossRef]
6. Hubbard, G.; Wolffram, S.; Lovegrove, J.; Gibbins, J.M. The role of polyphenolic compounds in the diet as inhibitors of platelet function. *Proc. Nutr. Soc.* **2003**, *62*, 469–478. [CrossRef]
7. Xiang, Y.; Zhao, R.; Lai, F.N.; Sun, X.; Sun, X.H.; Dai, H.Y.; Zhang, Y.G. Analysis of flavonoid fractions and antioxidant activity of red flesh apple peel. *J. Plant Physiol.* **2016**, *52*, 1353–1360.
8. Zhang, X.; Sun, X.H.; Bo, S.H.; Bo, H.X.; Hou, H.M.; Sun, X.; Zhang, Y.G. Anthocyanin content and in vitro antioxidant study of four red flesh apple extracts. *J. Qingdao Agric. Univ. (Nat. Sci. Ed.)* **2018**, *35*, 179–185+199.
9. Li, C.X.; Zhao, X.H.; Zuo, W.F.; Zhang, T.L.; Zhang, Z.Y.; Chen, X.S. Phytochemical profiles, antioxidant, and antiproliferative activities of four red-fleshed apple varieties in China. *J. Food Sci.* **2020**, *85*, 718–726. [CrossRef]
10. Chen, X.S.; Mao, Z.Q.; Wang, N.; Zhang, Z.Y.; Wang, Z.G.; Xu, Y.H.; Jiang, S.H.; Dong, M.X.; Li, J.M. Evaluation, mining and innovative utilization of 'Fuji' and Xinjiang red-fleshed apple (*Malus sieversii* f. niedzwetzkyana). *China Fruit* **2020**, *4*, 1–4.
11. Sun, X.H.; Bo, S.H.; Hou, H.M.; Sun, X.; Zhu, J.; Dai, H.Y.; Zhang, Y.G. A new red-fleshed apple cultivar 'Daihong'. *Acta Hortic. Sin.* **2019**, *46*, 2729–2730.
12. Zhang, L.Y.; Jiang, C.Y. Breeding of new red flesh apple variety 'Hongyun'. *China Fruit* **2023**, *11*, 108–109.
13. Würdig, J.; Flachowsky, H.; Hofer, M.; Peil, A.; Ali, M.A.M.S.E.; Hanke, M.V. Phenotypic and genetic analysis of the german *Malus* germplasm collection in terms of type 1 and type 2 red-fleshed apples. *Gene* **2014**, *544*, 198–207. [CrossRef]
14. Wang, N.; Jiang, S.H.; Zhang, Z.Y.; Fang, H.C.; Xu, H.F.; Wang, Y.C.; Chen, X.S. *Malus sieversii*: The origin, flavonoid synthesis mechanism, and breeding of red-skinned and red-fleshed apples. *Hortic. Res.* **2018**, *5*, 70. [CrossRef]
15. Wang, X.Q. Phenolic Metabolism of Red-Fleshed Apples and its Response to Stress. Ph.D. Thesis, Northwest Agriculture and Forestry Univsersity, Yangling, China, 2015.
16. Umemura, H.; Otagaki, S.; Wada, M.; Kondo, S.; Matsumoto, S. Expression and functional analysis of a novel MYB gene, *MdMYB110a_JP*, responsible for red flesh, not skin color in apple fruit. *Planta* **2013**, *238*, 65–76. [CrossRef]
17. Liu, R.; Wang, Y.X.; Wang, Z.; Shi, J.R.; Fan, Z.Z.; Zhang, Y.G.; Sun, X.H. Physicochemical analysis of the aging process of 'Hongxun 1' apple cider. *J. Qingdao Agric. Univ. (Nat. Sci.)* **2023**, *40*, 237–242.
18. Nie, J.Y.; Lv, D.G.; Li, J.; Liu, F.Z.; Li, H.F.; Wang, K. A preliminary study on the flavonoids in fruits of 22 apple germplasm resources. *Sci. Agric. Sin.* **2010**, *43*, 4455–4462.
19. Wang, D.J.; Wang, K.; Li, J.; Gao, Y.; Zhao, J.R.; Liu, L.J.; Gong, X.; Dong, X.G. Variation and correlation analysis of polyphenolic compounds in Malus germplasm. *J. Hortic. Sci. Biotechnol.* **2018**, *93*, 26–36. [CrossRef]
20. Kondo, S.; Tsuda, K.; Muto, N.; Ueda, J.E. Antioxidative activity of apple skin or flesh extracts associated with fruit development on selected apple cultivars. *Sci. Hortic.* **2002**, *96*, 177–185. [CrossRef]
21. Pang, W. The Separation Purification and Antioxidation Research of Apple Polyphenols. Ph.D. Thesis, Northwest University, Xi'an, China, 2007.
22. Wang, S.X.; Liu, J.C.; Jiao, Z.G.; Jiao, Z.G.; Zhang, S.N.; Yang, L. Changes of polyphenols during fruit development in apples. *J. Fruit Sci.* **2003**, *20*, 427–431.
23. Jiang, H.; Ji, B.P.; Liang, J.F.; Zhou, F.; Yang, Z.W.; Zhang, G.Z. Changes of contents and antioxidant activities of polyphenols during fruit development of four apple cultivars. *Eur. Food Res. Technol.* **2006**, *223*, 743–748. [CrossRef]

24. Kevers, C.; Pincemail, J.; Tabart, J.; Defraigne, J.O.; Dommes, J. Influence of cultivar, harvest time, storage conditions, and peeling on the antioxidant capacity and phenolic and ascorbic acid contents of apples and pears. *J. Agric. Food Chem.* **2011**, *59*, 6165–6171. [CrossRef] [PubMed]
25. Renard, C.M.G.C.; Dupont, N.; Guillermin, P. Concentrations and characteristics of procyanidins and other phenolics in apples during fruit growth. *Phytochemistry* **2007**, *68*, 1128–1138. [CrossRef]
26. Henry-Kirk, R.A.; Mcghie, T.K.; Ander, C.M.; Allan, A.C. Transcriptional analysis of apple fruit proanthocyanidin biosynthesis. *J. Exp. Bot.* **2012**, *63*, 5437–5450. [CrossRef]
27. Zhao, J.R.; Liu, G.J.; Chang, R.F.; Cao, K.; Shen, F.; Wu, T.; Wang, Y.; Han, Z.H.; Zhang, X.Z. Diversity of flesh polyphenols and their progressive dilution during fruit expansion in *Malus* germplasm. *Sci. Hortic.* **2015**, *197*, 461–469. [CrossRef]
28. Li, X.D.; Wang, F.; Tong, P.P.; Zhang, Y.R.; Liu, Y.J.; Jiang, Z.W.; Wang, H.B. Accumulation of anthocyanosides in the pulp of Xinjiang red-fleshed apple and the expression of their related gens. *Mol. Plant Breed.* **2024**, *22*, 4233–4239.
29. Wang, Y.L.; Zhang, Y.M.; Feng, S.Q.; Song, Y.; Xu, Y.T.; Zhang, Y.P.; Chen, X.S. The mechanism of red coloring different between skin and cortex in *Malus sieversii* f. neidzwetzkyana (Dieck). *Langenf. Sci. Agric. Sin.* **2012**, *45*, 2771–2778.
30. Wang, L.; Wang, F.; Tang, L.; Tong, P.P.; Zhang, Y.R.; Wang, J.B. Changes of anthocyanin content and expression of synthesis-related genes in peel of Xinjing red flesh apples in different periods. *Acta Agric. Jiangxi* **2021**, *33*, 6–10.
31. Wu, Q. Studies on the Phenolic Compounds by Widely Targeted Metabolomics and Flavonoid Extraction and Purification of 'hongmantang' Apple Fruit. Ph.D. Thesis, Shanxi Agricultural Univsersity, Taigu, China, 2023.
32. Wang, X.Q.; Li, C.Y.; Liang, D.; Zou, Y.J.; Li, P.M.; Ma, F.W. Phenolic compounds and antioxidant activity in red-fleshed apples. *J. Funct. Foods* **2015**, *18*, 1086–1094. [CrossRef]
33. Muthuswamy, S.; Vasantha, H.P. Fruit phenolics as natural antimicrobial agents: Selective antimicrobial activity of catechin, chlorogenic acid and phloridzin. *J. Food Agric. Environ.* **2007**, *5*, 81–85.
34. Kumar, S.; Deng, C.H.; Molloy, C.; Kirk, C.; Plunkett, B.; Wang, K.L.; Allan, A.; Espley, R. Extreme-phenotype GWAS unravels a complex nexus between apple (*Malus domestica*) red-flesh clour and internal flesh browning. *Fruit Res.* **2022**, *2*, 12. [CrossRef]
35. Zhou, L. Study on the Concentration Changes and Related Genes of Flavonoids in Apple. Ph.D. Thesis, Chinese Academy of Agricultural Sciences, Beijing, China, 2013.
36. Song, J.; Amyotte, B.; Yu, C.H.J.; Campbell-Palmer, L.; Vinqvist-Tymchuk, M.; Rupasinghe, H.P.V. Untargeted metabolomics analysis reveals the biochemical variations of polyphenols in a diverse apple population. *Fruit Res.* **2023**, *3*, 29. [CrossRef]
37. Wu, W.; Zhu, Q.G.; Wang, W.Q.; Grierson, D.; Yin, X.R. Molecular basis of the formation and removal of fruit astringency. *Food Chem.* **2022**, *372*, 131234. [CrossRef] [PubMed]
38. Xing, H.Y.; Wu, J.L.; Wang, L.R. Advances in the metabolism and regulation of astringent substances in fruits. *J. Fruit Sci.* **2023**, *40*, 1728–1740.
39. Wang, L.J.; Li, J.H.; Gao, J.J.; Feng, X.X.; Shi, Z.X.; Gao, F.Y.; Yang, L.Y. Inhibitory effect of chlorogenic acid on fruit russeting in 'Golden Delicious' apple. *Scientia Horticulturae* **2014**, *178*, 14–22. [CrossRef]
40. Podsedek, A.; Wilska-Jeszka, J.; Anders, B.; Markowski, J. Compositional characterisation of some apple varieties. *Eur. Food Res. Technol.* **2000**, *210*, 268–272. [CrossRef]
41. Li, S.J.; Chen, J.J.; Sarengaowa, C.C.; Hu, W.Z. Application of procyanidins from *Aronia melanocarpa* (Michx.) elliott in fresh-cut apple preservation. *Horticulturae* **2024**, *10*, 556. [CrossRef]
42. Feng, S.H.; Yi, J.Y.; Li, X.; Wu, X.Y.; Zhao, Y.Y.; Ma, Y.C.; Bi, J.F. Systematic review of phenolic compounds in apple fruits: Compositions, distribution, absorption, metabolism, and processing stability. *J. Agric. Food Chem.* **2021**, *69*, 7–27. [CrossRef]
43. Awad, M.A.; Jager, A.D.; Plas, L.H.W.V.D.; Krol, A.R.V.D. Flavonoid and chlorogenic acid changes in skin of 'Elstar' and 'Jonagold' apples during development and ripening. *Sci. Hortic.* **2001**, *90*, 69–83. [CrossRef]
44. Awad, M.A.; Jager, A.D.; Westing, L.M.V. Flavonoid and chlorogenic acid levels in apple fruit: Characterisation of variation. *Sci. Hortic.* **2000**, *83*, 249–263. [CrossRef]
45. Ju, Z.G.; Yuan, Y.B.; Liu, C.Q.; Wang, Y.Z.; Tian, X.P. Dihydroflavonol reductase activity and anthocyanin accumulation in 'Delicious', 'Golden Delicious' and 'Indo' apples. *Sci. Hortic.* **1997**, *70*, 31–43. [CrossRef]
46. Feng, F.J.; Li, M.J.; Ma, F.W.; Cheng, L.L. The effects of bagging and debagging on external fruit quality, metabolites, and the expression of anthocyanin biosynthetic genes in 'Jonagold' apple (*Malus domestica* Borkh.). *Sci. Hortic.* **2014**, *165*, 123–131. [CrossRef]
47. Yu, L.J.; Sun, Y.Y.; Zhang, X.; Chen, M.C.; Wu, T.; Zhang, J.; Xing, Y.F.; Tian, Y.; Yao, Y.C. ROS1 promotes low temperature-induced anthocyanin accumulation in apple by demethylating the promoter of anthocyanin-associated genes. *Hortic. Res.* **2022**, *9*, uhac007. [CrossRef] [PubMed]
48. Karl, A.D.; Peck, G.M. Great sunlight exposure during early fruit development increases polyphenol concentration, soluble solid concentration, and fruit mass of cider apples. *Horticulturae* **2022**, *8*, 99. [CrossRef]
49. Su, H.R. *Malology*; China Agriculture Press: Beijing, China, 1999.
50. Li, Y.N. *Researches of Germplasm Resources of Malus Mill*; China Agriculture Press: Beijing, China, 2001.
51. Lu, Q.N.; Jia, D.X. *China Fruit Tree-Apple*; China Forestry Publishing House: Beijing, China, 1999.

52. Yan, G.R.; Yu, W.W.; Yang, M.L.; Xu, Z. *The Malus sieversii in China*; China Forestry Publishing House: Beijing, China, 2020.
53. Williams, A.H. Chemical evidence from the flavonoids relevant to the classification of *Malus* species. *Bot. J. linn. Soc.* **1982**, *84*, 31–39. [CrossRef]

horticulturae

MDPI

Article

Isolation and Identification of Pear Ring Rot Fungus and Resistance Evaluation of Different Pear Varieties

Chen Wang [1,†], Yanjie Zhang [1,†], Chunqing Ou [1], Fei Wang [1], Li Ma [1], Caihong Wang [2,*] and Shuling Jiang [1,*]

1 Institute of Pomology, Chinese Academy of Agricultural Sciences, Xingcheng 125100, China; orange19991026@163.com (C.W.); zhangyanjie@caas.cn (Y.Z.); ouchunqing@caas.cn (C.O.); wangfei@caas.cn (F.W.); mali@caas.cn (L.M.)

2 College of Horticulture, Qingdao Agricultural University, Qingdao 266109, China

* Correspondence: chwangt@163.com (C.W.); jiangshuling@caas.cn (S.J.)

† These authors contributed equally to this work.

Abstract: *Botryosphaeria dothidea* is a significant plant pathogen responsible for causing ulcers, wilt, and fruit decay across a wide range of host plants. One notable fungal disease attributed to *B. dothidea* is pear tree ring rot, which currently ranks among the most severe diseases affecting pear trees in China. This pathogen primarily targets branches and fruits, occasionally impacting leaves as well, leading to tree weakening, fruit rot, and leaf drop. The annual repercussions of this disease severely affect both the yield and quality of pear fruits, thereby impeding the healthy development of the pear industry. Recent studies have indicated that other species within the *B. dothidea* complex can also induce pear ring rot; however, specific physiological strains of *B. dothidea* remain unreported. Consequently, this study collected tissues from pear trees infected with ring rot from orchards located in Liaoning, Hebei, Shandong, and other regions throughout China. Through morphological characterization combined with pathogenicity assessments and DNA sequence comparisons involving partial internal transcribed spacer (ITS), translation elongation factor (TEF), and β-tubulin (TUB) genes, 21 strains belonging to the *Botryosphaeria* spp. were identified. These 21 strains served as research subjects for inoculating dormant annual branches from 30 germplasm resources of pear trees in vitro. The results demonstrated that all tested strains could induce lesions on the branches which were characterized by dark brown spots. Furthermore, inoculation experiments involving these 21 strains were conducted to evaluate the resistance levels of various pear varieties against ring rot disease. The resistance was assessed by inoculating different isolates onto distinct pear varieties; this approach established the criteria for evaluating resistance while minimizing identification errors stemming from the variable responses exhibited by certain varieties towards individual strains. Ultimately, this study aims to provide a theoretical foundation for effective prevention and treatment strategies against pear ring rot.

Keywords: pear; ring rot disease; *Botryosphaeria dothidea*; pathogen isolation and identification; resistance evaluation

1. Introduction

The cultivated area and pear yield mean it is ranked third among fruit trees in China. According to the National Bureau of Statistics, the cultivated area in 2021 was 940,700 hm^2, while the output in 2022 reached 19,265,300 t. Disease resistance is a crucial characteristic of pear fruit and represents one of the primary objectives in pear breeding. *Pyrus* have very rich germplasm resources—at least 22 major pear species have been reported—but only *P. bretschneideri*, *P. pyrifolia*, *P. communis*, *P. sinkiangensis*, and *P. ussuriensis* are cultivated for actual fruit production [1]. Despite these numerous varieties with attractive fruit characteristics and favorable taste profiles, their poor disease resistance leads to increased costs associated with chemical disease control measures and contributes to environmental pollution. There is an urgent need for non-chemical methods to manage these diseases effectively. Through resistance identification in different pear varieties, screening out the resistant germplasm for disease-resistance breeding is the most effective measure for disease control [2].

Pear ring rot is one of the major diseases affecting pear trees and is caused by the *Botryosphaeria* spp. This disease is prevalent in all pear-producing regions of China. It damages the branches, fruits, and even the leaves of pear trees [3], leading to premature aging and resulting in a significant loss of leaves and fruit. It has been estimated that pear ring rot can reduce production by approximately 25% annually [4]. In severe cases, it can affect up to 80% of the fruit in an entire orchard [5]. When a tree becomes infected it exhibits two types of symptoms. The first occurs when the infection affects the branches, fruits, and leaves of the pear tree, resulting in a ring pattern symptom known as ring rot [6]. The second manifests as an ulceration of infected branches and is referred to as dry rot. Zhai [7] identified four pathogens responsible for these symptoms: *B. dothidea*, *B. rhodina*, *B. obtusa*, and *B. parva*, with *B. dothidea* being the dominant species causing ring rot disease. Xiao [8] demonstrated that the pathogen responsible for pear ring rot closely resembles that of apple ring rot. In addition to *B. dothidea*, *B. kuwatsukai* was also present, and its virulence was significantly stronger than that of *B. dothidea*. Pathogens were isolated from samples exhibiting pear wheel pattern symptoms and dry rot symptoms. Only *B. kuwatsukai* was isolated from the samples with pear wheel pattern symptoms, while both *B. dothidea* and *B. kuwatsukai* were isolated from the samples showing dry rot symptoms.

In "Descriptors and Data Standard for Pear (*Pyrus* spp.)" [9], the methodology employed for evaluating the resistance of pear fruit to ring rot involved the spray inoculation of the fruit with conidia from the pathogenic fungus. However, it is noteworthy that no established criteria exist for assessing the resistance of leaves and stems to ring rot. According to various scholarly sources [10–12], the incidence rate and disease index associated with the inoculation of mycelial blocks using punctured branches were higher than those resulting from the conidia inoculation method. This alternative method is advantageous as it does not require the continuous cultivation of the fungus under light conditions for approximately ten days to produce conidia, thereby rendering it a more rapid and straightforward approach for evaluating disease resistance. Furthermore, some researchers assessed the resistance of different pear germplasm resources through physiological indicators, such as measuring the accumulation of primary phenolic compounds in pear fruits and observing the alterations in these compounds following inoculation with *Botryosphaeria* spp. [13,14]. Currently, there is no consensus on the evaluation criteria for pear ring rot resistance. Therefore, establishing a rapid and comprehensive evaluation system for branch ring rot resistance is of considerable importance for the breeding of resistant pear varieties.

This research aimed to (i) gather and isolate the pathogenic fungi implicated in ring rot disease across the provinces of Shandong, Hebei, and Liaoning; (ii) identify the strains through morphological examination and multilocus analysis; (iii) evaluate the virulence of the isolated strains and formulate grading criteria for the virulence of *B. dothidea*; and (iv) investigate the resistance of different pear cultivars to ring rot disease, while also establishing the criteria for assessing resistance to this affliction.

2. Materials and Methods

2.1. Isolation and Purification of Pathogenic Fungi

Commencing in August 2022, tissue samples were collected from various regions, including Shandong, Hebei, and Liaoning. These samples, which comprised leaves, fruits, and branches, were individually stored in ziplock bags at a temperature of 4 °C within a refrigerator. Following the separation of tissues, infected samples were thoroughly washed and dried using sterile water, and subsequently disinfected with 75% ethanol. For the leaf or fruit samples, a sterilized scalpel was utilized to excise a tissue block approximately 5 mm in diameter from the interface of the diseased and healthy tissue. In the case of the branch samples, the tumor was carefully excised with a scalpel and then bisected along its diameter using sterilized scissors to obtain the requisite tissue block. The tissue blocks were immersed in a 0.1% mercuric chloride solution for 30 s, followed by a 30 s rinse with sterile water. They were then soaked in 75% ethanol for an additional 30 s and subsequently rinsed twice with sterile water for 30 s each [15]. Following these procedures, the tissues were cultured in Potato Dextrose Agar (PDA) medium at a temperature of 28 °C. After a growth period of three days, a pure single colony was selected and transferred to a new PDA medium for further cultivation.

2.2. Morphological and Molecular Identification

2.2.1. Observation of Mycelial Growth Status and Induction of Conidia

The characteristics of the colony and the growth state of the mycelia were critical criteria for identification. The aerated mycelia of this genus, when cultured on PDA medium, displayed a progressive color change from gray to black as the fungus matured. Simultaneously, the reverse side of the colony transitioned from white to yellow, followed by a change to dark green, and ultimately to black. To facilitate the observation of colony morphology, three distinct media were utilized for culturing the strains.

The colony morphology of each strain was meticulously documented. Hyphal blocks, measuring 5 mm in diameter, were excised from the periphery of the purified colonies and cultured on PDA medium for seven days. Following this incubation, the growth characteristics of the colonies on both sides of the Petri dish were evaluated. Mycelia cultured on PDA for seven days were subsequently analyzed under an Olympus IX51 inverted microscope to assess their morphological features. The colony was cultured on PDA medium for three days, and the mycelium pieces with a diameter of 5 mm were selected and placed on PDA, OA (Oatmeal Agar), and WA (Water Agar) plates (each plate with a diameter of 90 mm and a medium volume of 25 mL; all culture medium allocation methods are shown in Table S1) and incubated at a constant temperature and in darkness at 28 °C [16]. The process was repeated three times with each strain. The growth morphology of the mycelium was observed every 48 h.

Moreover, each strain was inoculated onto PDA, OA, and WA. Once the colonies had fully colonized the culture dish, the mycelium was carefully scraped using a sterile bamboo skewer. The conidial morphology was then examined microscopically while subjected to continuous ultraviolet irradiation within a constant-temperature incubator maintained at 28 °C for 45 days [17].

2.2.2. Molecular Identification and Phylogenetic Analysis

Different strains were activated on PDA medium for three days, and white mycelia with edges were selected and transferred to a new PDA medium for propagation. After five days, a small amount of mycelia were selected and the total genomic DNA was extracted by the Chelex-100 method [18]. The internal transcribed spacer (ITS) [19], translation elongation factor (TEF) [20], and β-tubulin (TUB) [21] were amplified using the primers in Table S2.

The PCR system was 50 μL, including 25 μL of premix Taq (TaKaRa, Biotechnology, Dalian, China), 1 μL of each primer, 1 μL of DNA template solution, and 22 μL of dd H_2O. The procedure was as follows: initial denaturation at 94 °C for 5 min; followed by 30 cycles at 94 °C for 30 s, primer annealing at a suitable temperature for 30 s (58 °C for ITS, 60 °C for TEF and TUB), and primer extension at 72 °C for 45 s; and extension at 72 °C for 10 min. The PCR product was observed under UV illumination on a 1% agarose gel containing ethidium bromide (0.5 mg/mL) and amplicons were sequenced by Taihe Biotechnology (Beijing, China). The sequencing results were compared and homologously analyzed with the known sequences in the GenBank database using the BLAST program. The sequences were spliced according to the ITS-TEF-TUB sequence, and the sequences were spliced and aligned using ACOPTloos. Using *Neofusicoccum parvum* as the outgroup, MEGA7.0 [22] was applied to construct the phylogenetic tree of the three-gene join by the neighbor-joining method [23], and 1000 repeats were performed by the bootstrap method [24].

2.3. Pathogenicity Determination of Pathogenic Fungi

The pathogenicity of the strains was tested according to Koch's rule. Initially, the surfaces of the dormant annual branches from randomly selected pear tree varieties were disinfected using 75% alcohol, followed by two rinses with sterile water. After allowing the branches to dry, they were cut into segments measuring 120 mm, with both ends sealed using paraffin wax. Wounds were created in the middle of the branches, extending to the xylem, while avoiding the flower buds. The isolated strains, cultured on PDA medium for five days, were then formed into disks using a 5 mm diameter hole punch. The mycelial surface of these disks was subsequently inoculated at the designated points on the branches, with each strain being tested in ten replicates. A blank PDA medium served as a control. The inoculated branches were then incubated in an artificial-climate chamber maintained at 28 °C, where the incidence of infection was monitored and recorded. Pathogenic fungi were isolated and purified from the re-infected branches using the previously described method, ensuring that the isolates were consistent with the originally inoculated strains.

To assess the differences in virulence among the strains, the spread of lesions caused by different strains on the 'Hongxiangsu' variety was analyzed through cluster analysis. The 21 strains were categorized into three virulence levels: strong, medium, and weak. Following the aforementioned inoculation method, the virulence of each strain was observed across different pear tree varieties. According to the lesion extension length at 21 d, Origin v2022 software was used for a systematic clustering analysis, and combined with the distribution of branches of different disease grades, the classification standard of virulence was determined.

2.4. Identification of Resistance of Different Pear Germplasm Resources

A total of 30 pear germplasm resources (Table 1) were selected from the experimental orchard of the Pear Breeding Research Group of the Research Institute of Pomology of the CAAS (Xingcheng) in January 2024. The strains were inoculated on the annually dormant branches of the 30 pear germplasm resources. The treatment method was as referred to in Section 2.3, with the process repeated ten times for each variety and a blank PDA medium inoculated as the control. The length of lesions on the branches were measured at 14, 21, and 28 days after inoculation.

Table 1. Pear germplasm resources tested in this study.

P. bretschneideri	*P. pyrifolia*	*P. ussuriensis*	*P. communis*	*P. sinkiangensis*	Stock
Zaosu	Huobali	Jingbaili	Wujiuxiang	Kuerlexiangli	Duli
Jinfeng	Lvxiu	Nanguoli	Jinxiang		Zhongai 1
Hongxiangsu		Huagaili	Zhongaihongli		Zhongai 3
Yuluxiang			Bayuehong		
Huangguan			Zaohong Comice		
Pingguoli			Fenlao		
Zaojinsu			Zhongjia 1		
Jinhua			Bali		
Huasu			Conference		
Dangshansuli			Zaojinxiang		
Zaosuhong					

2.5. Statistical Analysis

The experimental data were compared by means of a one-way analysis of variance (ANOVA) and SPSS statistical software 26.0 ($p < 0.05$). The homogeneity test of variance was conducted before ANOVA.

A cluster analysis was carried out using Origin v2022's systematic clustering.

3. Results

3.1. Isolation of Pathogenic Fungi

A comprehensive collection of 65 samples exhibiting ring rot was conducted, encompassing leaves, fruits, and branches. The manifestations of the disease are illustrated in Figure 1. Infected leaves and fruits displayed dark brown lesions, while branches affected by the fungus exhibited cracking of the bark and the formation of warty growths.

Figure 1. Ring rot disease symptoms on pears. (**A**,**B**) leaf; (**C**,**D**) fruit; and (**E**,**F**) trunk symptoms in the field.

3.2. Morphological Analysis of Pathogenic Fungi

The purified strains were inoculated onto PDA medium. After 7 days, the entire surface of the plates was colonized, with strains LN-ZJ4, LN-ZJ5, and HB-ZJZ3 exhibiting nearly complete white coverage, while the majority of the other colonies displayed gray and white coloration. The growth patterns across these plates were relatively uniform; however, strain HZ-SZ2162 demonstrated a highly irregular growth, with colonies aging rapidly and distinct color demarcations becoming apparent on the plate. Notably, while the peripheral mycelium remained white, the central mycelium exhibited significant blackening. Upon examining the reverse side of the plate, it was observed that colonies with a white coloration corresponded to a yellowish-gray hue on the back, whereas the off-white colonies were associated with a grayish-black appearance.

Microscopic examination of the mycelium after 7 days of culture at a magnification of 400× revealed that the mycelium was colorless and transparent, exhibiting a flat and straight morphology, with branching structures, including thin branches, and a consistent thickness throughout. It could be seen that the mycelium was associated with the state of the colony (Figure 2).

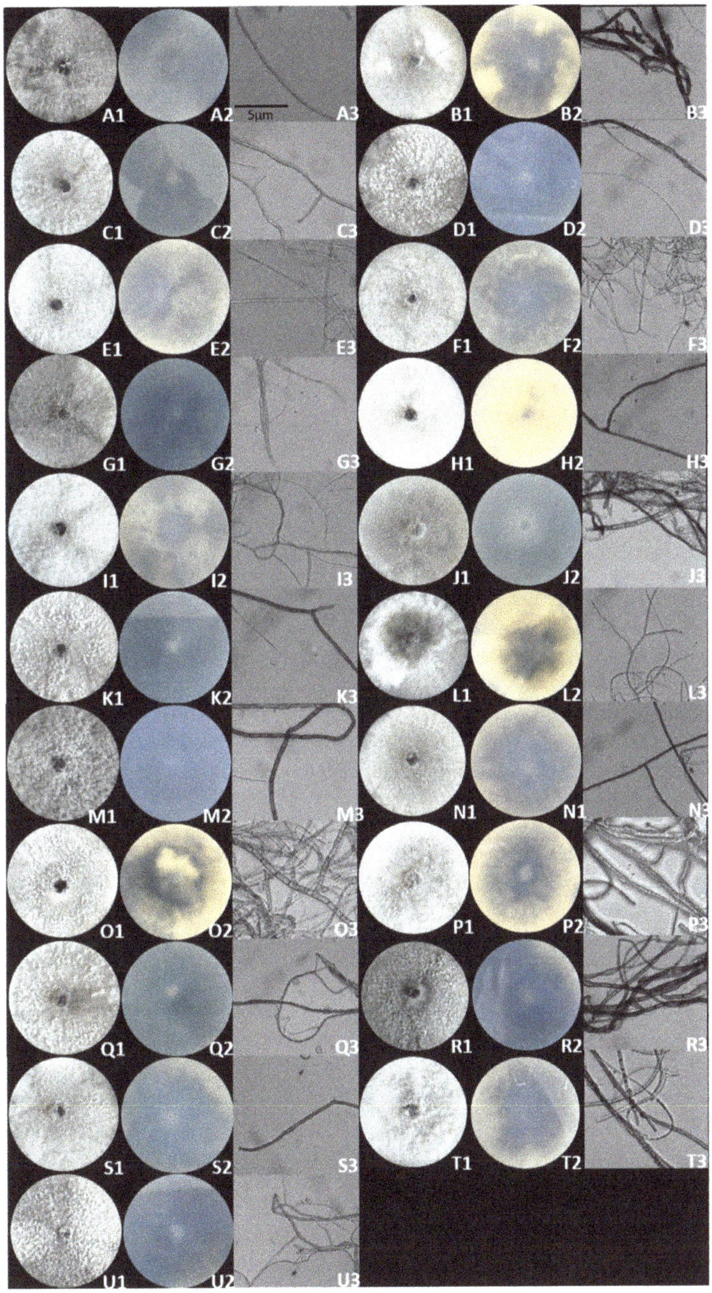

Figure 2. Colony and mycelial morphology on PDA medium: (**A**) LN-ZJ1, (**B**) HB-CZ6, (**C**) LN-ZJ2, (**D**) HB-SJZ2, (**E**) LN-ZJ4, (**F**) HB-ZJZ3, (**G**) LN-ZJ5, (**H**) GGS-K1, (**I**) LN-ZJ6, (**J**) HZ-MJ231, (**K**) LN-ZJ8, (**L**) HZ-SZ2162, (**M**) LN-SDGZ, (**N**) HB-CZ3, (**O**) LN-HS, (**P**) HB-HDWX, (**Q**) HB-CZ1, (**R**) HB-ZX, (**S**) HB-CZ2, (**T**) SD-YM, and (**U**) HB-CZ5 isolates; 1—fronts of colonies cultured on PDA, 2—backs of colonies cultured on PDA, and 3—mycelial morphology observed via a microscope at 400×.

The strains were inoculated on the media of PDA, OA, and WA for morphological observation. The results showed that all colonies on the PDA medium were white at the beginning of the period and showed radial growth, irregular edges, and grew outward layer by layer. The entire colony was snowflake-shaped, and some strains developed yellow pigments at 3 d, especially LN-ZJ5, LN-ZJ6, LN-ZJ8, HB-CZ2, HB-CZ5, and HB-ZX, which turned olive green in the middle of the plate. This showed that these strains aged quickly. The growth rate of different strains was also different. LN-ZJ4 and LN-HS had covered the whole plate by three days, while GGS-K1 and HZ-MJ231 only expanded by about 3 cm. All strains had overgrown the entire plate at 5 d, and the mycelium was dense and gradually changed to white and green. On the PDA medium, over time, the mycelium gradually changed from white to olive green, and finally completely changed to pure black. There are two kinds of air mycelia. The first kind are LN-ZJ4, LN-HS, LN-ZJ1, HB-ZJZ3, HB-ZX, and SD-YM, and the mycelia grew vertically and reached the top of the Petri dish. The second kind grew close to the underside of the plate. In OA medium, the colony expanded faster, the mycelia were dense, and the edges were neat, but the mycelia did not grow in the middle of the plate, and the interior showed concentric circles of black loss. The airborne mycelium was very abundant at the beginning, but with time, the number of mycelia gradually decreased, and more mycelia were present only in the periphery of the plate at 9 d.

The strain grew slowly on WA medium, only a small amount of mycelia grew closely to the medium, the stratification was not obvious, and the aging rate of the mycelia was slow. Only HB-CZ6 colonies turned black at 5 d, the other strains only had melanin precipitation at 7 or 9 d, and the colonies did not overgrow the entire plate at 9 d (Figure S1).

The strains cultivated on PDA, OA, and WA failed to produce conidia, despite the mechanical disruption of the mycelium using a bamboo skewer and the continuous exposure to ultraviolet light for 45 days. The induction of conidia in the context of pathogens proved challenging under the experimental conditions employed. Furthermore, alternative methods aiming to enhance fungal conidia production were found to be ineffective for these pathogens.

3.3. Molecular Identification and Phylogenetic Analysis

The DNA from 21 strains was extracted by the Chelex-100 method, and the ITS, TEF, and TUB of the strains were amplified by primers. Bands of 568, 277, and 437 bp were obtained, respectively, and then compared by Blast on NCBI. The homology with *B. kuwatsukai* and *B. dothidea* was more than 99%, which indicated that the scientific classification status of the new isolates was clear. The sequence of strains with high homology and 58 strains of *Botryosphaericeae* and other species were selected as the reference objects (Table S3: Gene bank used in constructing the phylogenetic tree). Through phylogenetic analysis, the sequences of each strain were analyzed in tandem according to ITS–TEF–TUB, and a combined phylogenetic tree of multiple genes was constructed (Figure 3). The 21 isolates were clustered in the same clade as *B. kuwatsukai*. At the same time, they all had a close kinship with *B. dothidea*.

3.4. Strain Virulence Test

Twenty-one isolates were inoculated onto healthy annual dormant branches of pear trees, which could cause the host branches to develop disease at 14 d, producing obvious reddish-brown to black-brown concentric circle disease spots. Subsequently, these lesions expanded from the inoculation site towards both ends of the branches in an oval configuration, leading to the formation of epidermal folds and even cracks in the areas where the lesions proliferated. The boundary between health and sickness was clear (Figure 4). At 21 d there was a marked toroidal lesion.

Figure 3. Phylogram of *Botryosphaeria* based on three combined genes: partial internal transcribed spacer (ITS), translation elongation factor (TEF), and β-tubulin (TUB). *Neofusicoccum parvum* was taken as the outgroup. The tree was constructed using concatenated sequences of ITS, TEF, and TUB genes. The node is marked with bootstrap values. The colored areas represent the branches that contain the isolates in this study.

Figure 4. Symptoms of disease spots produced by each isolated strain on pear branches.

In the virulence test of the branches inoculated on 'Hongxiangsu', HB-SJZ2 strains showed the strongest virulence, with an average spot diameter of 98.8 mm at 21 d, while HZ-MJ231 strains showed the weakest virulence, with an average spot diameter of 12.1 mm (Figure 5). There were significant differences in virulence between the 21 strains at 21 d (Figure 6), which indicated that there was a differentiation of virulence.

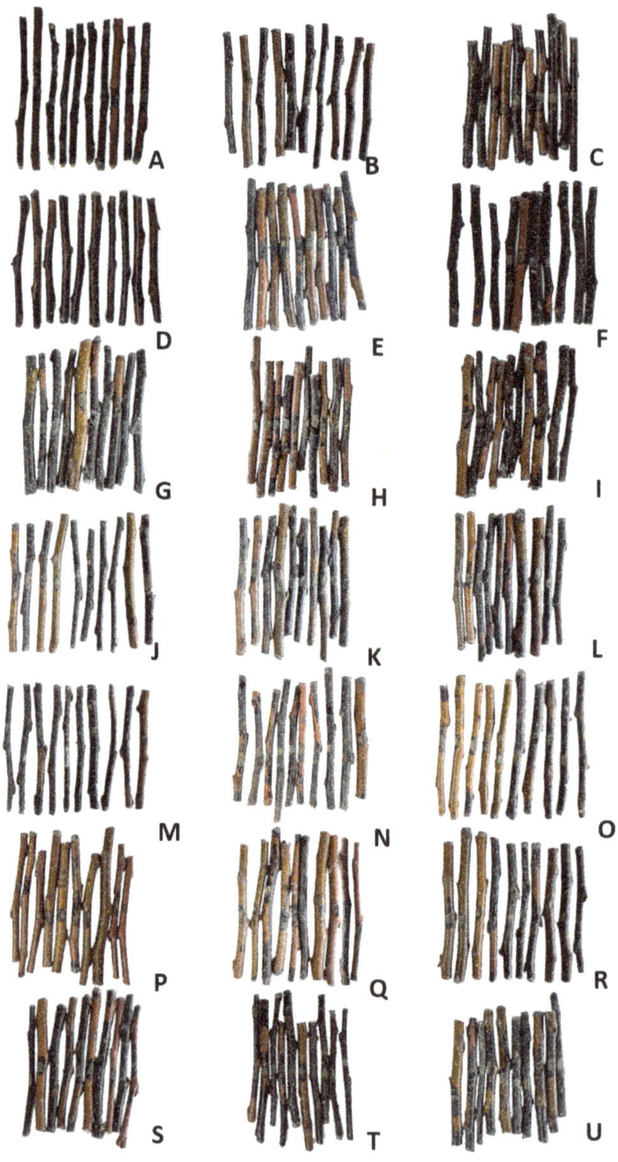

Figure 5. Symptoms of isolated fungi inoculated on 'Hongxiangsu' isolated branches. (**A**) LN-ZJ1, (**B**) LN-HS, (**C**) GGS-K1, (**D**) LN-ZJ2, (**E**) HB-CZ1, (**F**) HZ-MJ231, (**G**) LN-ZJ4, (**H**) HB-CZ2, (**I**) HZ-SZ2162, (**J**) LN-ZJ5, (**K**) HB-CZ5, (**L**) HZ-CZ3, (**M**) LN-ZJ6, (**N**) HB-CZ6, (**O**) HB-HDWX, (**P**) LN-ZJ8, (**Q**) HB-SJZ2, (**R**) HB-ZX, (**S**) LN-SDGZ, (**T**) HB-ZJZ3, and (**U**) SD-YM isolates.

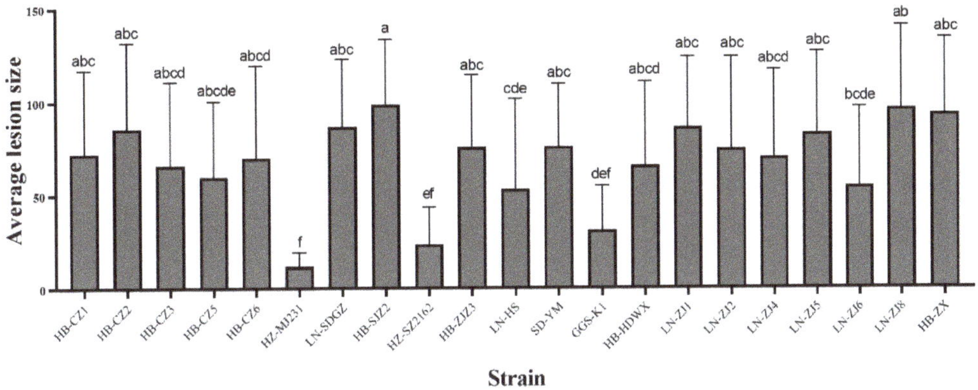

Figure 6. Differences in average damage length of isolated branches inoculated with different pathogenic strains on 'Hongxiangsu'. On the bar chart, the same letters indicate no significant difference, while different letters indicate a significant difference ($p < 0.05$).

The average diameter of lesions (ADLs) of the disease spots produced by each strain on the branches of 'Hongxiangsu' were selected for a systematic clustering analysis (Figure 7), and the virulence of the strains was determined according to the clustering results. The ADL scale was measured at 21 days of inoculation; when ADL $\geq$ 80 mm the strains were classified with strong virulence; when 40 mm $\leq$ ADL < 80 mm they were classified with medium virulence; and when ADL < 40 mm they were classified with weak virulence.

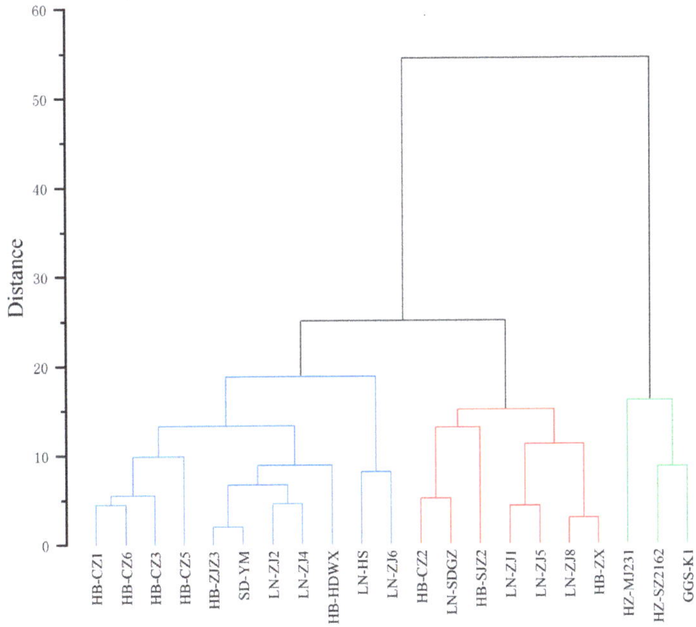

Figure 7. The average lesion length of each strain on the red fragrant crisp branches was analyzed by systematic clustering. All isolates were grouped into three categories. In the figure, the red category is the strong-virulence strain, the blue category is the moderate-virulence strain, and the green category is the weak-virulence strain.

Twenty-one strains were inoculated on the annual isolated branches of thirty germplasm resources, and according to the above criteria for the division of strong- and weak-virulence strains, the number of strains exhibiting strong and weak virulence in different germplasm resources was counted (Table 2).

Table 2. Each strain shows a different degree of virulence in the tested strains.

	Strong Virulence	Medium Virulence	Weak Virulence
GGS-K1	0	2	28
HB-CZ1	0	10	20
HB-CZ2	1	5	24
HB-CZ3	1	13	16
HB-CZ5	1	10	19
HB-CZ6	3	6	21
HB-HDWX	3	8	19
HB-SJZ2	4	8	18
HB-ZJZ3	1	12	17
HB-ZX	3	10	17
HZ-MJ231	0	1	29
HZ-SZ2162	0	2	28
LN-HS	0	8	22
LN-SDGZ	2	5	23
LN-ZJ1	4	7	19
LN-ZJ2	3	10	17
LN-ZJ4	1	9	20
LN-ZJ5	3	2	22
LN-ZJ6	2	8	20
LN-ZJ8	1	7	22
SD-YM	1	7	22

Most strains showed a weak virulence, among which HZ-MJ231 showed the weakest virulence. HB-CZ6, HB-HDWX, HB-SJZ2, HB-ZX, LN-ZJ1, LN-ZJ2, and LN-ZJ5 showed a strong virulence to three or more germplasm resources, and HB-CZ3 and HB-ZJZ3 showed a moderate virulence to more than ten germplasm resources.

3.5. Resistance Evaluation of Different Pear Germplasm Resources to Ring Rot Disease

By cluster analysis of the lesion size of different germplasm resources inoculated with HB-SJZ2, the resistance evaluation system of isolated branches was established as follows: after 21 days of trauma inoculation, ADL = 5 mm was an immune individual plant; 5 mm < ADL $\leq$ 15 mm was a high-resistance individual plant; 15 mm < ADL $\leq$ 50 mm was a medium-resistance individual plant; and 50 mm < ADL $\leq$ 80 mm was a moderately susceptible single plant. The single plant with ADL > 80 mm was highly susceptible (Figure 8).

Similarly, 21 strains were inoculated on the annual isolated branches of 30 germplasm resources, and the resistance performance of each germplasm was calculated according to the number of germplasm resources that showed resistance to different strains according to the above criteria for dividing resistant individual strains (Table 3).

The results of the resistance identification showed that most of the germplasm resources were resistant to the fungus. Among them, 12 pear germplasm resources, including 'Zaohong Comice', 'Pingguoli', 'Jinxiang', 'Wujiuxiang', 'Zaosuhong', 'Bali', 'Zaosu', 'Duli', 'Jinhua', 'Conference', 'Zhongjia 1', and 'Zhongaihongli', showed resistance to all strains. They were called ring-rot-resistant germplasm resources. In addition, 'Jingbai', 'Hongxiangsu', 'Huobali', and 'Yuluxiang' were all susceptible to the vast majority of strains, so they were defined as susceptible germplasm resources.

From the point of view of different pear series, *P. communis* showed the best resistance to ring rot disease, followed by *P. bretschneideri*, *P. ussuriensis*, *P. pyrifolia*, and *P. sinkiangensis*.

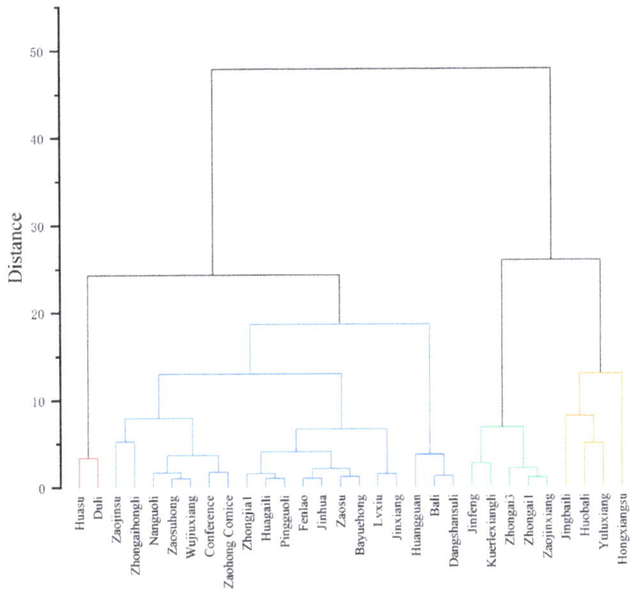

Figure 8. The average lesion size of HB-SJZ2 inoculated on different germplasm resources was analyzed utilizing systematic clustering. All germplasm resources are divided into 4 categories. In the figure, the red category is highly resistant, the blue category is moderately resistant, the green category is moderately susceptible, and the yellow category is highly susceptible.

Table 3. Resistance evaluation of different germplasm resources to strains.

	Highly Resistant	Moderately Resistant	Moderately Susceptible	Highly Susceptible
Huasu	8	10	3	0
Zaohong Comice	3	18	0	0
Pingguoli	1	20	0	0
Huangguan	1	13	7	0
Jinxiang	4	17	0	0
Zaojinsu	4	16	1	0
Nanguoli	1	19	1	0
Wujiuxiang	4	17	0	0
Bali	6	15	0	0
Zaosuhong	12	9	0	0
Jingbaili	1	6	6	8
Kuerlexiangli	1	10	10	0
Hongxiangsu	1	2	9	9
Zaosu	10	11	0	0
Duli	20	1	0	0
Jinfeng	2	18	1	0
Huagaili	4	16	1	1
Jinhua	2	19	0	0
Conference	2	19	0	0
Huobali	1	8	4	8
Lvxiu	1	12	6	2
Zhongjia 1	11	10	0	0
Bayuehong	7	13	1	0
Yuluxiang	1	2	9	9
Zhongaihongli	7	14	0	0
Dangshansuli	3	11	6	1
Zhongai 1	1	17	3	0
Fenlao	7	13	1	0
Zhongai 3	3	17	1	0
Zaojinxiang	4	15	1	1

4. Discussion

In this research, leaf, fruit, and branch samples were collected from pear-producing regions across various provinces in northern China, and the isolated strains were subjected to morphological and polygenic molecular identification [25]. The results showed that there were two pathogenic species of ring rot disease collected, *B. dothidea* and *B. kuwatsukai* [26,27]. There exists a debate among researchers regarding the etiology of ring rot disease; some posit that it results from the synergistic action of both pathogens, while others argue that one species is predominant, leading to a distinct symptomatology associated with each pathogen [8]. They believed that different species of pathogens are isolated from different plaque samples, but in this study, all of the pathogens were isolated from the samples with pear wheel pattern symptoms.

Furthermore, this study compared the cultural characteristics and virulence of the isolated *Botryosphaeria* spp. across three different culture media, revealing that the choice of medium significantly influences the colony morphology.

B. dothidea exhibits growth across various media, although the rate of growth varies significantly. PDA serves as a standard medium for fungal cultivation, with each strain demonstrating consistent and vigorous growth on this substrate. At 5 d, the mycelium typically covers the entire surface of a 90 mm Petri dish. Conversely, while strains exhibit accelerated mycelial development on OA, the aging of the mycelium occurs at an expedited rate. The growth on WA is comparatively slower for all strains, and this medium appears to inhibit mycelial proliferation, which could theoretically facilitate the germination of conidia. However, the current study did not observe any induction of conidium production, suggesting that the conditions provided were not conducive to sporulation in this fungus. Consequently, alternative strategies to enhance conidia production were deemed necessary, as the methods employed were ineffective for this particular fungal species. Further research is warranted to identify more effective sporulation techniques.

The virulence of strains isolated from various geographical regions exhibits a degree of molecular differentiation [28]. Consequently, a classification system for the 21 strains was established based on virulence standards, specifically for pear ring rot disease. This classification was determined by measuring the ADL following a 21-day trauma inoculation period. Strains were categorized as follows: those with an ADL of 80 mm or greater were classified as exhibiting strong virulence; between 40 and 80 mm were classified as exhibiting medium virulence; and less than 40 mm were classified as weak virulence. Among the strains analyzed, seven, including HB-CZ2, LN-SDGZ, HB-SJZ2, LN-ZJ1, LN-ZJ5, LN-ZJ8, and HB-ZX, were identified as exhibiting strong virulence, while three strains, namely HZ-SZ2162, HZ-MJ231, and GGS-K1, were classified with weak virulence.

In terms of the growth state, most *B. dothidea* grew close to the medium in PDA medium, while the strains belonging to *B. kuwatsukai* tended to grow upright against the lid of the Petri dish. In terms of virulence, the combination analysis of different virulence strains and phylogeny showed that there was no obvious correlation between the strength of virulence and the type of fungi. There were both strong virulence strains and weak virulence strains in *B. dothidea*. This is also the case in *B. kuwatsukai*.

The assessment of virulence across different strains significantly influences our understanding of their virulence. Given the observed differentiation in virulence among the isolates and the absence of published physiological subspecies for *B. othidea*, inoculating various strains onto different plant varieties can elucidate the resistance of those varieties to specific differentiated strains, thereby providing clearer insights into their resistance profiles. Ideally, resistant germplasm resources should demonstrate resilience against the majority of strains. Therefore, employing a diverse array of strains for inoculation can yield a more accurate representation of a variety's resistance level, mitigating the risk of misjudging resistance based on a variety's performance against a singular strain. This approach will facilitate the selection of more targeted strains for subsequent resistance evaluations.

The findings from the resistance assessment indicated that the majority of the evaluated germplasm resources exhibited resistance to ring rot disease. Notably, the germplasm resources 'Zaohong Comice', 'Pingguoli', 'Jinxiang', 'Wujiuxiang', 'Zaosuhong', 'Bali', 'Zaosu', 'Duli', 'Jinhua', 'Conference', 'Zhongjia 1', and 'Zhongaihongli' demonstrated resistance to all tested strains. Conversely, the varieties 'Yuluxiang', 'Jingbai', 'Huobali', and 'Hongxiangsu' displayed a high degree of susceptibility, being affected by nine, eight, eight, and nine strains, respectively, and exhibited greater vulnerability to *B. dothidea* in comparison to other varieties. In terms of the different pear species examined, *P. communis* emerged as the most resistant to ring rot disease, with the *P. communis* varieties in this study showing significant resistance. In contrast, certain varieties of *P. sinkiangensis* exhibited heightened susceptibility to the disease.

Currently, there is no consensus regarding the identification of resistance to pear ring rot disease, and various inoculation sites present distinct advantages and disadvantages depending on the experimental context. While leaves are convenient for use as identification materials, it is noteworthy that pear ring rot predominantly affects fruits and branches rather than leaves, raising questions about the representativeness of the results obtained from leaf-based assessments. In contrast, using fruits as inoculation materials allows for rapid disease onset and shorter experimental durations; however, the presence of latent diseases and pests within the fruits, which may not be detectable at the outset of the experiment, can significantly influence the outcomes. Furthermore, the inconsistency in fruit maturity among different individual plants, as well as the variability in the maturity of individual fruits, can also impact the experimental results. Utilizing branches as inoculation materials may provide a more accurate reflection of the field conditions, yet this approach is characterized by longer experimental durations, with disease symptoms in most individual plants typically beginning to manifest after 14 days, and significant differences becoming apparent only after 21 to 28 days. This extended timeline poses a challenge for rapid resistance identification. Overall, a comprehensive evaluation suggests that branches are more suitable for resistance identification. If an inoculation method can be developed to expedite disease progression in branches, the resistance evaluation framework could be further enhanced.

5. Conclusions

In this study, 21 strains were characterized using morphological and molecular biological techniques, with phylogenetic analysis revealing that all strains clustered closely with *B. kuwatsukai*. Examination of the colony morphology across three distinct media indicated that colonies grown on the PDA medium exhibited relative stability, with moderate rates of mycelial growth and aging. In contrast, mycelial growth and disappearance rates were more rapid on the OA medium, while growth was inhibited on the WA medium. The 21 strains were categorized into groups of strong, moderate, and weak virulence based on inoculation experiments conducted on isolated branches from various cultivars. There was no significant relationship between virulence and the geographical location of the pathogenic fungi. Additionally, the resistance of thirty cultivars were assessed, resulting in the identification of twelve resistant germplasm resources and four susceptible germplasm resources.

Supplementary Materials: The following supporting information can be downloaded at: https://www.mdpi.com/article/10.3390/horticulturae10111152/s1, Figure S1: Morphological characteristics of each strain on different media; Table S1: All culture media involved in this study; Table S2: All primer sequences involved in this study; Table S3: Gene bank used in constructing the phylogenetic tree.

Author Contributions: Conceptualization, C.W. (Chen Wang) and Y.Z.; methodology, C.W. (Chen Wang); software, C.W. (Chen Wang); validation, C.W. (Chen Wang), Y.Z. and C.O.; formal analysis, C.W. (Chen Wang); investigation, F.W. and L.M.; resources, C.W. (Chen Wang); data curation, C.W. (Chen Wang); writing—original draft preparation, C.W. (Caihong Wang) and Y.Z.; writing—review and editing, S.J. and C.W. (Caihong Wang); visualization, Y.Z.; supervision, S.J.; project administration, Y.Z.; and funding acquisition, S.J. All authors have read and agreed to the published version of the manuscript.

Funding: This research was supported by the Special talents support program of the Chinese Academy of Agricultural Sciences—Leading Talents Program; China Agriculture Research System—Pear (CARS-28); and the Agricultural Science and Technology Innovation Program of Chinese Academy of Agricultural Sciences (CAAS-ASTIP-2021-RIP-01).

Data Availability Statement: The original contributions presented in this study are included in the article/Supplementary Materials; further inquiries can be directed to the corresponding authors.

Conflicts of Interest: The authors declare no conflicts of interest.

References

1. Wu, J.; Wang, Z.W.; Shi, Z.B.; Zhang, S.; Ming, R.; Zhu, S.L.; Khan, M.A.; Tao, S.T.; Korban, S.S.; Wang, H.; et al. The genome of the pear (*Pyrus bretschneideri* Rehd.). *Genome Res.* **2013**, *23*, 396–408. [PubMed]
2. Wang, Y.; Zhang, Y.; Fan, J.; Li, H.; Chen, Q.; Yin, H.; Qi, K.; Xie, Z.; Zhu, N.; Sun, X.; et al. Physiological and autophagy evaluation of different pear varieties (*Pyrus* spp.) in response to *Botryosphaeria dothidea* infection. *Tree Physiol.* **2024**, *44*, tpad139. [CrossRef] [PubMed]
3. Wang, Y.; Sun, X.; Zhang, Z.; Pan, B.; Xu, W.; Zhang, S. Revealing the early response of pear (*Pyrus bretschneideri* Rehd) leaves during *Botryosphaeria dothidea* infection by transcriptome analysis. *Plant Sci.* **2022**, *315*, 111146. [PubMed]
4. Chen, G.J. The occurrence and prevention of pear ring rot. *Southeast Horti.* **2001**, *2*, 36.
5. Li, M.T. Occurrence Laws and Prevention and Control Technology of Pear Ring Rot. *J. Agri. Cata.* **2013**, *3*, 25–28.
6. Sun, X.; Wang, Y.; Pan, B.; Xu, W.; Zhang, S. Transcriptome analysis of pear leaves in response to calcium treatment during *Botryosphaeria dothidea* infection. *Phytopathology* **2021**, *111*, 1638–1647.
7. Zhai, L.; Zhang, M.; Lv, G.; Chen, X.; Jia, N.; Hong, N.; Wang, G. Biological and Molecular Characterization of Four *Botryosphaeria* Species Isolated from Pear Plants Showing Stem Wart and Stem Canker in China. *Plant Dis.* **2014**, *98*, 716–726.
8. Xiao, F. Gene Function of Bklip1 and Identification *botryosphaeria* Causing Pear Ring Rot in China. Ph.D. Thesis, Huazhong Agricultural University, Wuhan, China, 2023.
9. Cao, Y.F. Specification for description of pear germplasm resources. In *Descriptors and Data Standard for Pear (Pyrus* spp.*)*; Dong, Y.C., Liu, X., Eds.; Springer: Berlin/Heidelberg, Germany; China Agriculture Press: Beijing, China, 2006; pp. 35–37.
10. Long, L.; Zhou, Z.K.; Jiang, J.X. Identification of the Causal Pathogen of Pear Ring Rot in Nanchang Suburb. *South China Fruit* **2012**, *41*, 36–38.
11. Zhou, Z.Q.; Hou, H.; Wang, L.; Zhu, F.L. Trunk apple ring rot artificial inoculation method and the identification of cultivar resistance. *J. Fruit Sci.* **2010**, *27*, 952–955.
12. Wei, S.W.; Zhang, Y.; Wang, S.M. Evaluation of resistance of 43 local pear germplasm resources in Shandong province to stem ring rot and rot. *Deci. Fruit* **2016**, *48*, 15–16.
13. Zhao, M. Study on the Resistance Physiology Against the Pathogens of Anthracnose and Ring Rot on Pear Fruit. Ph.D. Thesis, Nanjing Agricultural University, Nanjing, China, 2013.
14. Yin, L.; Zou, Y.; Ke, X.; Liang, D.; Du, X.; Zhao, Y.; Ma, F. Phenolic responses of resistant and susceptible Malus plants induced by *Diplocarpon mali*. *Sci. Hortic.* **2013**, *164*, 17–23. [CrossRef]
15. Guo, Y.S.; Crous, P.W.; Bai, Q.; Fu, M.; Yang, M.M.; Wang, X.H.; Du, Y.M.; Hong, N.; Xu, W.X.; Wang, G.P. High diversity of *Diaporthe* species associated with pear shoot canker in China. *Persoonia* **2020**, *45*, 132–162. [CrossRef]
16. De Angelis, G.; Menchinelli, G.; Torelli, R.; De Carolis, E.; Posteraro, P.; Sanguinetti, M.; Posteraro, B. Different detection capabilities by mycological media for Candida isolates from mono- or dual-species cultures. *PLoS ONE* **2020**, *15*, e0226467. [CrossRef] [PubMed]
17. Li, Z.J.; Li, W.J.; Mao, Y.Y.; He, L.S.; Li, S.J. Identification and biological characteristics of branch withered disease causing pathogen to Koelreuteria paniculata Laxm. *J. Northwest AF Univ.* **2020**, *48*, 53–63.
18. Walsh, P.S.; Mitzger, D.A.; Higuchi, R. Chelex-100as a medium for simple extraction of DNA for PCR-based typing from forensic material. *Biotechniques* **1991**, *10*, 506–513. [PubMed]
19. White, T.J.; Bruns, T.D.; Lee, S.B.; Taylor, J.W. Amplification and direct sequencing of fungal ribosomal RNA genes for phylogenetics. In *PCR Protocols*; Gelfand, D.H., Sninsky, J.J., White, T.J., Eds.; Academic Press: New York, NY, USA, 1990; pp. 315–322.
20. Carbone, I.; Kohn, L.M. A method for designing primer sets for speciation studies in filamentous ascomycetes. *Mycologia* **1999**, *91*, 553–556. [CrossRef]
21. Glass, N.L.; Donaldson, G.C. Development of primer sets designed for use with the PCR to amplify conserved genes from filamentous ascomycetes. *Appl. Environ. Microbiol.* **1995**, *61*, 1323–1330. [CrossRef]
22. Kumar, S.; Stecher, G.; Tamura, K. MEGA7: Molecular evolutionary genetics analysis version 7·0 for bigger datasets. *Mol. Biol. Evol.* **2016**, *33*, 1870–1874. [CrossRef]
23. Fu, M.; Crous, P.W.; Bai, Q.; Zhang, P.F.; Xiang, J.; Guo, Y.S.; Zhao, F.F.; Yang, M.M.; Hong, N.; Xu, W.X.; et al. Colletotrichum species associated with anthracnose of *Pyrus* spp. in China. *Persoonia* **2019**, *42*, 1–35. [CrossRef]
24. Felsenstein, J. Confidence limits on phylogenies: An approach using the bootstrap. *Evolution* **1985**, *39*, 783–791. [CrossRef]

25. Minnis, A.M.; Kennedy, A.H.; Grenier, D.B.; Palm, M.E.; Rossman, A.Y. Phylogeny and taxonomic revision of the *Planistromellaceae* including its coelomycetous anamorphs: Contributions towards a monograph of the genus *Kellermania*. *Persoonia* **2012**, *29*, 11–28. [CrossRef] [PubMed]
26. Xu, C.; Wang, C.S.; Ju, L.L.; Zhang, R.; Biggs, A.R.; Tanaka, E.; Li, B.Z.; Sun, G.Y. Multiple locus genealogies and phenotypic characters reappraise the causal agents of apple ring rot in China. *Fungal Divers.* **2015**, *71*, 215–231. [CrossRef]
27. Wang, B.; Liang, X.F.; Gleason, M.L.; Zhang, R.; Sun, G.Y. Comparative genomics of *Botryosphaeria dothidea* and *B. kuwatsukai*, causal agents of apple ring rot, reveals both species expansion of pathogenicity-related genes and variations in virulence gene content during speciation. *IMA Fungus* **2018**, *9*, 243–257. [CrossRef] [PubMed]
28. Marsberg, A.; Kemler, M.; Jami, F.; Nagel, J.H.; Postma-Smidt, A.; Naidoo, S.; Wingfield, M.J.; Crous, P.W.; Spatafora, J.W.; Hesse, C.N.; et al. *Botryosphaeria dothidea*: A latent pathogen of global importance to woody plant health. *Mol. Plant Pathol.* **2017**, *18*, 477–488. [CrossRef]

horticulturae

MDPI

Article

Effects of Different Tree Shapes on Canopy Characteristics of Zaosu Pear (*Pyrus* ssp. Va.) in the Northwest Arid Zone

Mingxin Zhao [1,*,†], Mingze Ma [2,†], Wei Wang [1], Sufang Cao [1], Gang Cao [1] and Hongxu Li [1,*]

[1] Institute of Fruit and Floriculture Research, Gansu Academy of Agricultural Sciences, Lanzhou 730070, China; wangwei@gsagr.cn (W.W.); caosufang1210@gsagr.cn (S.C.); caogang@gsagr.cn (G.C.)
[2] College of Horticulture, Gansu Agricultural University, Lanzhou 730070, China; mamingze0515@163.com
* Correspondence: zmx850312@163.com (M.Z.); lihongxu@gsagr.cn (H.L.)
† These authors contributed equally to this work.

Abstract: Zaosu pear (*Pyrus* ssp. Va., Zaosu) is the main variety of pear cultivated in Gansu Province. Tree shape selection is important in pear cultivation: reasonable tree shapes affect the growth and development of fruit trees and fruit quality, which, in turn, affect the economic benefits of orchards. In this study, Zaosu pear was used as a test material, and five different tree shapes—Y, trunk, single-arm, double-arm, and spindle shapes—were studied regarding three aspects, namely, branching and leafing parameters, light radiation parameters, and root distribution, in order to provide a theoretical basis for the selection of pear tree shapes in arid areas. The results showed that the single-arm and double-arm shapes had the highest fruit weight per fruit values, reaching 280.06 g and 278.27 g, respectively, and the difference in soluble solids—12.42 $mg \cdot g^{-1}$ and 12.91 $mg \cdot g^{-1}$, respectively—was not significant; the mu yield of the single-arm shape was 3282 kg; and the yield of a single tree was 39.54 kg, which was the highest for the five tree shapes. The trees mainly had short fruiting branches when they entered the fruiting stage, and the double-arm shape had the highest number of short fruiting branches, which was significantly higher than those of the other tree shapes. Correlation analysis revealed that the number of long fruiting branches was significantly and positively correlated with axillary flower buds. The distribution of light intensity for all five tree shapes showed a gradual increase from bottom to top and from inside to outside, and the single-arm and double-arm shapes had a uniform distribution of light intensity, which translated into better growth and fruiting potential. The trunk shape had the largest leaf area index, resulting in too much depression within the canopy and reducing light, while the single-arm and double-arm shapes had the smallest leaf area indexes, resulting in the light being evenly distributed inside and outside the canopy, and the percentage of light intensity was greater than 50% in all cases, reaching up to 80%. The double-arm form had the highest chlorophyll a content of 24.55 $mg \cdot g^{-1}$, implying the ability to convert more light energy into chemical energy, and the content of chlorophyll a + b was highest in the single-arm form and the double-arm form, with values of 31.13 $mg \cdot g^{-1}$ and 33.65 $mg \cdot g^{-1}$. The total length of the root system and the total surface area of the root system for the single-arm shape were the largest, i.e., 558,550.64 cm and 79,252.32 cm^2, respectively, indicating that it has stronger adaptability and resource acquisition ability compared to the other tree shapes. In conclusion, the single-arm form is more suitable for planting in arid areas.

Keywords: Zaosu pear; tree shape; photosynthetic properties; tree growth potential

1. Introduction

China's pear species are rich in resources, and there is a long history of pear cultivation in the country, which is the foremost producer in the world, with the largest cultivation area and the highest production levels [1]. Zaosu pear (*Pyrus* ssp. Va., Zaosu) is an important fruit resource in Gansu Province [2], with a high fruit yield, thin fruit skin, and crisp, sweet juice; it is also rich in amino acids, vitamins, phenolics, and other characteristics [3–5] and has a high nutritional and economic value [6]. Tree shape selection is the basis of quality pear cultivation and a necessary condition for the production of high-quality fruit [7]. A reasonable tree structure helps to improve photosynthetic efficiency such that the pear orchard can maintain good ventilation and light-transmission conditions during the full fruiting period, thus improving the yield and quality of fruits. Similarly, proper tree shape can also reduce the cost of pruning and pest control and reduce investment in production. It also makes mechanized operations easier and improves management efficiency, so economic benefits can be further improved by optimizing tree shape.

Since ancient times, the evolution of pear tree shape has been accompanied by the deepening of human understanding of fruit tree management and technological innovation. From the initial natural growth form to the elaborate design and later regulation, every change involved a deeper understanding of fruit tree growth. Since the 1950s, the cultivation of pear trees has gradually evolved from the sparse square planting of trees with large crowns and natural round head shapes to rectangular layouts in which a variety of tree shapes coexist. In the 1980s and 1990s, rectangular dense planting was the main mode of cultivation, and tree shapes became diversified. With the adjustment of the industrial structure, pear tree shapes are becoming light and simplified. On the one hand, the trend is to adapt to the actual situations of different ecological regions, improve management efficiency, reduce labor intensity, and reduce labor costs [8]; on the other hand, light and simplified tree shapes can also receive better light, especially in arid environments where water resources are scarce, and open trees have better ventilation and light transmission [9]. The light distribution in the tree crown is directly related to the composition and spatial distribution of tree branches and the tightness of the branches and leaves in the tree crown, which affect ventilation and light transmission, temperature and humidity changes, flower bud formation, fruit flowering, and fruit quality in the tree crown. Studies have shown that the content of soluble solids in apple fruits is affected by their location in the canopy. As the light intensity in the inner layer is weaker than that in the outer layer, the content of soluble solids in the inner layer is lower than that in the outer layer [10]. Kviklys [11] found that the anthocyanin content of the apple fruits at the top of the tree crown was much higher than that of the internal fruits. In studies on citrus, it was found that the CO_2 assimilation, stomatal conductance, and transpiration rates of leaves in the upper canopy were significantly higher than those in the middle layer, while those in the middle layer were significantly higher than those in the lower layer [12]. From this point of view, the adequacy of light conditions is the most critical factor for high and stable yields and the quality of fruit trees. Light and simplified tree shapes can make it easier for light to reach the interior of the tree crown, improving fruit quality and photosynthetic efficiency.

The trend towards labor-saving, lightweight, and simplified cultivation practices is an inevitable development in the fruit industry [13]. An appropriate tree structure can enhance spatial distribution and optimize light penetration within the canopy, which are crucial factors for achieving high quality yields in fruit production. Consequently, investigating suitable tree shapes is a vital aspect of enhancing both the quality and efficiency of pear production. This study aimed to elucidate the effects of various tree shapes on the growth and development, photosynthetic characteristics, yield formation, and fruit quality of Zaosu pear, thereby providing a theoretical foundation and practical guidance for selecting tree shapes in arid regions. The following hypotheses were formulated: (1) different tree shapes exhibit significant variations in canopy light distribution and photosynthetic capacity; (2) in arid environments, certain tree structures demonstrate superior light interception and response capabilities that contribute to enhanced fruit quality and improved economic returns.

2. Materials and Methods

2.1. Overview of the Test Site

The experimental park is located in Tiaoshan Farm, Jingtai County, Gansu Province, China (37°10′ N, 104°3′ E). located in the southern edge of Tengger Desert, with an altitude of 1619.5 m. It is a temperate arid climate with an average annual temperature of 8.2 °C, annual precipitation of 184.8 mm, annual sunshine duration of 2725 h, and annual frost-free period of 141 days. The orchard soil is sandy limestone soil, and the soil pH is 8.0, deep soil layer and organic matter content is 1.4% [7].

2.2. Test Materials

The experiment was conducted at the demonstration base of the Nongken Strip Hill Group in Jingtai County, Baiyin City, Gansu Province, for a period of five consecutive years, from 2018 to 2022. For each tree form, five trees exhibiting similar growth characteristics to the remainder of the garden were selected as test trees. Each tree was regarded as a distinct experimental unit, with a total of five replicates. As illustrated in Table 1, the test material comprises the existing varieties and trees in the test garden. All tree forms are five years old, the rootstock is Du Pear, the terrain is flat, irrigation is by drip, conventional management is employed, and fertilizer management is consistent. The trees are planted in a north–south row formation and cultivated to achieve a neat and standard shape.

Table 1. (*Pyrus* ssp. Va., Zaosu) tree species and plantings.

Cultivar	Cultivation Mode	Tree Shape	Row Space (m × m)	Tree Height (m)	Trunk Height (m)	Acres of Trees (Plants)
Zaosu pear	Vertical Cultivation	Trunk	1 × 4	3	0.6	164
		Spindle	2 × 4	3	0.6~0.7	83
	Shelf Cultivation	Y shape	2 × 4	3	0.5	83
		Single-arm	2 × 4	2~2.5	1.5~1.7	83
		Double-arm	4 × 4	2~2.5	1.5	42

2.3. Fruit Quality and Measurement of Tree Structure Indicators

The number of branches and axillary flower buds were enumerated manually. Following the measurement of branch types with a tape measure, the number of nutritive and fruit-bearing branches were recorded in accordance with the following criteria: long branches: >15 cm; medium branches: 5–15 cm; and short branches: <5 cm. Flower buds formed subsequently to the current year's branches were counted as axillary flower buds, and the number was recorded.

$$\text{Individual tree fruit yield} = \text{acre yield}/\text{acre of trees}$$

2.4. Determination of Canopy Light Distribution of Different Tree Shapes of Zaosu Pear

The light intensity was determined in accordance with the methodology proposed by He [14] et al. with a minor adaptation. One marker was positioned at a distance of 0.5 m in four directions—southeast, northwest, and north—with the central trunk serving as the center point. Each marker was used as the ground base point to divide the tree crown into a 0.5 m × 0.5 m × 0.5 m cube. Additionally, the vertical direction of the canopy was divided into three levels: the lower level (less than 1.0 m from the ground), the middle level (1.0 to 2.0 m), and the upper level (>2.0 m). In mid-July, a sunny and windless day was selected, and the TES-1332A digital illuminance meter (TES Electrical Electronic Corp., Taipei, China) was utilized to determine the light intensity of the canopy at varying levels and orientations at 11:00 and 15:00. The light intensity of an unshaded open space was employed as a control.

$$\text{Relative light intensity} = \text{light intensity within the canopy}/\text{control}$$

2.5. Measurement of Leaf Area Index, Radiant Transmission Coefficient and Extinction Coefficient

The leaf area index was determined through the use of the CI-110 canopy analyzer (CID Bio-Science, Camas, WA, USA). Canopy images were collected from the east, south, west, and north directions at a distance of 50 cm from the trunk under the canopy. The collected images were analyzed and processed using the canopy analysis software (Plant

Canopy Analysis System V6.0, CID Bio-Science,) to calculate the leaf area index, radiation transmission coefficients, and extinction coefficients.

2.6. Measurement of Chlorophyll Content

The chlorophyll content was determined in accordance with the methodology proposed by Ergun E [15] et al. with minor modifications. A total of 40 mature leaves from annual branches were selected from four locations at the periphery of the pear canopy, specifically the east, west, south, and north. The surface dirt was then removed from the leaves. Fresh leaves (0.1 g) were cut into 0.2 cm wide pieces and placed in 20 mL of ethanol–acetone (Sinopharm Chemical Reagent Co. Ltd., Shanghai, China) (1:1) extract. The absorbance was then measured at different wavelengths (663 nm and 645 nm) using a UV-2550 spectrophotometer (Shimadzu, Japan).

$$\text{Total chlorophyll (FW}\cdot\text{mg/g)} = (20.2\ \text{A645} + 8.02\ \text{A663}) \times \text{V}/(1000 \times \text{FW})$$

Here, FW represents fresh weight, and V represents the volume of the extract (mL).

2.7. Root Soil Collection

The profile excavation method was employed to obtain soil samples from a range of soil layers and horizontal distances within the root system. Samples of trees with different shapes were collected according to the planting density and spacing of the plant rows. The trunk and overall shape of the tree were sampled in a profile measuring 1.2 m long, 1.0 m wide, and 1.0 m deep. The spindle, Y, single-arm, and double-arm shapes were sampled in sections 2.0 m long, 1.0 m wide, and 1.0 m deep centered on the trunk. A homemade sampler (20 cm long, 20 cm wide, 20 cm height) was used to collect soil samples in the profile, including roots. The samples were collected in 20 cm × 20 cm × 20 cm cubes, with one cube for every 20 cm in the vertical and horizontal directions, for a total of 40 cubes per tree. The rooted soil samples were poured into a 0.1 mm sieve for soaking and rinsing and then scanned and analyzed using the WinRHIZO (2016 a, Regent Instruments Inc., Quebec, Canada) root analysis system. This system was used to classify pear roots into two categories: absorption roots (diameter < 2 mm) and transport roots (diameter < 10 mm for 2 mm ≤ diameter). This classification was based on the function and morphology of the root system. Additionally, the root lengths and surface areas of the root systems of different diameter classes in each sampling cube were analyzed. Finally, root volume indexes were calculated.

2.8. Data Analysis

Excel 2021 (Microsoft, Redmond, WA, USA) was used for data processing and charting, Origin 2024 (Origin Inc., San Francisco, CA, USA) software was used for bar charts and line graphs, and data were analyzed for significance using SPSS 26.0 (SPSS Inc., Chicago, IL, USA) for analysis of variance and multiple comparisons using one-way ANOVA (One Way ANOVA), with a significance level of $p < 0.05$.

3. Results

3.1. Effects of Different Tree Shapes on Fruit Quality, Yield and Branches of Zaosu Pear

The effects of different tree shapes on pear fruit quality and yield were studied. According to the results (Table 2), the single fruit weight of single-arm and double-arm shapes was the largest, i.e., 280.06 g and 278.27 g, which was significantly higher than that of other trees. There was no difference in the longitudinal diameter of fruit among the five types of trees. The transverse diameter of fruit of double-arm-shaped trees was the largest, reaching 7.68 cm; the fruit hardness of spindle-shaped trees was the largest, reaching 16.70 N, which was significantly higher than that of other trees. There was no difference between trunk and single-arm shape and Y and double-arm shape. There was no difference in soluble solids content between Y and spindle shape and the double-arm shape had the highest content, which was 12.91 mg·g^{-1}.

Table 2. Effect of different (*Pyrus* ssp. Va., Zaosu) tree shapes on fruit quality and yield of date palm pear.

Shape	Fruit Weight (g)	Vertical Diameter (cm)	Horizontal Diameter (cm)	Firmness (N)	Soluble Solid (mg·g^{-1})	Yield Per Mu (kg)	Yield Per Tree (kg)
Y shape	251.11 ± 4.90 b	8.51 ± 0.13 a	7.27 ± 0.25 ab	14.65 ± 0.24 c	12.07 ± 0.44 bc	2910	35.06
Trunk	243.47 ± 4.98 b	8.37 ± 0.19 a	7.17 ± 0.23 ab	15.39 ± 0.29 b	11.63 ± 0.24 c	2784	16.98
Single-arm	280.06 ± 5.87 a	8.85 ± 0.24 a	7.36 ± 0.18 ab	15.83 ± 0.17 b	12.42 ± 0.27 ab	3282	39.54
Double-arm	278.27 ± 5.10 a	8.77 ± 0.13 a	7.68 ± 0.16 a	14.67 ± 0.18 c	12.91 ± 0.27 a	1597	38.02
Spindle	214.95 ± 6.10 c	8.53 ± 0.30 a	7.10 ± 0.26 b	16.70 ± 0.08 a	11.82 ± 0.39 bc	2506	30.19

Note: The data are presented as mean ± standard error of three biological replicates. Different lowercase letters indicate significant differences ($p < 0.05$).

After statistics (Table 2), it was found that the fruit yield per mu of the single-arm shape was the highest, which was 3282 kg. Since the double-arm shape had fewer trees per mu, the yield per mu was the lowest, which was 1597 kg. However, in terms of the yield of a single tree, the double-arm shape was second only to the single-arm shape.

The vegetative branches of different tree shapes are mainly long branches and short branches (Figure 1A). Long branches are mostly concentrated in the middle and upper parts of the tree body, while short branches are mostly concentrated in the lower part of the tree body. The two-armed long branches and medium branches have the largest number, i.e., 74.67 and 10, respectively, which are significantly higher than other tree shapes. Because these two tree shapes are flat scaffolds, the top advantage leads to the development of too many upright branches on the back; short branches of the Y and trunk shapes are the most common, and the number of short branches of the trunk shape is relatively stable, indicating that this tree shape mainly maintains the stability of the size and yield of the crown through the proportion of short and medium fruit branches.

There were large differences in the total number of fruiting branches in different tree shapes (Figure 1B), especially in the number of short fruiting branches. The number of double-arm-shaped short fruit branches is the largest—597.33, which is significantly higher than other tree shapes, followed by Y and spindle-shaped, there is no difference in the number of short fruit branches between the Y and spindle-shaped. The number of trunk-shaped short fruit branches is the lowest, only 186. There is no difference in the number of long fruit branches between the trunk and single-arm shape.

Pearson correlation coefficient was used to analyze the relationship between the number of long fruiting branches, medium fruiting branches, and short fruiting branches and the number of axillary flower buds (Figure 1C). It was found that the number of long fruit branches was significantly positively correlated with the formation of axillary flower buds. As the number of long fruit branches increased, the number of axillary flower buds also increased. The correlation between the number of medium fruit branches and axillary flower buds was relatively weak.

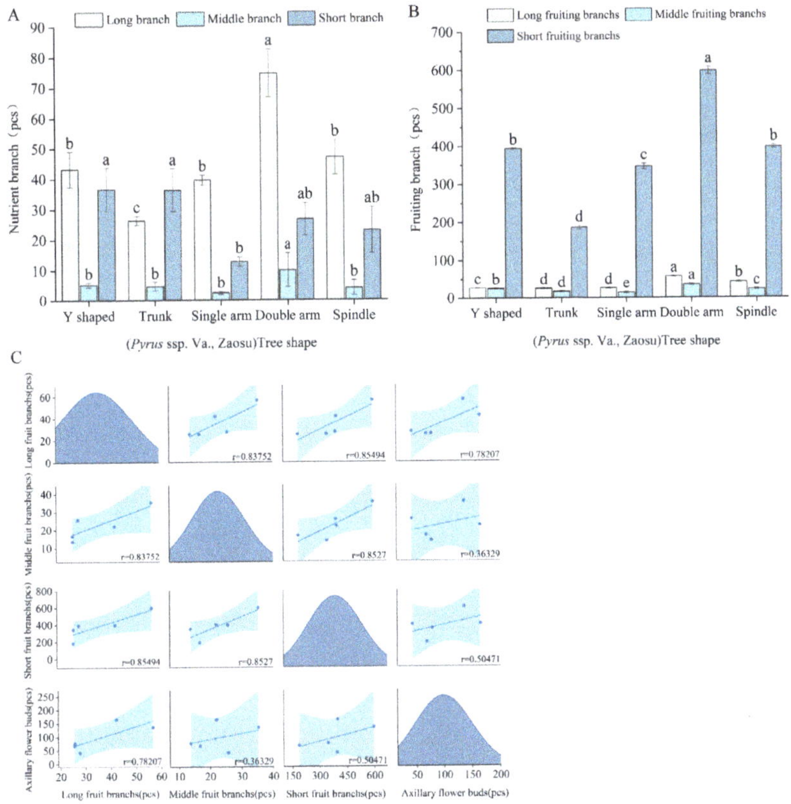

Figure 1. Effects of different (*Pyrus* ssp. Va., Zaosu) tree shapes on branches and flower buds: (**A**) number of nutrient branches; (**B**) number of fruiting branches; (**C**) relationship between the number of fruiting branches and axillary bud. (The dots represent the variables, the lines are the fit curves, the shaded areas represent the 95% confidence bands, and the diagonals are the distribution curves for each variable.) The data are presented as mean ± standard error of three biological replicates. Different lowercase letters indicate significant differences ($p < 0.05$).

3.2. *Effects of Different Tree Shapes on Light Distribution and Relative Light Intensity at Canopy Level*

Among the five tree shapes, the Y, single-arm, and double-arm shapes had the lowest proportion of inner-bore intercepted light radiation <30% (Table 3), indicating that there were few branches in the inner bore and most branches could see the light, which was conducive to receiving light energy; the spindle shape accounted for 65% of the 30–80% effective light area, and the entire crown had good lighting; the trunk shape accounted for 50%, while the Y, single-arm and double-arm shapes accounted for 35% each; the proportion of single-arm and double-arm shapes with light intensity greater than 80% reached 50%, which was higher than other tree shapes.

The light intensity distribution of the five tree shapes all shows a trend of gradually increasing from bottom to top and from inside to outside (Figure 2). The light intensity of the vertical cultivated tree structure in the center of the trunk is relatively low, which may be because the trunk and internal branches and leaves block the penetration of light. The light intensity of different tree structures is usually higher at 1 m on both sides, indicating that the light can be distributed more evenly to both sides of the tree. The Y-shape has a high proportion of light intensity in the center, mainly because the Y-shaped tree makes it easier for light to reach the center. The light distribution of the Y, single-arm, and double-arm tree shapes is significantly better than that of the main trunk and spindle shapes, which is mainly closely related to management measures and tree structure. The single-arm and double-arm shapes have a uniform distribution of light intensity and have better growth and fruiting potential.

Table 3. Proportion of relative light intensity in the canopy of different (*Pyrus* ssp. Va., Zaosu) tree shapes (%).

Shape	<30%	30–59%	60–80%	>80%
Y shape	15	25	10	40
Trunk	40	25	25	10
Single-arm	15	30	5	50
Double-arm	15	25	10	50
Spindle	25	40	25	10

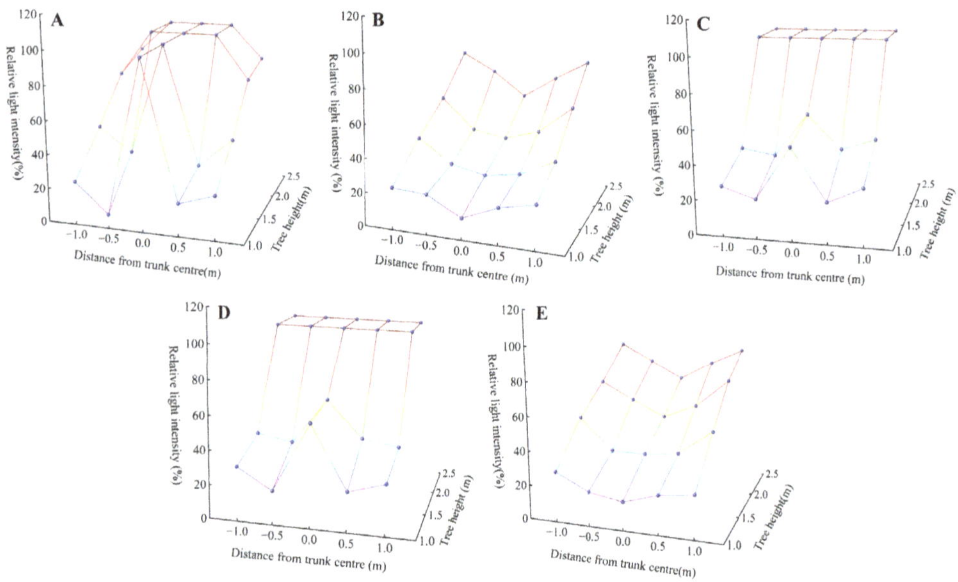

Figure 2. Relative light intensity in the canopy of different (*Pyrus* ssp. Va., Zaosu) tree shapes: (**A**) Y shape; (**B**) trunk; (**C**) single-arm; (**D**) double-arm; (**E**) spindle. Different colors represent different ranges of light intensity.

3.3. *Effect of Different Tree Shapes on Leaf Area Index and Light Radiation Characteristics*

Leaf area index (LAI) is a comprehensive index that represents the light energy utilization status of vegetation and canopy structure. In this study (Figure 3), the trunk-shaped leaf area index is the highest at 2.35, which is significantly higher than other tree shapes, followed by Y-shaped, with a leaf area index of 1.87, and single-arm, double-arm, and spindle-shaped leaf areas. The index is the lowest, and there is no difference between the three, which is consistent with the results of the tree structure investigation.

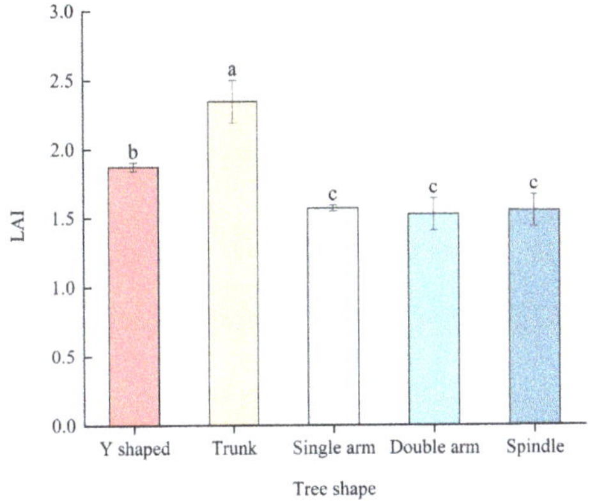

Figure 3. Effect of different (*Pyrus* ssp. Va., Zaosu) tree shapes on leaf area index. The data are presented as mean ± standard error of three biological replicates. Different lowercase letters indicate significant differences ($p < 0.05$).

3.4. Analysis of Radiation Transmission Coefficient and Extinction Coefficient of Different Tree Forms at Different Zenith Angles

Radiation transmission coefficients varied among tree shapes at different zenith angles (Table 4), with an overall trend of increasing and then decreasing transmission coefficients with increasing zenith angle, reflecting the effect of tree structure on light transmission and the change in the path of light through the canopy with the change in solar altitude angle. The Y-shape and single-arm leaf areas were exposed to the greatest radiation coefficient when the zenith angle was 29.75°; the trunk and double-arm were exposed to the greatest radiation coefficient when the zenith angle was 46.75°; and the spindle-shape was exposed to the greatest radiation coefficient at an angle of 55.25°.

Table 4. Radiation transmission coefficients of different (*Pyrus* ssp. Va., Zaosu) tree forms at different zenith angles.

Tree Shape	Radiative Transmittance									
	4.25	**12.75**	**21.25**	**29.75**	**38.25**	**46.75**	**55.25**	**63.75**	**72.25**	**80.75**
Y shape	0.103	0.095	0.175	0.285	0.278	0.243	0.200	0.158	0.090	0.090
Trunk	0.035	0.045	0.073	0.133	0.260	0.288	0.215	0.155	0.080	0.030
Single-arm	0.180	0.285	0.273	0.385	0.293	0.205	0.195	0.198	0.180	0.065
Double-arm	0.256	0.168	0.210	0.244	0.352	0.382	0.330	0.220	0.120	0.104
Spindle	0.328	0.135	0.130	0.195	0.245	0.300	0.378	0.338	0.280	0.138

The radiation transmittance coefficient of different tree shapes varies at different zenith angles (Table 4). The overall trend is that with the increase in the zenith angle, the transmittance coefficient first increases and then decreases, reflecting the influence of the tree structure on light transmittance and the change of the path of light passing through the tree crown with the change of the solar altitude angle. When the zenith angle is 29.75°, the Y and single-arm shapes receive the largest radiation coefficient; when the zenith angle is 46.75°, the trunk and double-arm shape receive the largest radiation coefficient; and the spindle shape receives the largest radiation coefficient at an angle of 55.25°.

The spindle shape has the largest extinction coefficient (K) at the zenith angle (4.25–72.25), followed by the Y and trunk shape. The single-arm shape has the smallest extinction coefficient between the zenith angle (4.25–72.25), and as the zenith angle increases, the extinction coefficients of different tree shapes all show an upward trend (Table 5), indicating that the longer the path length of light passing through the tree, the higher the extinction coefficient. A canopy structure with a large extinction coefficient will limit photosynthesis and lead to slow growth; conversely, the plant has sufficient photosynthesis and grows faster, which, in turn, affects the quality of the internal branches and buds.

Table 5. Extinction coefficients of different (*Pyrus* ssp. Va., Zaosu) tree forms at different zenith angles.

Tree Shape	Zenith Angle									
	4.25	12.75	21.25	29.75	38.25	46.75	55.25	63.75	72.25	80.75
Y shape	0.103	0.095	0.175	0.285	0.278	0.243	0.200	0.158	0.090	0.090
Trunk	0.035	0.045	0.073	0.133	0.260	0.288	0.215	0.155	0.080	0.030
Single-arm	0.180	0.285	0.273	0.385	0.293	0.205	0.195	0.198	0.180	0.065
Double-arm	0.256	0.168	0.210	0.244	0.352	0.382	0.330	0.220	0.120	0.104
Spindle	0.328	0.135	0.130	0.195	0.245	0.300	0.378	0.338	0.280	0.138

3.5. Effect of Different Tree Shapes on Leaf Chlorophyll Content

There are significant differences in the chlorophyll a content of different tree shapes (Figure 4). The chlorophyll a content of the double-arm tree shape is relatively high. The difference in chlorophyll a content among different tree shapes is the largest, which indicates that chlorophyll a is more sensitive to the physiological conditions and the growth environment of different tree shapes. The single-arm and double-arm shapes have higher chlorophyll b content, while the trunk and spindle shapes have lower chlorophyll b content. The double-armed tree shape has the highest chlorophyll a + b content, which indicates that it has higher photosynthetic capacity. Followed by Y and single-arm shape. The total amount of chlorophyll a + b is a more comprehensive indicator of the difference in photosynthetic ability of plants, and it can be seen that the different tree shapes of plants may affect their overall chlorophyll content, which may, in turn, affect their photosynthetic efficiency.

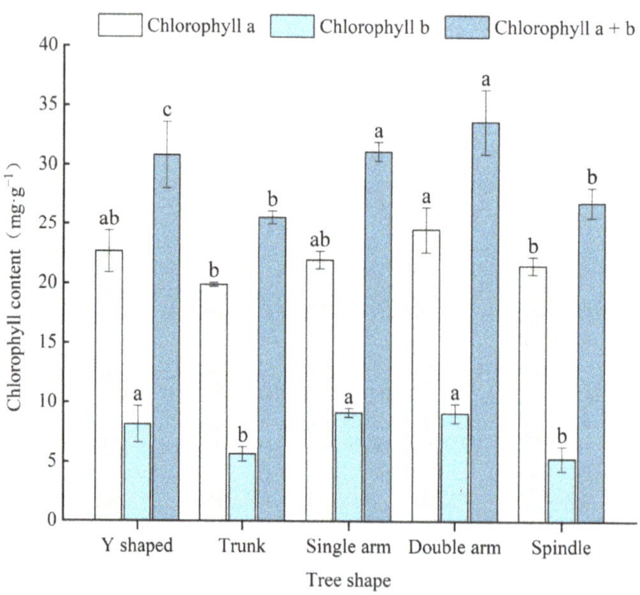

Figure 4. Effect of different (*Pyrus* ssp. Va., Zaosu) tree shapes on chlorophyll content. The data are presented as mean ± standard error of three biological replicates. Different lowercase letters indicate significant differences ($p < 0.05$).

3.6. Root Distribution and Growth of Different Tree Forms

The root lengths of different tree shapes vary significantly (Figure 5A). The single-arm shape has the longest absorptive root, reaching 540,674.32 cm, which is significantly higher than other tree shapes. The trunk-shaped root system has the shortest length, and the single-arm shape is 2.38 times longer. The tree shapes with the longest conductive roots are the single-arm and spindle-shaped at 17,876.32 cm and 18,210.51 cm, respectively. The root volumes of different tree shapes differed significantly (Figure 5B). The root volume of the single-arm absorptive root is still the largest at 1011.39 cm^3, which is significantly higher than other tree shapes. There is no significant difference between the Y and the spindle shape, and there is no significant difference between the main trunk and the double-arm shape. The largest conductive root volume is the spindle shape, reaching 3821.44 cm^3, while there are no differences among the Y, trunk, and single-arm shapes. The double-arm shape has the smallest root volume at only 1075.70 cm^3. The root surface areas of different tree shapes differed significantly (Figure 5C). The single-arm shape has the largest absorptive root surface area, which is significantly higher than other tree shapes, reaching 59,179.34 cm^2. There is no difference in the surface area of the Y- and trunk-shaped roots. The smallest is the trunk-shaped root, which is only 23,228.24 cm^2. The largest conductive root surface area is the spindle-shaped root at 24,563.83 cm^2, which is significantly higher than other tree shapes. The smallest ones are the trunk and the double-arm shape, which are 9762.35 cm^2 and 9105.90 cm^2, respectively, and there is no difference between them.

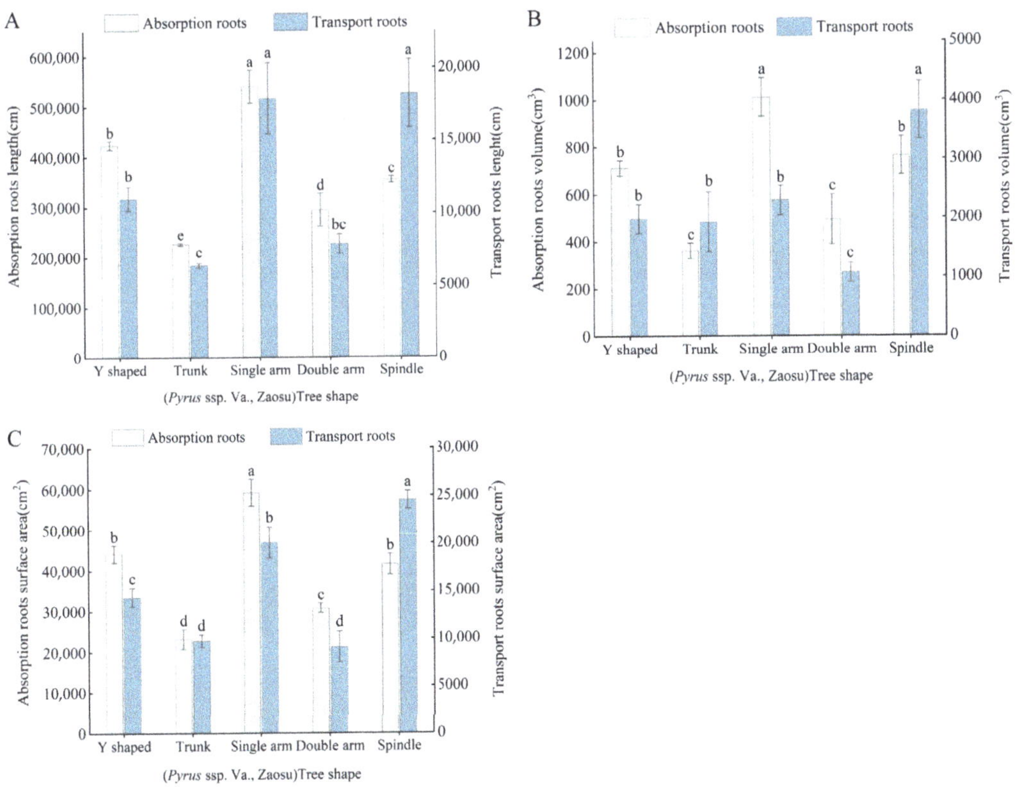

Figure 5. Root growth of different (*Pyrus* ssp. Va., Zaosu) tree shapes: (**A**) root length; (**B**) root volume; (**C**) root surface area. The data are presented as mean ± standard error of three biological replicates. Different lowercase letters indicate significant differences ($p < 0.05$).

When the absorbing roots and transporting roots are added together, the single-arm root system has the largest total length and total root surface area, which are 558,550.64 cm and 79,252.32 cm^2, respectively. The spindle-shaped root system has the largest total root volume, which reaches 4584.39 cm^3.

3.7. Correlation Analysis of Fruit Quality, Yield, and Photosynthetic Characteristics

Correlation analysis showed (Figure 6A) that the fruit horizontal diameter was positively correlated with soluble solids and chlorophyll a + b, there was a positive correlation between hardness and root volume, soluble solids and chlorophyll a + b, as well as between root length and root surface area; and there was a negative correlation between per mu yield and the number of nutrient branches.

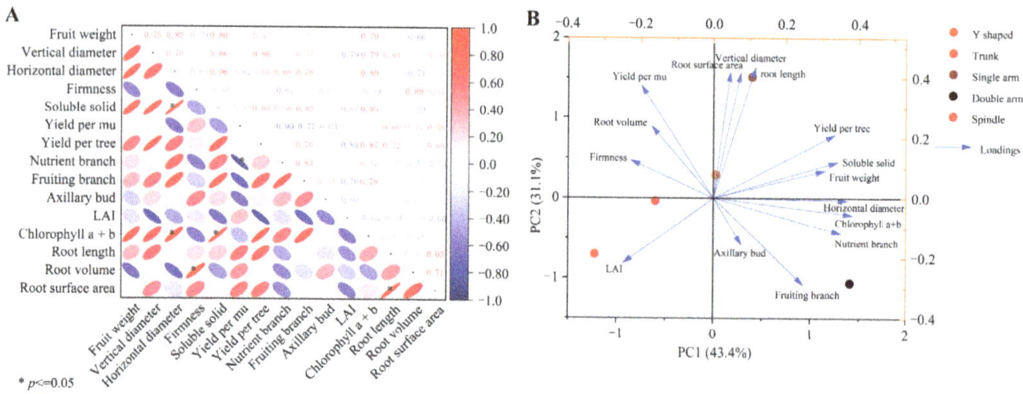

Figure 6. Pearson's correlation analysis (**A**) and principal component analysis (**B**) of fruit quality, yield, and photosynthetic characteristics of different (*Pyrus* ssp. Va., Zaosu) tree shapes. Data are expressed as average values. * represent significant correlations at the $p < 0.05$ levels (two-tailed).

The principal component analysis found (Figure 6B) that the sum of the first and second principal components reached 74.5%, explaining 43.4% and 31.1%, respectively. It can be seen from the figure that chlorophyll a + b and fruit diameter are the key indicators affecting the first principal component, while root length, root surface area, and fruit diameter are the key indicators affecting the second principal component. In addition, the acute angles between the five indicators, namely, soluble solids, single fruit weight, fruit diameter, chlorophyll a + b, and nutrient branches, can reflect the positive correlation between them.

4. Discussion

Appropriate cultivation mode and tree shape selection are the keys to the healthy development of the pear industry, reducing costs and increasing efficiency [16]. Tree shape directly affects fruit quality and yield in fruit tree cultivation. Different tree shape designs can optimize the photosynthetic efficiency, ventilation conditions, and nutrient distribution of fruit trees. These factors are crucial to fruit development. A high proportion of short branches can ensure the early high and stable yields of fruit-bearing trees [17]. The results of this study show that the number of long fruit branches is positively correlated with the formation of axillary flower buds, and the correlation between the number of medium fruit branches and axillary flower buds is relatively positive. Smaller, different tree shapes have a certain impact on the growth of branches, and the number of short fruit branches in the double-arm tree shape is significantly higher than that of other tree shapes. Wagenmakers [18] et al. found that the light distribution of the apple canopy has a significant impact on its yield and quality. Yoon [19] et al.'s research results showed that different tree shapes of cherries in South Korea will affect the increase in the number and decrease in length of new branches. At the same time, the cherry fruit yield of the elevated tree shape can reach 1.86~3.15 times that of other tree shapes. Zhao [20] et al. used six-year-old "Tiantaowang" peach as a test material to study the effects of different tree shapes on peach leaves, fruit quality, and yield. They found that tree shape affects chlorophyll, carotenoids, fruit shape index, fruit weight, and soluble sugar. The content has a significant impact, which is similar to the results of this study. Different tree shapes also have a significant impact on the fruit quality of pear. Similarly, differences in regional climate are also important factors affecting fruit quality. Lal [21] et al. studied the effects of different tree shapes on nectarine fruit quality in temperate regions and found that Y-shaped trees are better than other tree shapes.

A good tree structure has more reasonable canopy characteristics [22], more even distribution of branches in the canopy, and better light transmittance of the tree, which is conducive to the inner leaves receiving more light, thereby improving the photosynthetic efficiency of the canopy. This study found that among the five tree shapes, the proportion of single-arm and double-arm shapes with light intensity greater than 80% reached 50%, which was higher than other tree shapes. This may be because the leaf area index is small and the light inside the canopy is sufficient. The distribution of relative light intensity in the canopy is closely related to the tree shape and branch composition [23]. This study found that the distribution of light intensity in the five tree shapes all showed an increasing trend from bottom to top and from inside to outside, which is consistent with the results of Lu [24] et al. and Zhang [25] et al. The Y-shape has a high light intensity in the center, which may be because the Y-shaped structure allows light to reach the center more easily. The light distribution of the three tree shapes cultivated on the rack is significantly better than that of the two tree shapes cultivated on the vertical rack. This result is consistent with the results of Anthony [26] et al. in their study of light distribution in peach trees.

Leaf area index is used to evaluate the growth status of crop populations. Within a certain range, as the leaf area index increases, the crop yield will also increase. This is because the increase in leaf area brings more photosynthetic area, thereby increasing the efficiency of photosynthesis. However, an excessively large leaf area index will cause the crown to be too closed, reducing the amount of light received by the leaves, thereby reducing the photosynthetic efficiency. Leaf area index and diffuse radiation transmittance are important parameters for characterizing the structure of vegetation canopies [27]. They are closely related to the processes of light energy interception and utilization, yield, and quality formation in orchards and, to a certain extent, determine the production efficiency of orchards [28,29]. The photosynthetic rate of plants is related to the properties of leaves [30]. In addition to leaves, sufficient and high light efficiency are also conducive to photosynthesis, thereby improving fruit quality. In apples, LAI and K can be used as important parameters for evaluating the optical properties of the canopy [31]. Reasonable LAI is the main condition for fruit trees to fully utilize light energy. If LAI is too large or too small, it will affect photosynthetic production or reduce the accumulation of assimilated products of fruit trees. In this study, the leaf area index of different tree shapes was ranked from high to low as trunk > Y > single-arm > spindle > double-arm shape. This may be due to the different densities of leaves per unit area in different (*Pyrus* ssp. Va., Zaosu) tree shapes. This shows that single-arm, double-arm, and spindle-shaped tree shapes can effectively avoid canopy closure caused by dense leaves in the middle and lower parts of the crown. The trunk shape has a large number of main branches and often has a higher leaf density, which leads to more leaves growing in a relatively small space, thereby increasing the total leaf area, affecting the ventilation and light transmission of the tree, causing the lower branches to die and the fruit to move outward. Therefore, reasonable planting density is a key factor in orchard production.

The chlorophyll content of plant leaves is an important indicator of photosynthetic capacity and physiological activity [32]. Both chlorophyll a and chlorophyll b are important components of the light-harvesting pigment complex, affecting the absorption, transmission and conversion of light energy [33]. They are affected by the influence of light shading [34]. Different tree shapes have significant effects on chlorophyll content, which, in turn, affects photosynthetic efficiency and fruit quality. Tree shapes with better openness (such as the Y and single-arm shape) usually help to improve the light energy utilization of leaves. These structures can improve the light distribution within the canopy such that the lower leaves can also receive sufficient light, thus increasing chlorophyll. content. In this study, the chlorophyll a + b content of the double-arm shape was higher than that of the Y and single-arm shape, which may be due to the influence of the drought environment.

The root system of fruit trees is responsible for fixing the tree, absorbing and transporting water, mineral nutrients, and a small amount of organic matter, and also has the function of storing and synthesizing organic matter [35]. Root morphological configuration indicators such as root length, root surface area, root diameter, and root volume largely determine the root system's ability to absorb water and nutrients [36]. In this study, the single-arm root system had the longest total root length and the largest root surface area and root volume, indicating that the single-arm root system has a strong growth potential and absorption capacity, which helps plants obtain more nutrients and water to adapt to different growth conditions. The organization of the root system, including shape and position, especially the position of the functional root segment, has a direct impact on the plant's ability to explore soil nutrients [37–39]. In this study, the single-arm root system performed best in root growth, with a wider root distribution and larger root length, root surface area, and root volume.

5. Conclusions

In this study, different tree shapes were found to have an impact on the fruit quality, branch composition, canopy light, chlorophyll, and root system of pear trees. The single-arm and double-arm shapes had larger single fruit weights and higher soluble solids content. The single-arm shape had an acreage yield of 3282 kg, which was the highest among the five tree shapes. At the same time, the canopy light intensity of the two tree shapes was evenly distributed, and the proportion of strong light greater than 80% was the largest, and the chlorophyll a + b content was the highest. The single-arm shape performed best in root growth, with a wider root distribution, a larger root length, root surface area, and root volume, and better growth and fruiting potential. In summary, the single-arm shape can be used for the development and promotion of the Zaosu pear industry in the arid areas of Northwest China.

This study provides ideas and data support for the selection of pear tree shapes in arid areas. However, the study was conducted in an arid area, which limits the application of the research results to other climates or regions. In addition, the research content still needs to be deepened in terms of photosynthesis. The next step of research should be in this direction in order to better understand the complex relationships between tree shape, tree photosynthetic characteristics, growth, and development.

Author Contributions: Conceptualization, M.Z. and H.L.; methodology, S.C.; validation, G.C. and H.L.; formal analysis, M.Z. and M.M.; investigation, S.C.; resources, W.W. and S.C.; data curation, M.Z. and M.M.; writing—original draft preparation, M.Z. and M.M.; writing—review and editing, W.W. and H.L.; visualization, M.M.; supervision, G.C.; project administration, G.C.; funding acquisition, M.Z. and H.L. All authors have read and agreed to the published version of the manuscript.

Funding: This research was funded by National Modern Agricultural Industrial Technology System (CARS-28-47); Gansu Provincial Science and Technology Major Special Program (23ZDNA001); Gansu Provincial Academy of Agricultural Sciences Regional Collaborative Innovation Project (2024GAAS06).

Data Availability Statement: The original contributions presented in the study are included in the article, further inquiries can be directed to the corresponding authors.

Conflicts of Interest: The authors declare no conflicts of interest.

References

1. Teng, Y.W. The Pear industry and research in China. *Acta Hortic.* **2011**, *909*, 161–170. [CrossRef]
2. Huang, Y.; Li, Y.C.; Li, D.M.; Bi, Y.; Liu, Y.X.; Mao, R.Y.; Zhang, M.; Jiang, Q.Q.; Wang, X.J.; Prusky, D. Molecular Characterization of Phospholipase C in Infection Structure Differentiation Induced by Pear Fruit Surface Signals, Stress Responses, Secondary Metabolism, and Virulence of *Alternaria alternata*. *Phytopathology* **2022**, *112*, 2207–2217. [CrossRef] [PubMed]
3. Jiang, W.; Yan, P.; Zheng, Q.; Wang, Z.; Chen, Q.; Wang, Y. Changes in the Metabolome and Nutritional Quality of Pulp from Three Types of Korla Fragrant Pears with Different Appearances as Revealed by Widely Targeted Metabolomics. *Plants* **2023**, *12*, 3981. [CrossRef] [PubMed]
4. Coşkun Topuz, F.; Bakkalbaşı, E. Physical, chemical and bioactive properties of four different pears (*Pyrus communis* L.) varieties grown in Turkey. *Yüzüncü Yıl* **2022**, *27*, 303–314. [CrossRef]
5. Sun, H.L.; Wang, X.Q.; Cao, X.Y.; Liu, C.; Liu, S.Q.; Lyu, D.; Du, G. Chemical composition and biological activities of peels and flesh from ten pear cultivars (*Pyrus ussuriensis*). *Food Meas.* **2021**, *15*, 1509–1522. [CrossRef]
6. Li, X.; Li, X.J.; Wang, T.T.; Gao, W.Y. Chapter 24—Nutritional Composition of Pear Cultivars (*Pyrus* spp.). In *Nutritional Composition of Fruit Cultivars*; Academic Press: Cambridge, MA, USA, 2016; pp. 573–608.
7. Zhao, M.; Sun, W.; Li, H.; Wang, W.; Cao, G.; Wang, F. The Effects of the Tree Structure of Zaosu Pear on the Transport and Distribution of Photosynthetic Assimilates and Fruit Quality under Desert-Area Conditions. *Agronomy* **2022**, *12*, 2440. [CrossRef]
8. Hamada, N.A.; Moreira, R.R.; Nesi, C.N.; May De Mio, L.L. Pathogen dispersal and Glomerella leaf spot progress within apple canopy in Brazil. *Plant Dis.* **2019**, *103*, 3209–3217. [CrossRef]
9. Kiran Sagar, D.C.; Sathish, B.B.; Jyothsna, J. Canopy management and its influence on plant architecture to produce good yield and quality in fruit crops. *Adv. Hortic.* **2022**, *25*, 53–68.
10. Kaučić, M.; Vuković, M.; Gašpar, L.; Fruk, G.; Vidrih, R.; Nečemer, M.; Fruk, M.; Jatoi, M.A.; Fu, D.; Kobav, M.B.; et al. The Effect of Canopy Position on the Fruit Quality Parameters and Contents of Bioactive Compounds and Minerals in 'Braeburn'Apples. *Agronomy* **2023**, *13*, 2523. [CrossRef]
11. Kviklys, D.; Viškelis, J.; Liaudanskas, M.; Janulis, V.; Laužikė, K.; Samuolienė, G.; Uselis, N.; Lanauskas, J. Apple Fruit Growth and Quality Depend on the Position in Tree Canopy. *Plants* **2022**, *11*, 196. [CrossRef]
12. Dian, Y.Y.; Liu, X.Y.; Hu, L.; Zhang, J.Z.; Hu, C.G.; Liu, Y.Z.; Zhang, J.X.; Zhang, W.B.; Hu, Q.Q.; Zhang, Y.H.; et al. Characteristics of photosynthesis and vertical canopy architecture of citrus trees under two labor-saving cultivation modes using unmanned aerial vehicle (UAV)-based LiDAR data in citrus orchards. *Hortic. Res.* **2023**, *10*, 2052–7276. [CrossRef] [PubMed]
13. Lu, L.; Niu, Z.; Jiang, C.; Yu, L.; Hongning, W.; Qiao, M. Influences of open-central canopy on photosynthetic parameters and fruit quality of apples(*Malus* × *domestica*) in the Loess Plateau of China. *Hortic. Plant J.* **2022**, *8*, 133–142.
14. He, F.L.; Wang, F.; Wei, Q.P.; Wang, X.W.; Zhang, Q. Relationships Between the Distribution of Relative Canopy Light Intensity and the Peach Yield and Quality. *Agric. Sci. China* **2008**, *7*, 297–302. [CrossRef]
15. Ergun, E.; Demirata, B.; Gumus, G.; Apak, R. Simultaneous determination of chlorophyll a and chlorophyll b by derivative spectrophotometry. *Anal. Bioanal. Chem.* **2004**, *379*, 803–811. [CrossRef]
16. Hampson, C.R.; Quamme, H.A.; Brownlee, R.T. Canopy growth, yield, and fruit quality of'Royal Gala'apple trees grown for eight years in five tree training systems. *HortScience* **2002**, *37*, 627–631. [CrossRef]
17. Zhou, S.; Shen, Z.; Yin, B.; Liang, B.; Li, Z.; Zhang, X.; Xu, J. Effects of Dwarfing Interstock Length on the Growth and Fruit of Apple Tree. *Horticulturae* **2023**, *9*, 40. [CrossRef]
18. Wagenmakers, P.S.; Callesen, O. Light distribution in apple orchard systems in relation to production and fruit quality. *J. Hortic. Sci.* **1995**, *70*, 935–948. [CrossRef]
19. Yoon, L.K.; Nam, E.Y.; Yun, S.K.; Chung, K.H. Overhead arbor tree shape affects tree development and increases fruit yields in sweet cherry. *Acta Hortic.* **2017**, *1161*, 193–196. [CrossRef]
20. Zhao, C.P.; Wang, Q.X.; Han, M.Y.; Wang, A.Z.; Liu, H.K. Effects of tree shape on the quality of leaf and fruit and the yield in peach. *Agric. Food Sci.* **2010**, *38*, 160–164, 170.
21. Lal, S.; Sharma, O.C.; Singh, D.B. Effect of tree architecture on fruit quality and yield attributes of nectarine (*Prunus persica* var. nectarina) cv. Fantasia under temperate condition. *Indian J. Agric. Sci.* **2017**, *87*, 1008–1012. [CrossRef]
22. Nunes, L.; Pasalodos-Tato, M.; Alberdi, I.; Sequeira, A.C.; Vega, J.A.; Silva, V.; Vieira, P.; Rego, F.C. Bulk Density of Shrub Types and Tree Crowns to Use with Forest Inventories in the Iberian Peninsula. *Forests* **2022**, *13*, 555. [CrossRef]
23. Sharma, Y.; Singh, H.P.; Singh, S.P. Effect of light interception and penetration at different levels of fruit tree canopy on quality of peach. *Curr. Sci.* **2018**, *115*, 1562–1566. [CrossRef]
24. Lu, Y.; Si, Y.; Zhang, L.; Sun, Y.; Su, S. Effects of Canopy Position and Microclimate on Fruit Development and Quality of *Camellia oleifera*. *Agronomy* **2022**, *12*, 2158. [CrossRef]
25. Zhang, Y.; Wen, Y.; Bai, Q.; Ma, Z.; Ye, H.; Su, S. Spatio-temporal effects of canopy microclimate on fruit yield and quality of *Sapindus mukorossi* Gaertn. *Sci. Hortic.* **2019**, *251*, 136–149. [CrossRef]
26. Anthony, B.M.; Minas, I.S. Optimizing Peach Tree Canopy Architecture for Efficient Light Use, Increased Productivity and Improved Fruit Quality. *Agronomy* **2021**, *11*, 1961. [CrossRef]
27. Wei, B.; Ma, X.; Guan, H.; Yu, M.; Yang, C.; He, H.; Wang, F.; Shen, P. Dynamic simulation of leaf area index for the soybean canopy based on 3D reconstruction. *Ecol. Inform.* **2023**, *75*, 102070. [CrossRef]

28. Xiao, Z.; Lei, H.; Jin, C.; Pan, H.; Lian, Y. Relationship between the Dynamic Characteristics of Tomato Plant Height and Leaf Area Index with Yield, under Aerated Drip Irrigation and Nitrogen Application in Greenhouses. *Agronomy* **2023**, *13*, 116. [CrossRef]
29. Tustin, D.S.; Breen, K.C.; van Hooijdonk, B.M. Light utilisation, leaf canopy properties and fruiting responses of narrow-row, planar cordon apple orchard planting systems—A study of the productivity of apple. *Sci. Hortic.* **2022**, *294*, 110778. [CrossRef]
30. Wang, G.; Zeng, F.; Song, P.; Sun, B.; Wang, Q.; Wang, J. Effects of reduced chlorophyll content on photosystem functions and photosynthetic electron transport rate in rice leaves. *J. Plant Physiol.* **2022**, *272*, 153669. [CrossRef]
31. Knerl, A.; Anthony, B.; Serra, S.; Musacchi, S. Optimization of leaf area estimation in a high-density apple orchard using hemispherical photography. *HortScience* **2018**, *53*, 799–804. [CrossRef]
32. Hailemichael, G.; Catalina, A.; González, M.R.; Martín, P. Relationships between water status, leaf chlorophyll content and photosynthetic performance in Tempranillo vineyards. *S. Afr. J. Enol. Vitic.* **2016**, *37*, 149–156. [CrossRef]
33. Elango, T.; Jeyaraj, A.; Dayalan, H.; Arul, S.; Govindadamy, R.; Prathap, K.; Li, X. Influence of shading intensity on chlorophyll, carotenoid and metabolites biosynthesis to improve the quality of green tea: A review. *Energy Nexus* **2023**, *12*, 100241. [CrossRef]
34. Wang, Y.B.; Huang, R.D.; Zhou, Y.F. Effects of shading stress during the reproductive stages on photosynthetic physiology and yield characteristics of peanut (*Arachis hypogaea* Linn.). *J. Vol. Issue* **2021**, *20*, 1250–1265. [CrossRef]
35. Sun, M.; Liu, X.; Shi, K.; Peng, F.; Xiao, Y. Effects of Root Zone Aeration on Soil Microbes Species in a Peach Tree Rhizosphere and Root Growth. *Microorganisms* **2022**, *10*, 1879. [CrossRef] [PubMed]
36. Li, X.X.; Zeng, R.S.; Liao, H. Improving crop nutrient efficiency through root architecture modifications. *J. Integr. Plant Biol.* **2016**, *58*, 193–202. [CrossRef] [PubMed]
37. Chen, W.; Chen, Y.; Siddique, K.H.; Li, S. Root penetration ability and plant growth in agroecosystems. *Plant Physiol. Biochem.* **2022**, *183*, 160–168. [CrossRef]
38. Galindo-Castañeda, T.; Lynch, J.P.; Six, J.; Hartmann, M. Improving soil resource uptake by plants through capitalizing on synergies between root architecture and anatomy and root-associated microorganisms. *Front. Plant Sci.* **2022**, *13*, 827369. [CrossRef]
39. Fort, F. Grounding trait-based root functional ecology. *Funct. Ecol.* **2023**, *37*, 2159–2169. [CrossRef]

horticulturae

MDPI

Article

Impact of Aerated Irrigation Duration on the Growth of Greenhouse Grape Seedlings and Rhizosphere Soil Microorganisms

Yinchi Liang [1,2], Wei Wang [1,2], Jinshan Xi [1,2], Shiqiang Zhao [1,2], Yufan Zhou [1,2], Huanhuan Zhang [1,2], Kun Yu [1,2], Zhong Zheng [3,*] and Fengyun Zhao [1,2,*]

1 Department of Horticulture, College of Agriculture, Shihezi University, Shihezi 832003, China; liangyinchi@stu.shzu.edu.cn (Y.L.); wangweipatti@shzu.edu.cn (W.W.); 20222012045@stu.shzu.edu.cn (J.X.); zhaoshiqiang23@163.com (S.Z.); zhouyufan119@163.com (Y.Z.); 20212312221@stu.shzu.edu.cn (H.Z.); yukun410@163.com (K.Y.)

2 Key Laboratory of Special Fruits and Vegetables Cultivation Physiology and Germplasm Resources Utilization of Xinjiang Production and Construction Corps, Department of Horticulture, College of Agriculture, Shihezi University, Shihezi 832003, China

3 Department of Agronomy, College of Agriculture, Shihezi University, Shihezi 832003, China

* Correspondence: zhzh_agr@shzu.edu.cn (Z.Z.); zhaofyshz@163.com (F.Z.)

Abstract: This study investigates the effects of varying durations of aerated irrigation, administered at a consistent frequency, on the growth of greenhouse grape seedlings and the structure of the rhizosphere soil microbial community. Using two-year-old 'Flame Seedless' grape seedlings as the test material, we established a control group with no aeration (CK) and three treatment groups with aeration durations of 10 min (T1), 20 min (T2), and 30 min (T3), respectively. We determined grape seedling growth under different aerating durations. Additionally, changes in the rhizosphere soil microbial community of the plants were analyzed using 16S and ITS high-throughput genome sequencing to further explore the correlation between microbial diversity and plant growth. The results revealed that: (1) Aerated irrigation significantly enhanced plant growth, with the T2 treatment yielding superior increases in plant height, above-ground dry weight, below-ground dry weight, total root length, and root volume compared to T1 and T3 treatments. (2) Aeration treatments notably elevated the Shannon and Chao1 indices of the rhizosphere soil fungal community, with the T2 treatment exhibiting the most substantial effects, and the Shannon index of the bacterial community was also significantly higher under the T2 treatment. (3) The T2 treatment significantly increased the relative abundance of beneficial aerobic bacterial genera such as *Flavobacterium*, *Ellin6067*, and *Coniochaeta*, while decreasing the relative abundance of detrimental fungal genera like *Fusarium* and *Gibberella*. In conclusion, a 20 min aeration duration can effectively promote grape seedling growth, enhance the diversity of rhizosphere soil microbial communities, increase beneficial aerobic microorganisms, and reduce harmful ones. This study provides a theoretical basis for optimizing aerated irrigation practices in facility grape cultivation.

Keywords: grapes; aerated irrigation duration; growth; Illumina sequencing; microbial community

1. Introduction

Facility agriculture is a specialized cultivation technology designed to enhance the productivity of plants and animals under relatively controlled environmental conditions. It represents an important advancement in modern agriculture [1] and plays an increasingly vital role in the global food system [2,3]. In China, the facility grape industry has experienced rapid growth, with the area dedicated to facility grape cultivation reaching 230,700 hectares by the end of 2016, positioning the country as the second largest producer worldwide [4]. Soil, as an important resource in agricultural production systems, can provide the necessary water and nutrients for crop growth and development, and soil conditions such as nutrient effectiveness, porosity, and aeration can significantly affect crop root structure [5]. Key environmental factors, such as water, fertilizer, air, and heat, are crucial for plant growth and maintaining soil fertility [6]. However, in many facility agriculture systems, practices such as excessive irrigation; over-fertilization; prolonged mechanization; inadequate tillage; and imbalances in water, heat, and air can lead to a substantial decline in soil oxygen levels. Consequently, crop roots become increasingly sensitive to hypoxic stress [7].

Subsurface drip irrigation technology is widely adopted in agricultural production due to its high irrigation efficiency and effective water utilization [8]. However, traditional subsurface drip irrigation methods can lead to water saturation in soil pores [9], resulting in decreased soil aeration and stagnant water conditions [10]. This decline in soil oxygen can cause hypoxia in plant roots, leading to hypoxic stress [11] that adversely affects plant growth, development, yield, and quality [1]. Enhancing soil aeration is critical for improving soil fertility [12], which in turn promotes plant health. Aerated irrigation represents an innovative solution to address these challenges by integrating underground cavity storage with drip irrigation. This technique employs air pumps and ventilation devices to deliver air directly to the crop root zone. Additionally, Venturi aerating equipment can inject micro-bubbles or a water–gas mixture into the underground drip irrigation pipeline [13], ensuring that oxygen is continuously supplied to the root zone [14]. The micro-bubbles possess a large surface area and long lifespan, making them effective for mitigating soil hypoxia [15] and satisfying the oxygen demands of both the crop root system and the surrounding microbial community [16,17]. This can greatly promote crop growth and development and improve water utilization efficiency and overall yield [18]. Aerated irrigation technology has been widely implemented in crops such as cotton [19] and tomatoes [20]. Bhattarai et al. [21] demonstrated that tomatoes under aerated treatment exhibited a 21% increase in fresh weight compared to non-aerated treatments.

Microbial communities are vital components of soil ecosystems, playing a key role in regulating ecosystem functions and soil biogeochemistry [22]. They are integral to nutrient cycling and maintenance of soil structure, and their diversity can effectively respond to changes in soil quality [23,24]. Factors such as climatic conditions and irrigation practices can directly impact the structure of these microbial communities [25]. Variations in soil oxygen content significantly affect the diversity and composition of microbial communities [26]. For example, Chen et al. [27] showed that aerated irrigation increased the abundance and diversity of soil microorganisms in tomato rhizosphere soil, thereby enhancing soil respiration rates. Li Yuan et al. [9] found that aerated irrigation significantly affected microbial populations in melons. Research by Chen [28] and others indicated that increased oxygen levels could boost the relative abundance of aerobic microbes and facultative anaerobic microbes while reducing the prevalence of strictly anaerobic species. The rhizosphere is home to a diverse array of microorganisms that are essential parts of the soil–plant ecosystem, with the composition and diversity of these communities directly influencing plant growth and development [29].

Although aerated irrigation technology effectively addresses oxygen deficiency in the crop root zone, there is limited research on how different aerated irrigation durations affect plant growth and the structure of rhizosphere soil microbial communities. Our experiment focuses on two-year-old own-rooted 'Flame Seedless' grape seedlings, adopting aerated irrigation technology while using conventional non-aerated treatment as a control. The study aims to evaluate the effects of different aeration durations on grape seedling growth and the microbial community structure in the rhizosphere, ultimately identifying the optimal duration for aerated irrigation. This research seeks to provide a theoretical basis for the scientific and rational application of aerated irrigation in agricultural production, facilitating further optimization and improvement.

2. Materials and Methods

2.1. Plant Material and Experimental Design

Our experiment was carried out from April to December 2023 in a solar greenhouse at the College of Agriculture, Shihezi University, located in Shihezi City, New Uygur Autonomous Region (44°19′ N, 85°58′ E). During the experiment, greenhouse temperatures ranged from 16 to 32 °C, and relative humidity varied between 60% and 80%. A potting experiment was employed using topsoil collected from the vineyard of the Experimental Station at Shihezi University, specifically at a depth of 0–30 cm. The soil, classified as gray desert soil, was sieved through a 40-mesh screen prior to use. The soil's nutrient composition was analyzed, revealing a total nitrogen content of 1.22 g·kg^{-1}, ammonium nitrogen of 2.63 mg·kg^{-1}, nitrate nitrogen of 2.01 mg·kg^{-1}, quick-acting phosphorus of 43.6 mg·kg^{-1}, and quick-acting potassium of 305 mg·kg^{-1}. The soil's bulk density was measured at 1.40 g·cm^{-1}, the particle density at 2.65 g·cm^{-1}, and the pH was recorded at 6.56. The experimental pots used were cylindrical, with a mouth diameter of 30 cm, a base diameter of 25 cm, and a height of 45 cm.

Uniformly grown two-year-old own-rooted 'Flame Seedless' grape seedlings were planted in the center of each pot after acclimatization on 16 May 2023 (Figure 1), Inject oxygen gas that does not contain pollutants. A subsurface cavity storage drip irrigation system was used for drip irrigation and air injection treatment (Figure 1). The experimental design included a control group with no aeration (CK) and three treatments with different aeration durations: 10 min (T1), 20 min (T2), and 30 min (T3) per day. Each treatment consisted of nine plants in a single-plant replication. Aerated irrigation treatments commenced 30 days post-planting, with irrigation performed daily at 9 a.m. for a duration of 30 days. The irrigation water volume was kept consistent across treatments. An air compressor was used for aeration, maintaining a pressure of 0.04 MPa and a gas flow rate of 55 L·min^{-1}. The aeration volume was determined according to the formula $V = 0.0003\ (rR + r^2 + R^2)\ (1 - \rho_b/\rho_s)\ \pi h$ [30], where V represents the aeration volume (L), r is the radius of the pot mouth (15 cm), R is the radius of the pot bottom (12.5 cm), ρ_s is the soil density (2.65 g·cm^{-3}), ρ_b is the soil bulk density (1.40 g·cm^{-3}), and h is the pot height (45 cm). This resulted in an aeration volume of 11.38 L per pot per filling. A rigid PVC pipe, 10 cm in height and 5 cm in diameter, was used as the underground gas injection bucket for drip irrigation. This pipe was sealed at the top with openings at the bottom, featuring evenly distributed small holes (0.5 cm in diameter) along the pipe wall. The gas injection bucket was placed 30 cm deep, with a distance of 5 cm between the plant and the bucket. Valves connected the main tube to the branch tubes, and a gas solenoid valve linked to a control center was used to regulate the opening and closing of each valve, allowing for precise control over the duration of gas filling for each treatment.

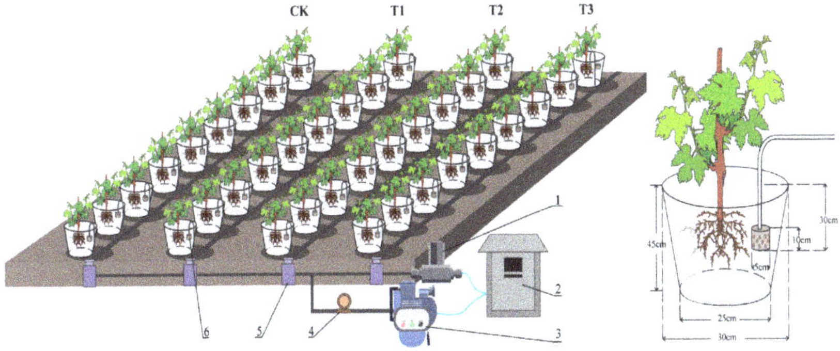

Figure 1. Schematic diagram of experimental design. Note: 1. water storage; 2. control center; 3. air compressor; 4. gas flow meter; 5. switch; 6. gas injection bucket for drip irrigation in underground cavity.

2.2. Measurement Items and Methods

2.2.1. Determination of Growth Indicators of Grape Seedlings

Growth indicators, including plant height, stem thickness, and SPAD values, were measured at 15, 18, 21, 24, and 27 days after the initiation of aerated irrigation treatments. Plant height was determined using a tape measure, stem thickness was assessed with a vernier caliper, and SPAD values were recorded using a SPAD-502 Plus chlorophyll meter.

2.2.2. Determination of Biomass and Root Activity in Grape Seedlings

After 30 days of aerated irrigation treatments, five uniformly growing grape seedlings were randomly selected from each treatment for destructive sampling.

Root activity was estimated using the modified triphenyl tetrazolium chloride (TTC) method following root removal [31]. The fresh roots were washed with dH_2O thrice, blotted on filter paper, and then stored at 4 °C to be used the same day. The standard curve was plotted based on the spectrophotometric absorption (λ = 485 nm) of different amounts of TTC (1 g·0.1 L^{-1}) solution with $Na_2S_2O_4$ and ethyl acetate. Roots were then sliced into one-centimeter pieces and immersed in 10 mL of an equally mixed solution of TTC (0.4%) and phosphate buffer (0.1 mol·L^{-1}, pH 7.0), then kept in the dark for 3 h at 37 °C. Subsequently, 2 mL of H_2SO_4 (1 mol·L^{-1}) was added to stop the reaction. The immersed root tips were dried with filter paper and extracted with ethyl acetate. The extracted solution was transferred into a tube with ethyl acetate cleaning solution to a total volume of 10 mL, and absorbance was read at 485 nm. Root activity was calculated from the standard curve and expressed as TTC reduction intensity: $\mu g \cdot g^{-1} \cdot h^{-1}$.

The dry weights (g) of both above-ground and below-ground plant parts were determined separately. The plant samples were dried in an oven preheated to 105 °C for 30 min, and then the temperature was adjusted to 80 °C until the weight was constant.

2.2.3. Determination of Morphological Indicators of Plant Roots

After 30 days of aerated irrigation treatments, three uniformly growing grape plants, free from pests and diseases, were selected for each treatment. The soil from the top 0–30 cm layer was collected, and the root systems were quickly extracted and rinsed with sterile water. The root systems were then scanned with a root scanner, and the scanned images were analyzed by WinRHIZO software purchased in Beijing (Regent, Vancouver, BC, Canada) [32]. This analysis provided morphological indicators of the root systems, including effective root surface area (cm^2), effective root volume (cm^3), total root length (cm), and the number of root tips. Following scanning, the roots were dried at 80 °C until a constant weight was achieved, and the dry weight of the root systems was subsequently recorded.

2.2.4. Rhizosphere Soil Sampling

Rhizosphere soil sampling was conducted using a soil auger method after 30 days of aerated irrigation treatments. For each cultivation pot, four sampling points were established evenly at a horizontal distance of 5 cm from the grape plant. Soil samples of 100 g were collected at a vertical depth of 20–30 cm from the soil surface. The soil from the four sampling points was thoroughly mixed, and this process was repeated five times. The collected soil samples were immediately sieved through a 40-mesh sieve, mixed again, and stored in a freezer at −80 °C for subsequent 16S and ITS high-throughput genome sequencing.

2.3. Soil DNA Extraction and Sequencing

Total soil DNA was extracted using the Fast DNATM SPIN Kit for Soil (MP Biomedicals, Irvine, CA, USA). The purity and concentration of the extracted DNA were assessed via 1.2% agarose gel electrophoresis. An appropriate volume of the sample taken in a centrifuge tube was diluted to 1 ng·μL^{-1} with sterile water and stored in a refrigerator at −80 °C for backup. The diluted DNA genome served as a template for PCR amplification, employing

Phusion® High-Fidelity PCR Master Mix with GC from New England Biolabs (Ipswich, MA, USA). Specific primers with Barcode Buffer and high-efficiency, high-fidelity enzymes targeting the V3-V4 regions of the bacterial 16S rRNA gene (primers: CCTAYGGGRBG-CASCAG and GGACTACNNGGGTATCTAAT) and the intra-fungal transcriptional spacer (ITS) genes (primers: CTTGGTCATTTAGAGGAAGTAA and TGCGTTCTTCATCGATGC) were used. The PCR amplification procedure included pre-denaturation at 98 °C for 1 min, followed by 30 cycles of denaturation at 98 °C for 10 s, annealing at 50 °C for 30 s, and extension at 72 °C for 30 s, concluding with a final extension at 72 °C for 5 min. The PCR products were then mixed, purified, and prepared for library construction and on-board sequencing.

2.4. Bioinformatics Analysis and Data Processing

The data for each sample were separated based on the Barcode sequences and PCR amplification primers. The sequences of the Barcode and primers were extracted using FLASH to splice the sequences for each sample, yielding high-quality raw data. These data were then filtered to remove chimeric sequences, resulting in valid sequences. Amplicon Sequence Variants (ASVs) were generated through noise reduction using DADA2 based on the validated data, and the representative sequences of each ASV underwent species annotation and abundance analysis, revealing the species composition and abundance distribution of the samples.

2.5. Statistical Analysis

Data organization was performed using Excel 2010. Statistical analyses were conducted using SPSS 27, Duncan's multiple comparisons was used as a post hoc test after a significant difference was found in ANOVA ($p < 0.05$). Graphs were created using Origin 2021 and Graphpad Pism 9.5 software. Pearson correlation coefficients were calculated to analyze the relationship between microorganisms and growth indicators, with a significance level set at $p < 0.05$.

3. Results

3.1. Effects of Aerated Irrigation Duration on Growth Indicators of Grape Seedlings

3.1.1. Impact on Plant Height, Stem Thickness, and SPAD Value of Grape Seedlings

As illustrated in Figure 2, a moderate increase in the duration of aerated irrigation significantly enhanced grape plant height. The heights of grape seedlings exhibited an initial increase followed by a decrease with prolonged aerated irrigation duration. Notably, 24 days after aerated irrigation, the height of the plants under T2 treatment was significantly greater than that under other treatments, showing increases of 78.3%, 27.7%, and 31.9% compared to CK, T1, and T3 treatments, respectively. At 27 days post-aeration, the plant height rankings were T2 > T3 > T1 > CK, with significant differences observed. In contrast, stem thickness did not exhibit significant variation across treatments. At each measurement period, the SPAD values under CK and T2 treatments were significantly higher than those under T1 and T3 treatments. Specifically, at 24 days post-aeration, the SPAD values were reduced by 6.2%, 22.8%, and 19.2% under CK, T1, and T3 treatments, respectively, in comparison to T2 treatment.

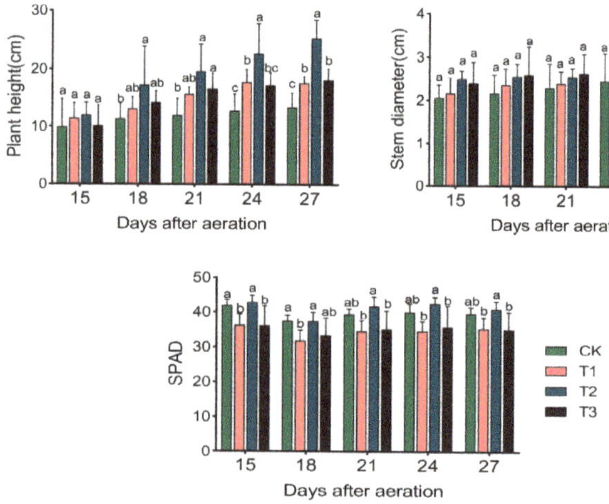

Figure 2. Impact of aerated irrigation duration on plant height, stem diameter, and SPAD value of grape seedlings, aeration durations are 0 min (CK), 10 min (T1), 20 min (T2), and 30 min (T3) per day. Note: Different lowercase letters indicate significant differences among treatments ($p < 0.05$).

3.1.2. Impact on Biomass and Root Activity of Grape Seedlings

As shown in Table 1, both above-ground and below-ground dry biomass exhibited increasing trends followed by decreases with increasing durations of aerated irrigation, with the highest values recorded under T2 treatment. There were no significant differences in above-ground dry biomass between T1 and T2 treatments; however, both were significantly greater than CK and T3. For below-ground dry biomass, T2 treatment resulted in increases of 49.0%, 34.9%, and 25.7% compared to CK, T1, and T3 treatments, respectively, with these differences being significant. Root activity under different treatments showed, in descending order, T2 (0.0520 $\mu g \cdot g^{-1} \cdot h^{-1}$) > CK (0.0482 $\mu g \cdot g^{-1} \cdot h^{-1}$) > T1 (0.0458 $\mu g \cdot g^{-1} \cdot h^{-1}$) > T3 (0.0455 $\mu g \cdot g^{-1} \cdot h^{-1}$), with the T2 treatment yielding the highest value, which was 7.9%, 13.8% and 13.5% higher than the CK, T1, and T3 treatments, respectively.

Table 1. Impact of aerated irrigation duration on dry biomass and root activity of grape seedlings, aeration durations are 0 min (CK), 10 min (T1), 20 min (T2), and 30 min (T3) per day.

Treatment	Above-Ground Dry Biomass (g)	Below-Ground Dry Biomass (g)	Root System Activity ($\mu g \cdot g^{-1} \cdot h^{-1}$)
CK	8.71 ± 0.13 b	4.69 ± 0.09 c	0.0482 ± 0.00 a
T1	10.28 ± 0.47 a	5.18 ± 0.27 bc	0.0458 ± 0.00 a
T2	10.38 ± 0.10 a	6.99 ± 0.38 a	0.0520 ± 0.01 a
T3	9.84 ± 0.52 ab	5.56 ± 0.13 b	0.0455 ± 0.00 a

Note: Different lowercase letters indicate significant differences among treatments ($p < 0.05$).

3.1.3. Impact on Root Morphology of Grape Seedlings

As shown in Figure 3, the T1, T2, and T3 treatments significantly increased the root volume and the number of root tips of grape seedlings compared to CK treatment, and positively affected the total root length. The T2 treatment exhibited the highest increases, with the total root length enhanced by 19.1%, 4.7%, and 11.9% compared to CK, T1, and T3 treatments, and the root volume increased by 16.6%, 12.5%, and 22.6%. Noteworthily, the number of root tips was significantly greater under T1 treatment compared to other treatments. While the root surface area did not show significant differences across treatments in comparison with CK, it displayed varying degrees of increase, peaking under T2 treatment. In summary, T2 treatment proved to be the most beneficial for the root growth of grape seedlings.

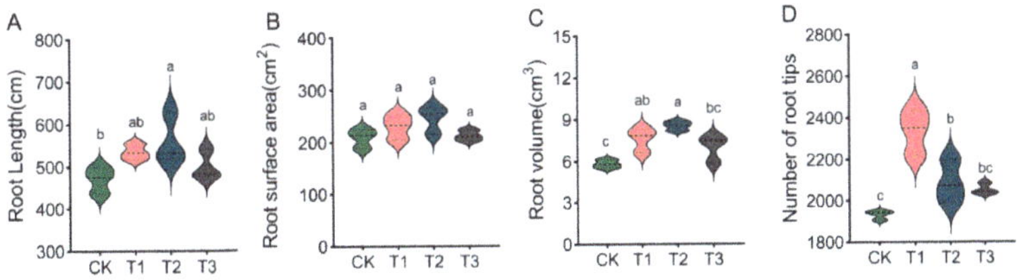

Figure 3. Impact of aerated irrigation duration on root morphological indices of grape seedlings, aeration durations are 0 min (CK), 10 min (T1), 20 min (T2), and 30 min (T3) per day. (**A**) Root length; (**B**) Root surface area; (**C**) Root volume; (**D**) Number of root tips. Note: Different lowercase letters indicate significant differences among treatments ($p < 0.05$).

3.2. Effects of Aerated Irrigation Duration on Microbial Diversity of the Rhizosphere Soil of Grape Seedlings

High-throughput sequencing yielded 3,632,356 valid sequences for fungi and 3,534,942 for bacteria. Each sample generated a minimum of 68,705 valid sequences, averaging 100,898 for bacteria and 98,193 for fungi. Following noise reduction with DADA2, 96,576 ASVs were identified for bacteria and 9164 for fungi. The sequencing depth index coverage for both bacterial and fungal samples exceeded 99.6%, indicating comprehensive representation of microbial composition. This suggests that the depth of sequencing essentially covered all species present, ensuring reliable sequencing results.

Alpha diversity indices can accurately reflect microbial richness and diversity. In this experiment, Chao1 and Shannon indices were used to assess the differences in the richness and diversity of microbial composition in the rhizosphere soil of grape seedlings. As shown in Table 2, the Shannon and Chao1 indices for rhizosphere soil bacteria under all treatments exhibited an increasing and then decreasing trend with longer aeration durations, with T2 treatment showing a significantly higher Shannon index compared to CK and T3 treatments, but no significant difference in the Chao1 index among treatments. For fungi, the Shannon index ranked T2 > T1 > CK > T3, with T2 treatment being 14.9%, 5.9%, and 19.2% higher than the CK, T1, and T3 treatments, respectively. Although T3 treatment did not significantly affect the Chao1 index of the fungal community compared to CK treatment, T2 treatment showed a 13.3% increase over CK treatment, which was significant. This suggests that the optimal aerated irrigation duration of 20 min resulted in the highest diversity and abundance of both bacterial and fungal communities in the rhizosphere soil of grape seedlings.

Table 2. Microbial richness and diversity indices in the rhizosphere soil of grape seedlings, aeration durations are 0 min (CK), 10 min (T1), 20 min (T2), and 30 min (T3) per day.

Treatment	Shannon		Chao1	
	Bacteria	**Fungi**	**Bacteria**	**Fungi**
CK	10.50 ± 0.12 b	4.20 ± 0.20 bc	2760.45 ± 15.47 a	276.06 ± 8.12 b
T1	12.31 ± 0.24 ab	4.56 ± 0.03 ab	3110.28 ± 53.54 a	298.59 ± 9.82 ab
T2	14.14 ± 0.91 a	4.83 ± 0.08 a	3006.08 ± 196.43 a	312.83 ± 10.54 a
T3	10.59 ± 0.58 b	4.05 ± 0.09 c	2874.71 ± 476.98 a	279.78 ± 9.38 b

Note: Different lowercase letters indicate significant differences among treatments ($p < 0.05$).

3.3. Effects of Aerated Irrigation Duration on the Rhizosphere Soil Microbial Community of Grape Seedlings

3.3.1. Impact on the Rhizosphere Soil Bacterial Community of Grape Seedlings

To assess the influence of different aerated irrigation durations on the structure of the soil bacterial community in the rhizosphere soil of grape seedlings, the species composition of each group was analyzed and compared. As presented in Figure 4A, the top five dominant bacterial phyla by relative abundance across treatments were Proteobacteria, Acidobacteriota, Gemmatimonadota, Bacteroidota, and Actinobacteriota, collectively accounting for 77.5% of the total. The relative abundance of Proteobacteria exhibited a gradual increase with longer aeration durations, peaking under T3 treatment, where it was 16.2%, 16.1%, and 11.3% higher than under CK, T1, and T2 treatments, respectively. In contrast, Acidobacteriota and Gemmatimonadota reached the highest relative abundances under T2 treatment, showing increases ranging from 4.5% to 48.0% and 21.7% to 91%, respectively, compared to the other treatments (CK, T1, and T3 treatments). Conversely, Bacteroidota and Actinobacteriota were more abundant under CK than under any of the aerated irrigation treatments. The dominant bacterial genera identified across the different aeration durations included *unidentified_Chloroplast*, *Sphingomonas*, *MND1*, *Clade Ia*, and *Planktomarina*. Notably, the relative abundance of *Sphingomonas* was significantly higher in T1 treatment, exceeding CK, T2, and T3 treatment by 8.6%, 26.2%, and 91.4%, respectively. The relative abundance of *MND1* increased and then decreased across treatments, peaking in T2 treatment and reaching its lowest in T3 treatment.

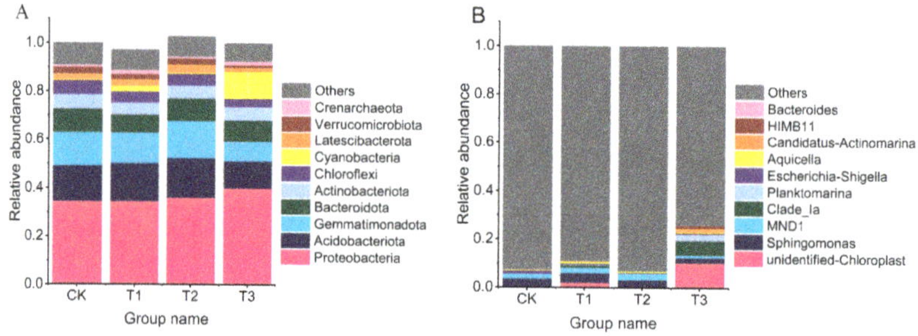

Figure 4. Community structure of bacterial phylum (**A**) and genus (**B**) in the rhizosphere soil of grape seedlings under different aerated irrigation durations.

3.3.2. Impact on the Rhizosphere Soil Fungal Community of Grape Seedlings

Fungal communities exhibited a simpler composition compared to bacterial communities (Figure 5). In terms of relative abundance, the top four dominant fungal phyla across all aeration duration treatments were Ascomycota, Chytridiomycota, Mortierellomycota, and Basidiomycota, collectively accounting for 92.7% of the total (Figure 5A). The dominant fungal genera identified under each treatment included *Pseudeurotium*, *Botryotrichum*, *Coniochaeta*, *Enterocarpus*, and *Rhizophlyctis*, which together comprised 65.3% of the total abundance (Figure 5B). The relative abundances of *Pseudeurotium*, *Coniochaeta*, and *Rhizophlyctis* were higher under T1 treatment compared to other treatments. In contrast, *Enterocarpus* displayed the highest relative abundance in T2 treatment, being 1.54, 0.55, and 1.22 times higher than in CK, T1, and T3 treatments, respectively.

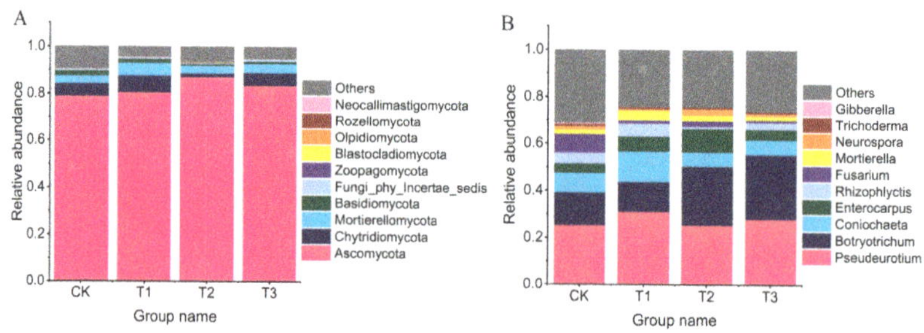

Figure 5. Community structure of fungal phylum (**A**) and genus (**B**) in the rhizosphere soil of grape seedlings under different aerated irrigation durations.

3.4. Effects of Aerated Irrigation Duration on Aerobic Microorganisms in the Rhizosphere Soil of Grape Seedlings

The impact of different aerated irrigation durations on the aerobic bacterial genera in the rhizosphere soil of grape seedlings is illustrated in Figure 6A. The relative abundance of *Flavobacterium* differed significantly among treatments, peaking under T2 treatment, which was 0.56 and 3.20 times significantly higher than under CK and T3 treatments, respectively, although no significant difference was noted when compared to T1 treatment. The relative abundance of *Lysobacter* followed the rankings of T2 > CK > T1 > T3. In contrast, the relative abundance of *Sphingomonas* was significantly higher under T1 treatment, being 1.04-fold greater than that under T3 treatment. Similarly, the relative abundance of *Gemmatimonas* increased and then decreased with increasing durations of aerated irrigation, reaching its highest level under T1 treatment, where it was 18.1% to 35.2% higher than in the other treatments. The relative abundance of *Ellin6067* was the highest under T2 treatment, being 27.9%, 43.1%, and 62.6% higher than under CK, T1, and T3 treatments, respectively.

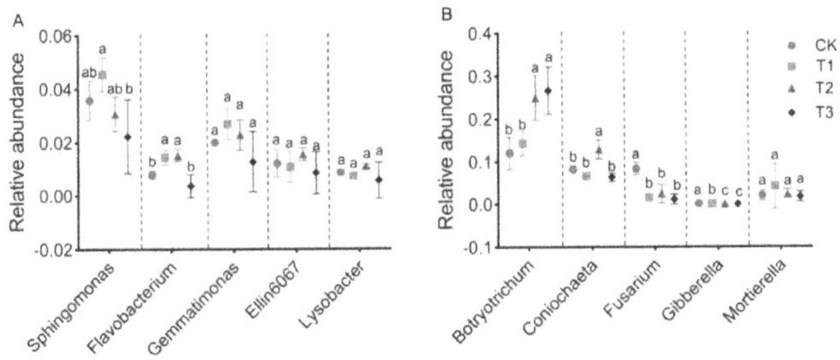

Figure 6. Effects of different aerated irrigation durations on aerobic microorganisms in the rhizosphere soil of grape seedlings, aeration durations are 0 min (CK), 10 min (T1), 20 min (T2), and 30 min (T3) per day. Note: (**A**). Aerobic bacteria of the genus. (**B**). Aerobic fungi of the genus. Different lowercase letters indicate significant differences among treatments ($p < 0.05$).

The aerated irrigation duration also significantly influenced the relative abundance of aerobic fungal species in the rhizosphere soil (Figure 6B). Compared to CK, the aerated irrigation treatments effectively increased the relative abundance of *Botryotrichum*, which was the highest under T3 treatment, being 1.2 and 0.85 times significantly greater than under CK and T1 treatments, respectively. The relative abundance of *Coniochaeta* peaked under T2 treatment, being 0.57, 0.94, and 1.05 times significantly higher than under CK, T1, and T3 treatments, respectively. Conversely, the relative abundances of *Fusarium* and *Gibberella* significantly decreased with longer aerated irrigation durations, with the highest values recorded under CK treatment. In addition, no significant effect was observed on the relative abundance of *Mortierella* across the different aerated irrigation durations.

3.5. Correlation Analysis of Grape Seedling Growth with the Genus of Aerobic Microorganisms in the Rhizosphere Soil

The results of Pearson's correlation analysis revealed significant relationships between the growth of both the root system and the above-ground parts of grape seedlings and the genera of aerobic microorganisms (Figure 7). Specifically, root activity exhibited a significant positive correlation with the relative abundances of *Coniochaeta*, *Lysobacter*, and *Ellin6067*. The SPAD value also showed a significant positive correlation with the relative abundance of *Coniochaeta*. Furthermore, the relative abundance of *Botryotrichum* was significantly positively correlated with plant stem thickness. Conversely, a significant negative correlation was observed between the relative abundance of *Gibberella* and below-ground biomass.

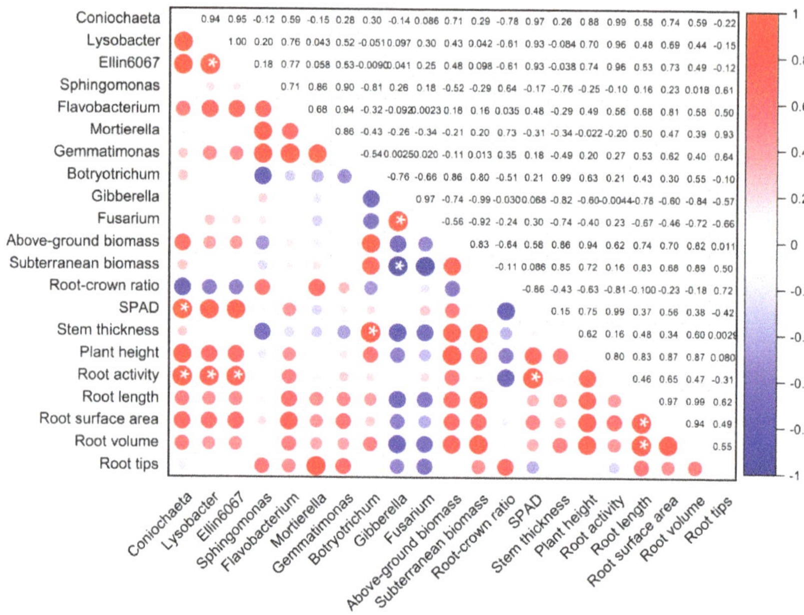

Figure 7. Correlation analysis between grape seedling growth indices and dominant microbial genera in the rhizosphere soil. Note: Red and blue colors indicate positive and negative correlations, respectively, and * indicates significant correlation ($p < 0.05$).

4. Discussion

4.1. Effects of Aerated Irrigation Duration on the Growth of Grape Seedlings

Aerated irrigation offers advantages over traditional irrigation methods by supplying optimal water and nutrients for crop growth while enhancing the soil microenvironment through improved aeration and the regulation of water, fertilizer, air, and heat ratios [33]. Previous studies, such as those by Qian et al. [34], demonstrated that aerated treatments can effectively promote plant growth, increasing above-ground dry matter and yield in crops like cucumber and melon. Similarly, Li et al. [18] reported significant increases in tomato plant height and stem thickness under aerated conditions. Our findings align with these results, showing that aerated irrigation significantly promoted grape seedling height, with the most pronounced effect observed at a duration of 20 min. While a positive effect on stem thickness was noted, it did not reach statistical significance (Figure 2). This discrepancy with previous studies may stem from variations in crop varieties, soil conditions, and climate. Li et al. [18] indicated that increasing the rhizosphere soil oxygen level positively influences crop canopy growth. Seridou et al. [14] concluded that inadequate soil aeration during irrigation is detrimental to above-ground plant growth. Our findings confirm that aerated irrigation treatments enhanced both above-ground and below-ground biomass, with the highest values recorded under T2 treatment (Table 1), aligning with the results of previous studies. Chlorophyll content is closely linked to a plant's photosynthetic capacity and directly affects plant biomass and yield, serving as a crucial indicator of nutritional growth [35]. Prior research has shown that aerated treatments significantly boost chlorophyll content in plant leaves [36]. The results of this study showed that T2 treatment effectively increased plant leaf SPAD values, and at 15 d after aerated irrigation, leaf SPAD values were significantly lower under T1 and T3 treatments compared to CK treatment. This suggests that a 20 min aerated irrigation duration could improve the rhizosphere soil aeration of grape seedlings, effectively promote crop growth, strengthen biomass accumulation, and positively affect overall crop health.

4.2. Effects of Aerated Irrigation Duration on Root Morphology of Grape Seedlings

The root systems of crops play a crucial role in exchanging materials, energy, and information with the external environment. The morphological characteristics and functionality of roots are affected by the soil's physical and chemical properties, as well as fertilizer application, all of which directly impact the growth and development of above-ground organs, the morphological composition of the plants, and, ultimately, the crop yield [37]. Aerated irrigation increases root length and effectively promotes root growth, thereby facilitating water and nutrient uptake by plant roots [21]. In this study, aerated irrigation significantly boosted the root system growth of grape seedlings. The total root length and root volume were notably higher with a 20 min irrigation aerated duration compared to other durations, while the number of root tips peaked at 10 min of aerated irrigation. The morphological indices of the root system showed a trend of initially increasing and then decreasing with longer aeration durations. This pattern suggests that, while aerated irrigation improves rhizosphere aeration and promotes root growth, excessive aeration could lead to elevated soil oxygen levels that may cause oxygen damage to root cells of grape seedlings, negatively affecting root growth [38]. In this experiment, the observed effects of different aerated irrigation durations on the growth of grape seedlings were consistent with the changes in root morphology, with the most favorable outcomes occurring at the T2 treatment duration. Nonetheless, when aeration was extended to T3 treatment, slight oxygen injury was noted, which may have suppressed plant growth. Overall, the growth of grape seedlings under aerated irrigation conditions remained superior to that observed under the un-aerated CK condition.

4.3. Effects of Aerated Irrigation Duration on the Diversity and Structure of the Rhizosphere Soil Microbial Community in Grape Seedlings

Microorganisms represent the most diverse and abundant biota on Earth, with soil microorganisms forming essential components of soil ecosystems. Their diversity reflects terrestrial biodiversity and can effectively regulate the biogeochemical cycling of nutrients, which is vital for the functioning of terrestrial ecosystems [39]. Rhizosphere soil microorganisms are an important bioindicator for assessing plant health and productivity [40]. Previous studies have shown that various agricultural practices can have a profound impact on microbial communities [41], and rhizosphere soil aeration significantly affects the abundance and diversity of soil microbial communities in the rhizosphere soil of crops [42,43]. And climatic conditions, soil type, aeration intensity, field management, and other environmental conditions can influence plant physiology and soil microbial activity, thereby affecting the structural composition of the rhizosphere microbial community [44]. The results of this study showed that the diversity and richness of bacterial species in the rhizosphere soil of grape seedlings were higher than those of fungi. Aerated irrigation treatments (T1, T2, and T3) promoted the richness and diversity of microbial communities compared to the non-aerated CK treatment, indicating that aerated irrigation could enhance rhizosphere microbial abundance [45]. Notably, the highest microbial diversity was recorded with the 20 min aerated irrigation duration. However, extending the duration of aerated irrigation to 30 min (T3) resulted in a decrease in bacterial and fungal diversity. This decline can be attributed to the fact that, while aerated irrigation relieved the soil from low-oxygen stress, thereby initially enhancing microbial diversity, prolonged aeration increased gas flow within the soil significantly. This heightened air flow can disturb microbial communities, ultimately leading to a reduction in both the number of microorganisms and their diversity [9].

The oxygen content in soil significantly affects the soil microbial community structure [42]. In recent years, various agricultural management practices, such as straw return [46], soil aeration [47], application of organic fertilizers [48], and the use of biochar [49], have been employed to enhance soil aeration, thereby affecting the composition of soil microbial communities. The dominant bacterial phyla in this study were Proteobacteria and Gemmatimonadota, while the dominant bacterial genera were *MND1* and *Clade_Ia*, which also included *Sphingomonas*, a genus known for promoting plant growth and strengthening plant resistance. Previous studies have demonstrated that variations in soil oxygen content can directly impact microbial biomass and community structure [26]. Our findings indicate that aerated irrigation treatments significantly increased the relative abundance of *Sphingomonas* in the rhizosphere of grape seedlings, particularly with a 10 min aeration duration. *Sphingomonas* are aerobic Gram-negative bacteria [50]; thus, the enhanced oxygen concentration resulting from aerated irrigation contributed to a substantial increase in their relative abundance. Moreover, the relative abundances of other dominant bacterial genera were also affected by different durations of aerated irrigation. The dominant fungal genera observed included *Pseudeurotium*, *Botryotrichum*, *Coniochaeta*, *Enterocarpus*, and *Rhizophlyctis*. Specifically, the relative abundances of *Pseudohyphae*, *Coniochaeta*, and *Rhizophlyctis* were significantly higher under the 10 min aeration condition, while *Botryotrichum* peaked at 20 min of aeration (Figure 5). This suggests that aerated irrigation improved the habitat conditions for these fungi, leading to increased abundance. However, prolonged aeration durations can disturb the growth of microorganisms due to frequent airflow, making their relative abundance decrease to some extent [9].

4.4. Effects of Aerated Irrigation Duration on Aerobic Microbial Genera in the Rhizosphere Soil of Grape Seedlings

Rhizosphere soil microbial communities are important regulators of plant growth, development, and stress tolerance [51], also serving as an ecological barrier to limit pathogen invasion [52]. Microorganisms directly and indirectly influence soil nutrient cycling and organic matter decomposition through their interactions with plant roots, thus effectively regulating crop growth and development [53]. For example, plant growth-promoting bacteria can promote nutrient uptake, stress tolerance, plant growth and development, and yield through symbiosis with plants (secretion of organic acids, enzymes, plant hormones, antioxidants, and cytokinins) [54]. In addition to this, plant growth-promoting bacteria fix atmospheric nitrogen and effectively improve soil organic carbon content, soil pH, and soil porosity [55], thus improving soil fertility as well as soil health. The ability of plants to absorb nutrients from the soil is closely related to root morphological indicators such as root length, root surface area, and root activity [1]. Suitable soil aeration conditions significantly impact both microbial communities and the metabolic activities of plant roots [42]. In this study, we observed a close relationship between the growth of both above-ground and root structures of grape seedlings and the presence of aerobic microorganisms. The results showed that aerated irrigation treatments could effectively increase the relative abundances of aerobic microorganisms, particularly *Coniochaeta*, *Lysobacter*, and *Ellin6067*, with the most pronounced effects seen in T2 treatment (Figure 6). Notably, a significant positive correlation was found between the relative abundances of *Coniochaeta*, *Lysobacter*, and *Ellin6067* and the root activity (Figure 7), which contributed positively to plant growth. There is mutually beneficial coexistence between plant rhizosphere soil microorganisms and their hosts or pathogenicity to their hosts as determinants of crop health and productivity [40]. Beneficial rhizosphere soil microorganisms can inhibit or directly kill pathogens in a variety of ways, leading to a reduction in soil-borne diseases [56]. *Coniochaeta* is recognized as a beneficial fungus with biocontrol properties that is capable of degrading hemicellulose in the soil and producing natural enzymes and bioactive compounds [57]. The genus *Lysobacter* is known to mitigate plant diseases; for instance, Ji Guanghai et al. [58] showed that *Lysobacter* can effectively inhibit a variety of pathogenic bacteria, including those causing rice leaf blight. Additionally, *Ellin6067*, an ammonia-oxidizing bacterium [59], plays a vital role in the nitrogen cycle, helping to maintain nitrogen balance and create favorable soil conditions for root growth, thus promoting overall plant health. By comparison, the relative abundances of the aerobic fungal genera *Gibberella* and *Fusarium* both decreased as the duration of aerated irrigation increased (Figure 6B). Significant negative correlations were observed between the relative abundances of *Gibberella* and *Fusarium* and the dry weight of the plant's underground portions, indicating adverse effects on grape growth

(Figure 7). In the process of agricultural production, *Gibberella* and *Fusarium* are known to be destructive phytopathogenic fungi that cause plant diseases. For example, *Gibberella* is responsible for wheat blast and potato tuber dry rot [60], while *Fusarium* can cause round spot root rot in grapes. These fungi can inhibit root growth and reduce nutrient uptake efficiency, thus adversely affecting the growth of grapes [61]. Overall, aerated irrigation could effectively increase the relative abundance of beneficial microorganisms such as *Coniochaeta*, *Lysobacter*, and *Ellin6067* in the rhizosphere soil while reducing pathogenic fungi like *Gibberella* and *Fusarium*. The most favorable effects on grape seedling growth were achieved with a 20 min aeration duration (T2).

5. Conclusions

(1) An aerated irrigation duration of 20 min significantly increased plant height, above-ground and below-ground biomass, total root length, and root volume in greenhouse grapes compared to other treatments. Additionally, a duration of 10 min notably increased the number of root tips.

(2) The Shannon index and Chao1 index of the rhizosphere soil fungal community were significantly enhanced after aeration, with the most pronounced effects observed at a 20 min aeration duration. The Shannon index of the bacterial community was also significantly higher at this duration; however, no significant differences were found in the Chao1 index across treatments, indicating that 20 min of aeration could effectively promote the diversity and richness of microbial communities in the rhizosphere soil.

(3) An aeration duration of 10 min significantly increased the relative abundance of *Sphingomonas*. The highest relative abundances of beneficial genera, including *Flavobacterium*, *Ellin6067*, and *Coniochaeta*, were observed at an aeration duration of 20 min. Conversely, the phytopathogenic fungi *Fusarium* and *Gibberella* were significantly more abundant under the non-aerated condition compared to all aerated treatments.

(4) Under aerated irrigation conditions, the relative abundances of *Coniochaeta*, *Lysobacter* and *Ellin6067* showed significant positive correlations with the root activity of grape seedlings. In contrast, a significant negative correlation was found between the relative abundance of *Gibberella* and below-ground biomass.

Author Contributions: Writing—original draft, visualization, software, methodology, investigation, formal analysis, Y.L.; visualization, methodology, W.W.; software, investigation, J.X.; methodology, investigation, S.Z.; investigation, Y.Z.; formal analysis, H.Z.; supervision, resources, K.Y.; writing—review and editing, supervision, project administration, methodology, investigation, funding acquisition, Z.Z. and F.Z. All authors have read and agreed to the published version of the manuscript.

Funding: This work was funded by the National Natural Science Foundation of China (32360718), the Research, Development and Demonstration of Key Equipment for the Integration of "Water, Fertilizer and Gas" Technology for Fruit Tree Burrow Storage and Drip Irrigation (2024AB038) and the Research Initiation Program for High-level Talents of Shihezi University (RCZK201925).

Data Availability Statement: Data have been provided in the manuscript as images or are available upon request from the corresponding author.

Conflicts of Interest: The authors declare no conflicts of interest.

References

1. Xiao, Z.Y.; Lei, H.J.; Lian, Y.J.; Zhang, Z.H.; Pan, H.W.; Yin, C.; Dong, Y.C. Impact of Aerated Drip Irrigation and Nitrogen Application on Soil Properties, Soil Bacterial Communities and Agronomic Traits of Cucumber in a Greenhouse System. *Plants* **2023**, *12*, 3834. [CrossRef]
2. Ma, Y.; Liu, Z.H.; Xi, B.D.; He, X.S.; Li, Q.L.; Qi, Y.J.; Jin, M.Y.; Guo, Y. Characteristics of Groundwater Pollution in a Vegetable Cultivation Area of Typical Facility Agriculture in a Developed City. *Ecol. Indic.* **2019**, *105*, 709–716. [CrossRef]
3. Pineda, I.T.; Lee, Y.D.; Kim, Y.S.; Lee, S.M.; Park, K.S. Review of Inventory Data in Life Cycle Assessment Applied in Production of Fresh Tomato in Greenhouse. *J. Clean. Prod.* **2021**, *282*, 124395. [CrossRef]
4. Wang, S.P.; Li, B. Overview of Facility Grape Development in China. *Deciduous Fruits* **2019**, *51*, 1–5.

5. Luo, Z.K.; Zhang, S.; Zhao, Z.G.; Minasny, B.; Chang, J.F.; Huang, J.Y.; Li, B.H.; Shi, Z.; Wang, M.M.; Wu, Y.S.; et al. Soil-smart cropping for climate-smart production. *Geoderma* **2024**, *451*, 117061. [CrossRef]
6. Lei, H.J.; Hu, S.G.; Pan, H.W.; Zang, M.; Liu, X. Research Progress on Soil Aeration and Oxygen Irrigation. *Acta Pedol. Sin.* **2017**, *54*, 297–308. [CrossRef]
7. Ma, J.W.; Rukh, G.; Ruan, Z.Q.; Xie, X.C.; Ye, Z.Q.; Liu, D. Effects of Hypoxia Stress on Growth, Root Respiration, and Metabolism of Phyllostachys praecox. *Life* **2022**, *12*, 808. [CrossRef]
8. Lei, H.J.; Lian, Y.L.; Du, J.; Pan, H.W.; Li, X.H.; Li, D.X.; Jin, C.C.; Xiao, Z.Y.; Hou, Y.R. Dynamic Optimization of Greenhouse Tomato Irrigation Schedule Based on Water, Fertilizer and Air Coupled Production Function. *Agronomy* **2023**, *13*, 776. [CrossRef]
9. Li, Y.; Niu, W.Q.; Zhang, M.Z.; Xue, L.; Wang, J.W. The Effect of Aerated Irrigation on Soil Enzyme Activities and Microbial Quantity of Greenhouse Muskmelon. *Trans. Chin. Soc. Agric. Mach.* **2015**, *46*, 121–129.
10. Sun, Y.N.; Duan, L.B.; Zhong, H.Y.; Cai, H.J.; Xu, J.T.; Li, Z.J. Effects of Irrigation-Fertilization-Aeration Coupling on Yield and Quality of Greenhouse Tomatoes. *Agric. Water Manag.* **2024**, *299*, 108893. [CrossRef]
11. Xiao, R.; Sun, K.P.; Lei, H.J.; Zhang, T.Y.; Chen, J. Study on the Growth, Physiological Characteristics, Nutrient Absorption, and Yield Relationship of Greenhouse Peppers under Aerated Irrigation. *J. North China Univ. Water Resour. Electr. Power* **2023**, *44*, 78–86.
12. Baram, S.; Evans, J.F.; Berezkin, A.; Ben-Hur, M. Irrigation with Treated Wastewater Containing Nanobubbles to Aerate Soils and Reduce Nitrous Oxide Emissions. *J. Clean. Prod.* **2021**, *280*, 124509. [CrossRef]
13. Yu, Z.Z.; Wang, H.X.; Zou, H.F.; Sun, H.T.; Wang, H.Y. Changes in Respiration Rates and Relationships with Soil Water and Oxygen under Aerated Irrigation in Red Soils. *Chin. J. Trop. Crops* **2022**, *43*, 110–118. [CrossRef]
14. Seridou, P.; Kalogerakis, N. Disinfection applications of ozone micro- and nanobubbles. *Environ. Sci. Nano* **2021**, *8*, 3493–3510. [CrossRef]
15. Zhou, Y.P.; Li, Y.K.; Liu, X.J.; Wang, K.Y.; Muhammad, T. Synergistic improvement in spring maize yield and quality with micro/nanobubbles water oxygation. *Sci. Rep.* **2019**, *9*, 1–10. [CrossRef]
16. Gao, L.L.; Li, J.; Huang, H.Y.; Xiang, B.; Li, S.F. The Effect of Aerated Irrigation under Drip Irrigation on Root Zone Habitat Factors and Yield of Winter Potato in Yunnan. *Agric. Res. Arid. Areas* **2022**, *40*, 108–115. [CrossRef]
17. Palencia, P.; Martinez, F.; Padua, D.; Oliveira, J.A. Effects of Oxyfertigation on Strawberry Plant Growth and Fruit Quality in a Soilless Growing System. *Acta Hortic.* **2019**, *1256*, 512–519. [CrossRef]
18. Li, Y.; Niu, W.Q.; Xu, J.; Wang, J.W.; Zhang, M.Z.; Lv, W. Root Morphology of Greenhouse Produced Muskmelon under Sub-surface Drip Irrigation with Supplemental Soil Aeration. *Sci. Hortic.* **2016**, *201*, 287–294. [CrossRef]
19. Rao, X.J.; Fu, Y.B.; Huang, J.; Feng, Y.Z.; Wang, Z.G. Effect of Oxygen Irrigation on Cotton Nutritional Characteristics and Soil Fertility. *Acta Pedol. Sin.* **2018**, *55*, 797–803.
20. Lu, Z.H.; Cai, H.J.; Wang, J.; Li, Z.J. Effects of Rhizosphere Aeration at Different Growth Stages on Growth and Yield of Greenhouse Tomatoes. *Sci. Agric. Sin.* **2012**, *45*, 1330–1337.
21. Bhattarai, S.P.; Su, N.H.; Midmore, D.J.; Sparks, D.L. Oxygation Unlocks Yield Potentials of Crops in Oxygen-limited Soil Environments. *Adv. Agron.* **2005**, *88*, 313–377. [CrossRef]
22. Shigyo, N.; Umeki, K.; Hirao, T. Seasonal Dynamics of Soil Fungal and Bacterial Communities in Cool-temperate Montane Forests. *Front. Microbiol.* **2019**, *10*, 1944. [CrossRef]
23. Zhang, M.Q.; Wang, X.Q.; Guan, H.B.; Yu, J.; Zhang, Y. The Effects of Salt Stress on the Growth of Balloonflower Seedlings and the Number of Rhizosphere Microorganisms. *West China J. Pharm. Sci.* **2024**, *39*, 89–93.
24. Velmurugan, A.; Swarnam, T.P.; Jaisankar, I.; Swain, S.; Subramani, T. Vegetation–Soil–Microbial Diversity Influences Ecosystem Multifunctionality across Different Tropical Coastal Ecosystem Types. *Trop. Ecol.* **2021**, *63*, 1–13. [CrossRef]
25. Zeglin, L.H.; Dahm, C.N.; Barrett, J.E.; Gooseff, M.N.; Fitpatrick, S.K.; Takacs-Vesbach, C.D. Bacterial Community Structure along Moisture Gradients in the Parafluvial Sediments of Two Ephemeral Desert Streams. *Microb. Ecol.* **2011**, *61*, 543–556. [CrossRef]
26. Biggs-Weber, E.; Aigle, A.; Prosser, J.I.; Gubry-Rangin, C. Oxygen Preference of Deeply-rooted Mesophilic Thaumarchaeota in Forest Soil. *Soil Biol. Biochem.* **2020**, *148*, 107848. [CrossRef]
27. Chen, H.; Shang, Z.H.; Cai, H.J.; Zhu, Y. Irrigation Combined with Aeration Promoted Soil Respiration through Increasing Soil Microbes, Enzymes, and Crop Growth in Tomato Fields. *Catalysts* **2019**, *9*, 945. [CrossRef]
28. Chen, Z.; Maltz, M.R.; Cao, J.X.; Yu, H.; Shang, H.; Aronson, E. Elevated O_3 Alters Soil Bacterial and Fungal Communities and the Dynamics of Carbon and Nitrogen. *Sci. Total Environ.* **2019**, *677*, 272–280. [CrossRef]
29. Coats, V.C.; Rumpho, M.E. The Rhizosphere Microbiota of Plant Invaders: An Overview of Recent Advances in the Microbiomics of Invasive Plants. *Front. Microbiol.* **2014**, *5*, 368. [CrossRef]
30. Xie, H.X.; Cai, H.J.; Zhang, Z.H. Comprehensive Benefit Evaluation of Aerated Irrigation for Greenhouse Muskmelon. *Trans. Chin. Soc. Agric. Mach.* **2010**, *41*, 79–83.
31. Richter, A.K.; Frossard, E.; Brunner, I. Polyphenols in the woody roots of Norway spruce and European beech reduce TTC. *Tree Physiol.* **2007**, *27*, 155–160. [CrossRef]
32. Anand, S.; Kundan, D.; Marcus, G.; Harchao, C.; Freschet, G.T.; York, L.M. RhizoVision Explorer: Open-source Software for Root Image Analysis and Measurement Standardization. *AoB PLANTS* **2021**, *13*, plab056. [CrossRef]
33. Wang, H.Y.; Wang, C.; Wang, F.; Wang, H.X.; Ma, G.Q.; Yu, Z.Z. Effects of Aerated Irrigation Technology on Soil and Crop Growth. *Res. Agric. Mech.* **2024**, *46*, 17–24.

34. Zhang, Q.; Du, Y.D.; Cui, B.J.; Sun, J.; Wang, J.; Wu, M.L.; Niu, W.Q. Aerated Irrigation Offsets the Negative Effects of Nitrogen Reduction on Crop Growth and Water-nitrogen Utilization. *J. Clean. Prod.* **2021**, *313*, 127945. [CrossRef]
35. Al-Gaadi, K.A.; Tola, E.; Madugundu, R.; Zeyeda, A.M.; Alameen, A.A.; Edrris, M.K.; Edress, H.F.; Mahjoop, O. Response of Leaf Photosynthesis, Chlorophyll Content and Yield of Hydroponic Tomatoes to Different Water Salinity Levels. *PLoS ONE* **2024**, *19*, 0293098. [CrossRef]
36. Li, Y.; Niu, W.Q.; Lv, W.; Gu, J.; Zou, X.Y.; Wang, J.W.; Liu, L.; Zhang, M.Z.; Xu, J. Aerated Irrigation Improving Photosynthesis Characteristics and Dry Matter Accumulation of Greenhouse Tomato. *Trans. Chin. Soc. Agric. Eng.* **2016**, *32*, 125–132.
37. Huang, J.; Hu, T.S.; Yasir, M.; Gao, Y.; Chen, C.; Zhu, R.; Wang, X.; Yuan, H.W.; Yang, J.W. Root Growth Dynamics and Yield Responses of Rice (*Oryza sativa* L.) under Drought-Flood Abrupt Alternating Conditions. *Environ. Exp. Bot.* **2019**, *157*, 11–25. [CrossRef]
38. Li, J.; Pan, Y.C.; Jiao, X.Y.; Hu, W.Y.; Liu, Y. Effects of Aerated Irrigation on Soil Reducibility and Rice Growth after Wheat Straw Incorporation. *Trans. Chin. Soc. Agric. Mach.* **2021**, *52*, 250–259.
39. Wu, B.H.; Luo, H.Y.; Wang, X.T.; Liu, H.K.; Peng, H.; Sheng, M.P.; Xu, F.; Xu, H. Effects of Environmental Factors on Soil Bacterial Community Structure and Diversity in Different Contaminated Districts of Southwest China Mine Tailings. *Sci. Total Environ.* **2022**, *802*, 149899. [CrossRef]
40. Darriaut, R.; Antonielli, L.; Martins, G.; Ballestra, P.; Vivin, P.; Marguerit, E.; Mitter, B.; Masneuf-Pomarede, I.; Compant, S.; Ollat, N. Soil composition and rootstock genotype drive the root associated microbial communities in young grapevines. *Front. Microbiol.* **2022**, *13*, 1031064. [CrossRef]
41. Finkel, O.M.; Castrillo, G.; Paredes, S.H.; Gonzalez, I.S.; Dangl, J.L. Understanding and exploiting plant beneficial microbes. *Curr. Opin. Plant Biol.* **2017**, *38*, 155–163. [CrossRef]
42. Qian, Z.Z.; Zhuang, S.Y.; Gao, J.S.; Tang, L.Z.; Harindintwali, J.D.; Wang, F. Aeration Increases Soil Bacterial Diversity and Nutrient Transformation under Mulching-induced Hypoxic Conditions. *Sci. Total Environ.* **2022**, *817*, 152659. [CrossRef]
43. Li, Y.; Niu, W.Q.; Zhang, M.Z.; Wang, J.W.; Zhang, Z.X. Artificial Soil Aeration Increases Soil Bacterial Diversity and Tomato Root Performance under Greenhouse Conditions. *Land Degrad. Dev.* **2020**, *31*, 1443–1461. [CrossRef]
44. Bamba, M.; Akyol, T.K.; Azuma, Y.; Quilbe, J.; Andersen, S.U.; Sato, S. Synergistic effects of plant genotype and soil microbiome on growth in Lotus japonicus. *FEMS Microbiol. Ecol.* **2024**, *100*, 56. [CrossRef]
45. Zhu, J.J.; Niu, W.Q.; Zhang, Z.H.; Siddique, K.; Sun, D.; Yang, R.Y. Distinct Roles for Soil Bacterial and Fungal Communities Associated with the Availability of Carbon and Phosphorus under Aerated Drip Irrigation. *Agric. Water Manag.* **2022**, *274*, 107925. [CrossRef]
46. Liu, D.T.; Song, C.C.; Xin, Z.H.; Fang, C.; Liu, Z.H.; Xu, Y.P. Agricultural Management Strategies for Balancing Yield Increase, Carbon Sequestration, and Emission Reduction after Straw Return for Three Major Grain Crops in China: A Meta-analysis. *J. Environ. Manag.* **2023**, *340*, 117965. [CrossRef]
47. Zhu, J.J.; Xu, N.; Siddique, K.H.M.; Zhang, Z.H.; Niu, W.Q. Aerated Drip Irrigation Improves Water and Nitrogen Uptake Efficiencies of Tomato Roots with Associated Changes in the Antioxidant System. *Sci. Hortic.* **2022**, *306*, 111471. [CrossRef]
48. Liu, J.A.; Shu, A.P.; Song, W.F.; Shi, W.C.; Li, M.C.; Zhang, W.X.; Li, Z.Z.; Liu, G.R.; Yuan, F.S.; Zhang, S.X.; et al. Long-term Organic Fertilizer Substitution Increases Rice Yield by Improving Soil Properties and Regulating Soil Bacteria. *Geoderma* **2021**, *404*, 115287. [CrossRef]
49. Zhang, Y.F.; Wang, J.M.; Feng, Y. The Effects of Biochar Addition on Soil Physicochemical Properties: A Review. *Catena* **2021**, *202*, 105284. [CrossRef]
50. Liu, H.; Wei, L.L.; Zhu, L.F.; Wei, H.; Bai, Y.X.; Liu, X.L.; Li, S.B. Advances in the Study of *Sphingomonas*. *Microbiol. Bull.* **2023**, *50*, 2738–2752.
51. Salas-González, I.; Reyt, G.; Flis, P.; Custódio, V.; Gopaulchan, D.; Bakhoum, N.; Dew, T.P.; Suresh, K.; Franke, R.B.; Dangl, J.L.; et al. Coordination between microbiota and root endodermis supports plant mineral nutrient homeostasis. *Science* **2020**, *371*, 6265. [CrossRef] [PubMed]
52. Hacquard, S.; Spaepen, S.; Garrido-Oter, R.; Schulze-Lefert, P. Interplay Between Innate Immunity and the Plant Microbiota. *Annu. Rev. Phytopathol.* **2017**, *55*, 565–589. [CrossRef] [PubMed]
53. Wang, R.Q.; Xiao, Y.P.; Lv, F.J.; Hu, L.Y.; Wei, L.G.; Yuan, Z.Q.; Lin, H.X. Bacterial Community Structure and Functional Potential of Rhizosphere Soils as Influenced by Nitrogen Addition and Bacterial Wilt Disease under Continuous Sesame Cropping. *Appl. Soil Ecol.* **2018**, *125*, 117–127. [CrossRef]
54. Gupta, P.; Kumar, V.; Usmani, Z.; Rani, R.; Chandra, A.; Gupta, V.K. A comparative evaluation towards the potential of Klebsiella sp. and Enterobacter sp. in plant growth promotion, oxidative stress tolerance and chromium uptake in *Helianthus annuus* (L.). *J. Hazard. Mater.* **2019**, *377*, 391–398. [CrossRef] [PubMed]
55. Fasusi, O.A.; Cruz, C.; Babalola, O.O. Agricultural Sustainability: Microbial Biofertilizers in Rhizosphere Management. *Agriculture* **2021**, *11*, 163. [CrossRef]
56. Ratnadass, A.; Fernandes, P.; Avelino, J.; Habib, R. Plant species diversity for sustainable management of crop pests and diseases in agroecosystems: A review. *Agron. Sustain. Dev.* **2011**, *32*, 273–303. [CrossRef]
57. Gao, H.X.; Li, Y.; Zhu, Z.X. Two New Asian Records of *Coniochaeta* from China. *Mycosystema* **2022**, *20*, 96–108.
58. Ji, G.H. Research Progress on *Lysobacter* and Its Role in Plant Disease Control. *J. Yunnan Agric. Univ.* **2011**, *26*, 124–130.

59. Li, X.; Lu, Y.Z.; Chen, Y.; Zhu, G.C.; Zeng, R.J. Constraining Nitrification by Intermittent Aeration to Achieve Methane-driven Ammonia Recovery of the Mainstream Anaerobic Effluent. *J. Environ. Manag.* **2021**, *295*, 113103. [CrossRef]
60. Cui, Y.F.; Huang, Y.; Jiang, L.H. Advances in the Study of *Gibberella* spp. in Agricultural Production. *Chin. Agric. Sci. Bull.* **2007**, *7*, 441–446. [CrossRef]
61. Zhang, H.H.; Xi, J.S.; Lv, Q.; Wang, J.W.; Yu, K.; Zhao, F.Y. Effect of Aerated Irrigation on the Growth and Rhizosphere Soil Fungal Community Structure of Greenhouse Grape Seedlings. *Sustainability* **2022**, *14*, 12719. [CrossRef]

horticulturae

Article

Identification and Evaluation of Flesh Texture of Crisp Pear Fruit Based on Penetration Test Using Texture Analyzer

Yulu Mou [1], Xingguang Dong [1,*], Ying Zhang [1], Luming Tian [1], Hongliang Huo [1], Dan Qi [1], Jiayu Xu [1], Chao Liu [1], Niman Li [2], Chen Yin [1] and Xiang Yang [1]

[1] Research Institute of Pomology, Chinese Academy of Agricultural Sciences (CAAS), Key Laboratory of Horticulture Crops Germplasm Resources Utilization, Ministry of Agriculture and Rural Affairs, Xingcheng 125100, China; mouyulude@163.com (Y.M.); wodeying1314@163.com (Y.Z.); tianluming@caas.cn (L.T.); huohongliang@caas.cn (H.H.); qidan@caas.cn (D.Q.); xujiayu01@caas.cn (J.X.); renshengsaniianshi@126.com (C.L.); yinchen4869@163.com (C.Y.); y18864836605@163.com (X.Y.)

[2] School of Software, Liaoning Technical University, Huludao 125105, China; liniman0518@163.com

* Correspondence: dongxingguangde@126.com

Abstract: Flesh texture is an important quality trait and is related to people's preference for fruit, especially for crisp pears. Puncture tests were carried out on 156 crisp pear fruit germplasm samples to analyze the diversity level of texture traits, to clarify the correlation between sensory description evaluation and instrumental traits, and to explore the effects of fruit ripening, size, and shelf life on the change in flesh texture. The results showed that puncture parameters were significantly different between crisp pear cultivars, and the work associated with the flesh limit compression force had the highest coefficient of variation (0.281). There was a significant correlation between puncture parameters and sensory evaluation scores. The correlation between sensory score and flesh firmness was the highest, with a correlation coefficient of 0.708, indicating that hardness can significantly influence the sensory evaluation of texture. Cluster analysis based on sensory evaluation and puncture determination could divide the germplasm resources of crisp pear into five texture categories: loosen, crunchy, crisp, tight–crisp, and dense. A comprehensive texture score model, constructed by principal component analysis, showed consistency with sensory evaluation scores and proved that the combination of a puncture test and sensory evaluation is the best way to identify and evaluate the texture of crisp pear. Further analysis of the influencing factors of flesh texture showed that fruit maturity and shelf life had significant effects on flesh quality. This study provides an important reference for the standardization, evaluation, and utilization of crisp pear variety resources.

Keywords: pear; germplasm resources; flesh texture; sensory evaluation; puncture method

1. Introduction

Pear (*Pyrus* spp.) is a globally significant fruit and one of the most extensively cultivated species in temperate regions. As the primary center of origin for pears, China boasts an exceptionally rich diversity of pear germplasm resources, encompassing 13 native species. Notably, the predominant cultivated species, including *P. bretschneideri*, *P. pyrifolia*, and *P. ussuriensis*, all originated from China. Pears demonstrate a wide geographical distribution and have diversified into numerous locally adapted cultivars in response to diverse environmental conditions [1,2]. Texture is a critical quality attribute of fresh fruits, alongside appearance and flavor; some people like soft pears but others like hard pears, and some pears that need after-ripening have a better flavor when they are

 https://doi.org/10.3390/horticulturae11040359

fully after-ripened, with texture significantly influencing consumer acceptance [3]. The diversity in flesh texture among pear varieties is considerable, and consumer preferences for specific textures play a pivotal role in the fruit industry's development [4–7]. Changes in texture are intricately associated with physiological processes, profoundly influencing crucial quality parameters such as sensory characteristics, storage potential, and processing suitability [8,9]. Therefore, studying the changes in flesh texture can help in improving the storage technology of pear fruit and diversifying its processing methods.

The flesh texture of pear can be divided into soft-type and crisp-type based on changes in fruit hardness during ripening. Cultivated species such as *P. ussuriensis* and *P. communis* belong to the soft type, which is further divided into melting and soft types; cultivated species such as *P. bretschneideri* and *P. pyrifolia* belong to the crisp type, which is further divided into crisp, loose–crisp, and tight–crisp texture types according to the tightness of the flesh [10]. According to the statistical analysis of 562 pear germplasm resources preserved in the National Germplasm Repository of pear in China, the crunchy type has the most resources, accounting for 40.57%. Soft-type pear cultivars account for 26.51%, while crisp-type pear cultivars account for more than 60% [11]. Crisp pears occupy a relatively high proportion in the germplasm resources of pears and have rich texture types, which plays an important role in research on the flesh texture of pears.

At present, the evaluation of fresh texture varieties mainly includes two methods: qualitative sensory evaluation and quantitative instrument determination [12,13]. Studies have shown that there is a good correlation between sensory properties such as hardness and crispness and hardness tester measurements [14]. Instrumental analysis offers a rapid and objective approach for evaluating fruit textural properties; however, the complexity and multidimensional nature of texture as a quality attribute cannot be fully captured by instrumental measurements alone. Consequently, while instrumental methods provide valuable quantitative data, they should be complemented with sensory evaluation, which serves as an essential reference standard for calibrating and validating instrumental readings. [15,16]. Sensory evaluation is more complex and subjective and can truly reflect a human's instant information on texture, and instrumental measurement is more economical, objective, and stable [17]. Therefore, the combination of sensory evaluation and instrument measurement is the best method to identify and evaluate texture properties.

A texture analyzer, also known as a physical property analyzer, is a commonly used instrument in the detection of fruit and vegetable texture characteristics. Puncture tests can be performed on unpeeled or peeled fruit samples to quantitatively determine the texture characteristics of fruits, and they are accurate, objective, and simple [18–20]. With the popularity of the texture analyzer, relevant texture evaluation systems have been established for many fruits, and the best test conditions for texture testing have been obtained [21–23]. Through puncture testing and texture profile analysis (TPA), the fruit texture traits of apples [24,25], pears [26,27], peaches [28,29], kiwifruits [30], and grapes [31,32] were studied, and most of the research contents focused on the differences and changes in fruit texture traits during storage after harvest. In kiwifruits [33], cucumbers [34], and watermelons [35], there are studies on the identification and evaluation of multiple germplasm resources by a texture analyzer and the comparison of the changes in texture parameters among different germplasm resources. However, there have been no studies on the systematic identification and evaluation of crisp pear varieties based on a texture analyzer and sensory evaluation.

In the literature on instrumental measurement and sensory evaluation, there are no studies focusing on the texture evaluation of crisp pears. Sensory texture descriptors are also more inclined to use general descriptors. However, compared with other common fruits, pears have more abundant texture types, so it is necessary to establish a texture evaluation system for pears. In this study, 156 germplasm resources of crisp pear germplasm were taken as the research object. The correlation between fruit puncture determination parameters was analyzed, and the varieties were classified based on sensory evaluation and texture determination. Then, the comprehensive score formula of principal components of was established through principal component analysis of puncture determination parameters, and the factors affecting the texture of pear flesh were analyzed so as to provide a theoretical basis for the establishment of accurate identification standards for the importance traits of crisp pear germplasm resources and the selection and utilization of resources.

2. Materials and Methods

2.1. Plant Material

In the National Germplasm Repository of Pear and Apple (xingcheng, China), 156 crisp pear fruit germplasm resources of Asian pear were selected from several systems, covering all the crisp-flesh types of *P. bretschneideri*, *P. pyrifolia*, *P. sinkiangensis*, and *P. ussuriensis* pears. For each variety, ≥20 fruits were harvested and stored at 4 °C. In addition, six representative crisp pear varieties, namely 'Korla pear' (*P. sinkiangensis*), 'Shuihong Xiao' (*P. bretschneideri*), 'Xuehua' (*P. bretschneideri*), 'Qiubai' (*P. bretschneideri*), 'Baozhu' (*P. pyrifolia*), and 'Chili' (*P. bretschneideri*), were picked for the determination of shelf life puncture parameters. The number of pears picked for each variety was 200. They were stored at 4 °C. Harvest timing for all 156 varieties followed the maturity standards documented in *Chinese Pear Genetic Resources*, [11], with exact harvest dates recorded.

2.2. Puncture Test and Parameter Setting

The Texture Measurement System–Professional Food Texture Analyzer (Food Technology Corporation, Sterling, VA, USA) was used to assess the fruits with no surface damage and no pests and diseases. Each fruit was punctured once from the front and back, ensuring that the puncture position was located in the central area of the fruit. When fruits that have reached harvestable maturity are stored at room temperature for a week after harvest, the flesh is fully ripe and has the best flavor and texture [36,37]. Accordingly, after cold storage at 4 °C, pear samples were transferred to room-temperature (25 °C) conditions. Texture measurements were conducted on day 7, using 10 fruits per variety. For shelf life puncture tests, 200 fruits per variety were maintained in a product observation chamber (25 °C). The first puncture tests were performed immediately after removal from cold storage (day 0), followed by subsequent measurements every 4 days using 20 randomly selected fruits per time point, and the results were averaged.

The puncture test was conducted to determine the texture of pear flesh and followed an existing pear flesh puncture test [27]; the parameter settings were slightly modified. Before the test, the pear fruit was peeled on both sides of the equator and placed into the puncture fixation device of the texture analyzer. The diameter of the probe was 6 mm, and the parameters were set as follows: trigger force 2 $kg \cdot m/s^2$ (N), test speed 60 $mm \cdot min^{-1}$, puncture distance 10 mm, and return speed 100 $mm \cdot min^{-1}$ (Figure 1). The measurement parameters and definitions are shown in Table 1. The fruit diameter was automatically output by the texture analyzer.

2.3. Sensory Evaluation of Flesh Texture

For each variety, 10 fruits with no surface damage and no pests and diseases were peeled, cut into evenly sized pieces, and mixed together for sensory texture evaluation. Sensory texture evaluation was carried out by experts trained in pear germplasm resources who compiled 10 classifications of flesh texture types in the book *Chinese Pear Genetic Resources* [11], which covers all texture categories, from melting (low hardness, high viscosity) to dense (high hardness, low porosity). The crispy flesh texture descriptors and their quantitative assignment are as follows: 6. loose; 7. crunchy; 8. crisp; 9. tight and crisp; 10. dense.

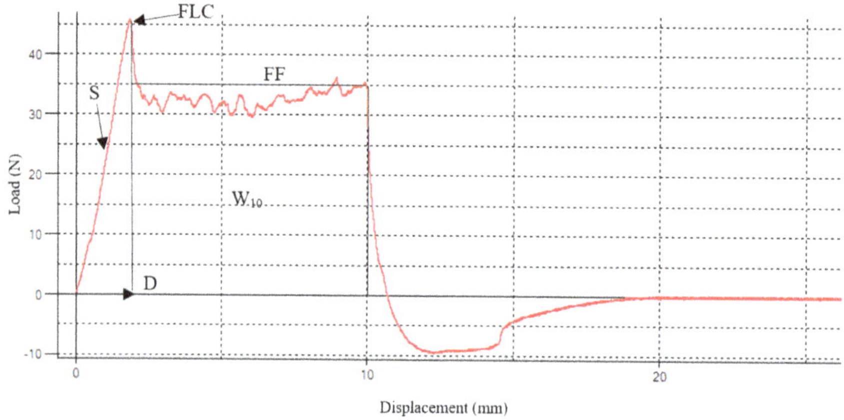

Figure 1. Load–displacement curve obtained by puncture test of crisp pear with texture analyzer.

Table 1. Definitions of flesh puncture test parameters.

Parameter	Calculation Method	Description
Flesh limit compression force (FLC, N)	Maximum force through the flesh	Represents the ultimate elasticity of flesh
Deformation associated with flesh limit compression force (D, mm)	Distance at which breaking force is displaced	Deformation of flesh with breaking force
Slope of force deformation curve (S, N·mm^{-1})	Gradient slope of force from 0 to FLC	Gradient measurement of puncture hardness
Work associated with FLC (WFLC, N·mm)	Area under the curve from 0 to D	The work required for the flesh to break
Work required to attain a flesh deformation of 10 mm (W_{10}, N·mm)	Displacement from 0 to 10mm area under the curve	The work required when the flesh is deformed to 10 mm
Flesh firmness (FF, N)	Mean of force required to shift from D to 10 mm	The average force required to penetrate the flesh to 10 mm

Note: The FLC and FF units were Newton (N), 1N = kg·m/s^2. During the puncture test, the probe was penetrated at a speed of 60 mm·min^{-1} to record flesh limit compression force and flesh firmness. The unit of D is mm, and it represents the displacement of the probe from contact with the flesh to penetration into the flesh. The unit of S is N·mm^{-1}, and it represents the force required per 1 mm for the probe to penetrate the flesh before reaching the FLC. The units of WFLC and W_{10} are N·mm, which is the unit of work, representing the product of the force felt by the probe and its displacement in the direction of motion.

2.4. Statistical Analysis

Microsoft Excel 2016 (https://www.microsoft.com/en-us/microsoft-365/excel, accessed on 17 March 2025) was used for data sorting, SPSS27.0 (https://www.ibm.com/products/spss-statistics, accessed on 17 March 2025) software was used for correlation analysis, principal component analysis, and cluster analysis, and Origin 2022 (https://www.originlab.com, accessed on 17 March 2025) software was used for plotting.

3. Results

3.1. Variation and Correlation Analysis of Puncture Measurement Parameters

The averages, extreme values (Min and Max), standard deviations, and coefficients of variation (CV) of six puncture measurement parameters were analyzed (Table 2). WFLC (N·mm) had the highest coefficient of variation (0.281), with a range of 11.40~53.63 N·mm, followed by FLC (N), FF (N), S (N·mm^{-1}), and W_{10} (N·mm). The coefficient of variation was 0.242, 0.232, 0.226, and 0.219, and the range of variation was 10.71~42.41 N, 11.88~38.59 N, 5.29~21.81 N·mm^{-1}, and 96.11~290.85 N·mm. The coefficient of variation for D (mm) was the smallest (0.136), and the range of variation was 1.24~2.45 mm. Most penetration parameters exhibited coefficients of variation exceeding 15%, demonstrating significant varietal differences in puncture characteristics. The parameters of the puncture test could be used to distinguish the texture characteristics of varieties.

Table 2. Differences in puncture test parameters of 156 crisp pear germplasm resources.

Indicators	FLC/N	D/mm	WFLC/N·mm	FF/N	W_{10}/N·mm	S/N·mm^{-1}
Average	26.15	1.83	28.49	23.03	187.30	14.55
Min	10.71	1.24	11.40	11.88	96.11	5.29
Max	42.41	2.45	53.63	38.59	290.85	21.81
Standard deviation	6.32	0.25	8.02	5.35	41.05	3.29
CV/%	24.20	13.60	28.1	23.2	21.9	22.6

Correlation analysis showed that there was a significant correlation among all parameters (Figure 2). FF (N) and W_{10} (N·mm) had the most significant correlation, and the correlation coefficient was the largest (0.992). FLC (N) was positively correlated with W_{10} (N·mm), FF (N), and WFLC (N·mm) ($p < 0.01$), and the correlation coefficients were 0.921, 0.92, and 0.865, respectively. Puncture parameters directly related to force had a high positive correlation. However, D (mm) and S (N·mm^{-1}) showed an extremely significant negative correlation ($p < 0.01$), and the correlation coefficient was −0.213. D (mm) was the deformation of flesh with breaking force, while S (N·mm^{-1}) was the gradient measurement of puncture hardness, and the two indicators showed a negative correlation, which may be related to the greater variation in FLC and S.

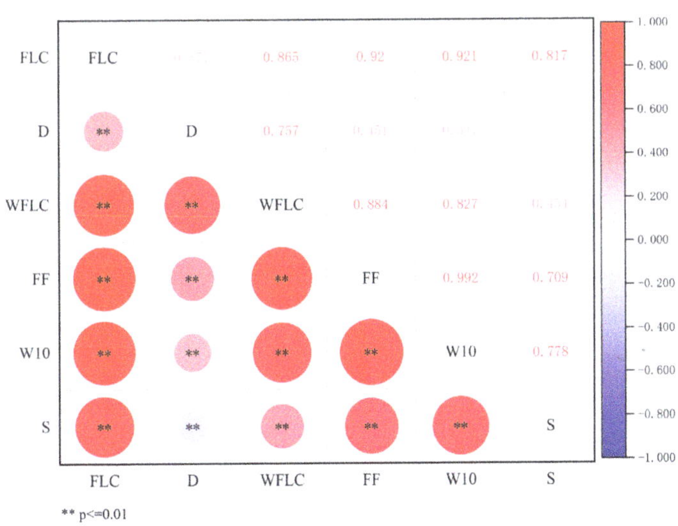

Figure 2. Correlation of puncture test parameters of 156 crisp pear germplasm resources.

3.2. Cluster Analysis Based on Sensory Texture Evaluation and Puncture Measurement

To investigate the associations between sensory texture evaluations and instrumental measurements, we performed correlation analyses between these parameters. The analysis revealed highly significant positive correlations between sensory texture scores and all six puncture measurement parameters. The strongest correlation (Table 3) occurred between sensory scores and FF (N), demonstrating that hardness represents the most influential texture parameter in sensory evaluation. According to the sensory texture score and puncture measurement parameters, systematic clustering and the gradual aggregation method were used to perform cluster analysis, and the approximate matrix was obtained by square Euclidean distance. At a clustering distance threshold of 5, the 156 crisp pear germplasm accessions were segregated into five distinct clusters (A-E). Cluster A contained 73 varieties, followed by cluster B, which contained 29 varieties, cluster C, which contained 24 varieties, cluster D, which contained 20 varieties, and cluster E, which contained 10 varieties (Figure 3). The five clusters showed strong correspondence with the five recognized crisp pear texture types in standard pear resource classifications.

Table 3. Correlation between sensory texture scores and puncture test parameters of 156 crisp pear germplasm resources.

	FLC/N	D/mm	WFLC/N·mm	FF/N	W_{10}/N·mm	S/N·mm^{-1}
Sensory texture score	0.698 **	0.294 **	0.648 **	0.708 **	0.696 **	0.529 **

Note: ** is extremely significant (1% significant level).

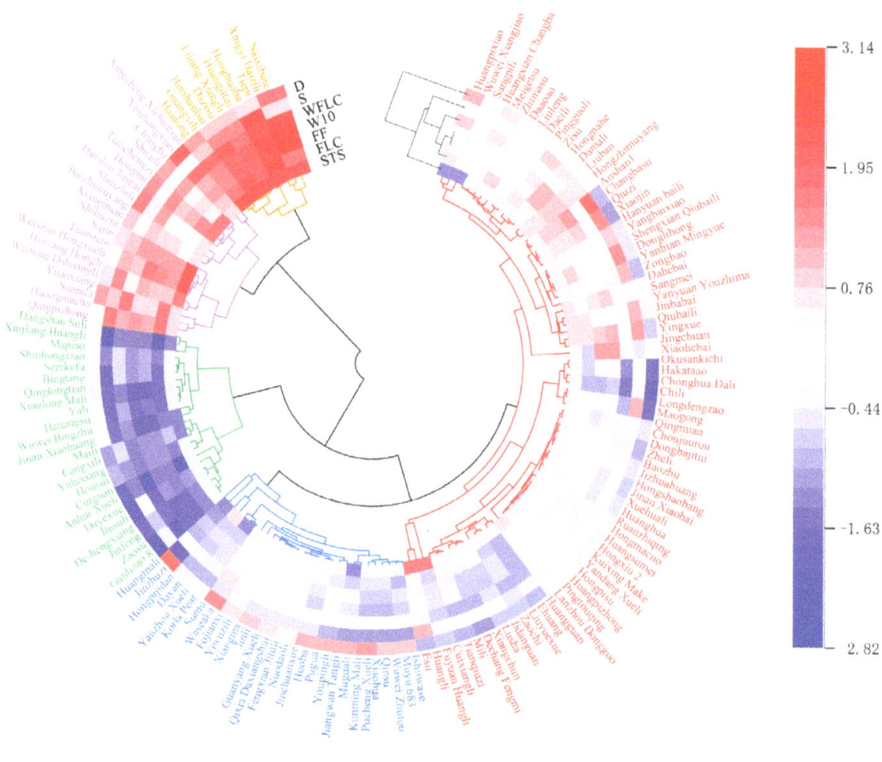

Figure 3. Circular heatmap of puncture test parameters and sensory texture score of 156 crispy pear germplasm resources. Note: The red varieties are labeled as cluster A; the blue varieties are labeled as cluster B; the green varieties are labeled as cluster C; the purple varieties are labeled as cluster D; the yellow varieties are labeled as cluster E; STS represents sensory texture score.

The six puncture measurement parameters of cluster E resources were higher than those of the other four types of resources (Figure 4a). The mean values of FLC (N), D (mm), WFLC (N·mm), FF (N), W_{10} (N·mm), and S (N·mm^{-1}) were 38.93 N, 2.14 mm, 45.74 N·mm, 34.28 N, 268.68 N·mm, and 18.57 N·mm^{-1}, respectively (Table 4), and the sensory texture scores of the 10 varieties of cluster E resources were all 10 (Figure 4b). Cluster E resources belong to dense flesh type. The six puncture measurement parameters in cluster D ranked second among the five clusters of resources, second only to cluster E resources (Figure 4a). The mean values of FLC (N), D (mm), WFLC (N·mm), FF (N), W_{10} (N·mm), and S (N·mm^{-1}) were 32.63 N, 2.04 mm, 37.86 N·mm, 28.76 N, 228.73 N·mm, and 16.42 N·mm^{-1}, respectively (Table 4). The varieties with sensory texture scores of 9 were mostly in cluster D resources (Figure 4b). Cluster D resources belong to the tight-and-crisp flesh type. The six parameters of cluster C resources are significantly lower than those of the other four types of resources (Figure 4a). The mean values of FLC (N), D (mm), WFLC (N·mm), FF (N), W_{10} (N·mm), and S (N·mm^{-1}) are 16.71 N, 1.67 mm, 17.93 N·mm, 15.40 N, 128.05 N·mm, and 10.35 N·mm^{-1}, respectively (Table 4). Most of the varieties with a sensory texture score of 6 were in cluster C resources (Figure 4b), and cluster C resources belonged to the loose flesh type. The two resource indicators of cluster A and cluster B are located in the middle level (Figure 4a). The mean values of FLC (N), D (mm), WFLC (N·mm), FF (N), W_{10} (N·mm), and S (N·mm^{-1}) of cluster A resources are 26.75 N, 1.70 mm, 26.74 N·mm, 23.26 N, 192.50 N·mm, and 16.03 N·mm^{-1}, respectively. The mean values of FLC (N), D (mm), WFLC (N·mm), FF (N), W_{10} (N·mm), and S (N·mm^{-1}) for cluster B resources are 23.58 N, 2.04 mm, 29.23 N·mm, 20.96 N, 166.63 N·mm, and 11.63 N·mm^{-1}, respectively (Table 4). The sensory texture scores were mostly 7 and 8 (Figure 4b), belonging to the crunchy and crisp-flesh types, and these two descriptions are difficult to distinguish by sensory evaluation. Overall, the above results indicate that there is good agreement between sensory evaluation and texture instrument puncture determination, both of which divide the variety resources of crisp pear into five germplasm types, and flesh hardness is the main factor affecting texture evaluation. These findings provide valuable scientific support for establishing standardized protocols for crisp pear texture assessment and varietal classification.

3.3. Principal Component Analysis of Puncture Measurement Parameters

According to the analysis results (Table 5), the parameters are mainly divided into two principal components. The characteristic values are 4.475 and 1.360. The weights of FF (N), FLC (N), W_{10} (N·mm), and WFLC (N·mm) for the first principal component are relatively large; FF (N) has the largest eigenvalue, followed by FLC (N). The pairings of the four parameters are all higher than 0.8, showing a significant positive correlation (Figure 2). Moreover, these four measurement parameters represent the elasticity and hardness of flesh. They reflect the masticatory characteristics of flesh. The second principal component S (N·mm^{-1}) has the largest negative weight, and D (mm) has the largest positive weight. These two parameters have an extremely significant negative correlation, which indicates that the degree of fragmentation and fracture of flesh sections reflects the cracking characteristics of flesh. The variance contribution rates of the first and second principal components are 74.591% and 22.672%, respectively, and the cumulative variance contribution rates are 74.591% and 97.262%, respectively. And the cumulative variance contribution rate of these two principal components was 97.262%, which basically reflected the flesh elasticity, hardness, and crispness of the 156 crisp pear germplasm resources and was highly representative of the texture and quality information of crisp pear.

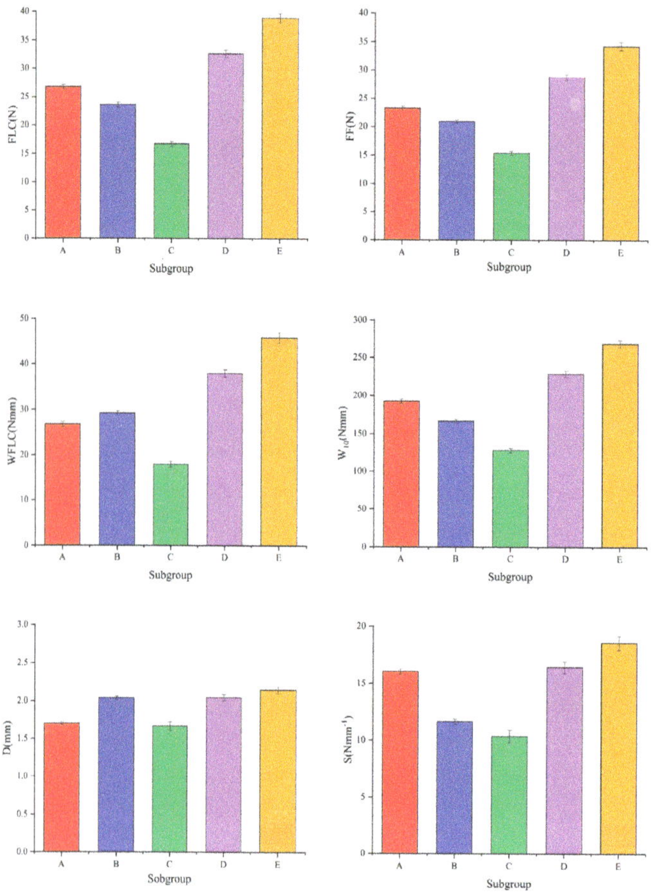

(a)

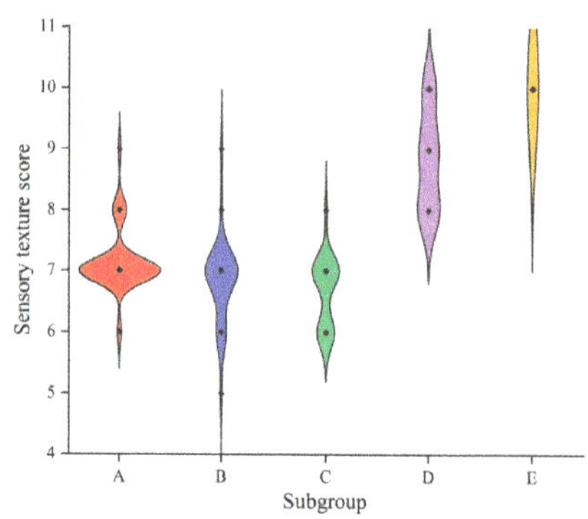

(b)

Figure 4. (**a**) Histogram of statistical analysis of six puncture parameters in different subgroups; (**b**) violin charts with sensory texture scores in different subgroups.

Table 4. Comparison of the mean values of six puncture test parameters between different subgroups.

Subgroups	FLC (N) ± SD	D (mm) ± SD	WFLC (N·mm) ± SD	FF (N) ± SD	W10 (N·mm) ± SD	S (N·mm^{-1}) ± SD
A	26.75 ± 0.4	1.70 ± 0.02	26.74 ± 0.54	23.26 ± 0.37	192.50 ± 2.9	16.03 ± 2.9
B	23.58 ± 0.45	2.04 ± 0.02	29.23 ± 0.49	20.96 ± 0.3	166.63 ± 2.35	11.63 ± 2.35
C	16.71 ± 0.47	1.67 ± 0.05	17.93 ± 0.65	15.40 ± 0.33	128.05 ± 2.73	10.35 ± 2.73
D	32.63 ± 0.69	2.04 ± 0.04	37.86 ± 0.79	28.76 ± 0.5	228.73 ± 4.34	16.42 ± 4.34
E	38.93 ± 0.73	2.14 ± 0.05	45.74 ± 1.16	34.28 ± 0.71	268.68 ± 4.75	18.57 ± 4.75

3.4. Principal Component Analysis of Puncture Measurement Parameters

According to the analysis results (Table 5), the parameters are mainly divided into two principal components. The characteristic values are 4.475 and 1.360. The weights of FF (N), FLC (N), W_{10} (N·mm), and WFLC (N·mm) for the first principal component are relatively large; FF (N) has the largest eigenvalue, followed by FLC (N). The pairings of the four parameters are all higher than 0.8, showing a significant positive correlation (Figure 2). Moreover, these four measurement parameters represent the elasticity and hardness of flesh. They reflect the masticatory characteristics of flesh. The second principal component S (N·mm^{-1}) has the largest negative weight, and D (mm) has the largest positive weight. These two parameters have an extremely significant negative correlation, which indicates that the degree of fragmentation and fracture of flesh sections reflects the cracking characteristics of flesh. The variance contribution rates of the first and second principal components are 74.591% and 22.672%, respectively, and the cumulative variance contribution rates are 74.591% and 97.262%, respectively. And the cumulative variance contribution rate of these two principal components was 97.262%, which basically reflected the flesh elasticity, hardness, and crispness of the 156 crisp pear germplasm resources and was highly representative of the texture and quality information of crisp pear.

Table 5. Principal component analysis of puncture test parameters of 156 crisp pear germplasm resources.

Parameter	Feature Vector	
	First Principal Component	Second Principal Component
FF (N)	0.982	−0.015
FLC (N)	0.971	−0.123
W10 (N·mm)	0.970	−0.134
WFLC (N·mm)	0.923	0.359
S (N·mm^{-1})	0.737	−0.662
D (mm)	0.482	0.872
Characteristic value	4.475	1.360
Contribution rate/%	74.591	22.672
Cumulative contribution rate/%	74.591	97.262

Note: The extraction method is principal component analysis, and 2 components have been extracted.

Principal component analysis was performed to calculate component score coefficients (Table 4), yielding the following two principal component expressions:

$$Y1 = 0.464 \times ZFF + 0.459 \times ZFLC + 0.459 \times ZW_{10} + 0.436 \times ZWFLC + 0.348 \times ZS + 0.228 \times ZD;$$
$$Y2 = -0.0129 \times ZFF - 0.1055 \times ZFLC - 0.1149 \times ZW_{10} + 0.3078 \times ZWFLC - 0.5677 \times ZS + 0.7477 \times ZD.$$

ZFF, ZFLC, ZW10, ZWFLC, ZS and ZD represent values normalized by standard deviation (Z-score) for FF (N), FLC (N), W10 (N·mm), WFLC (N·mm), S (N·mm^{-1}), and D (mm), respectively. The comprehensive model of principal component analysis was $Y = 0.74591 \times Y1 + 0.22672 \times Y2$.

The Z-score values of the texture parameters of 156 crisp pear germplasm resources were substituted into the two formulas Y1 and Y2, respectively, to obtain the scores of principal component 1 and principal component 2 of each variety. Then, the comprehensive score of principal component analysis of each variety was calculated by the comprehensive model of principal component analysis, which could also be regarded as the texture score of puncture determination (Figure 5). The results were sorted according to the comprehensive score of principal component analysis, and the texture scores measured by puncture were compared with those evaluated by sensory texture (Table S1). For example, 'Naixibute', 'Tiepi', and 'Xingyi Haizili' were the top 3 varieties among the 156 varieties, with overall texture scores of 4.66, 3.86, and 3.70, respectively. Their sensory texture scores were also the highest, at 10 for all of these, and the texture type of the flesh was dense. 'Ganlizao 8', 'Mapiao', and 'Anhui Xueli' were the three varieties with the lowest overall texture scores, which were −3.13, −3.15 and −3.24, respectively. Their sensory texture scores were also relatively low, at 6 or 7.

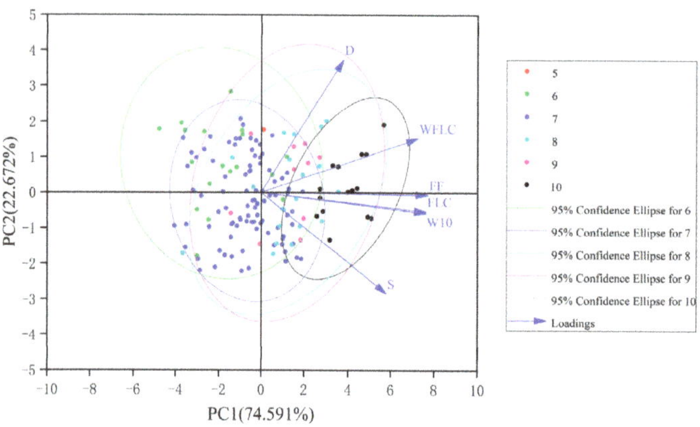

Figure 5. Principal component analysis biplot of puncture test parameters based on sensory texture score. Note: The sensory texture score was used as the grouping category.

Among the 36 varieties with texture scores greater than 1, all the varieties with cluster E and cluster D resources were included, and the texture parameters of cluster E and cluster D resources were higher than those of cluster A, cluster B, and cluster C resources. The varieties belonging to cluster E with dense-type flesh include 'Naixibute', 'Tiepi', and 'Xingyi Haizili', and so on. The varieties belonging to cluster D with tight- and crisp-type flesh include 'Xingcheng Xiehuatian', 'Huangmacha', 'Youhongxiao', etc. The range with scores greater than −1.36 and less than 1 has 90 varieties, including most of the intermediate cluster A resources and all cluster B resources. There are 30 varieties with scores less than −1.36, including all cluster C resources, such as 'Dangshan suli', 'Yuluxiang', 'Cuiguan', etc. The sensory texture scores and instrumental texture scores exhibited consistent trends, where higher values corresponded to more compact tissue structures, while lower values indicated looser textures. This agreement demonstrates that both evaluation methods effectively capture textural differences among crisp pear germplasm resources.

3.5. Analysis of Influencing Factors of Fruit Texture

Maturity and fruit size represent crucial agronomic traits primarily governed by genetic background. The analysis of their correlation with flesh texture is of great significance for screening important genetic resources and carrying out directional breeding for important traits. Correlation analysis (Table 6) showed that fruit maturity was significantly positively correlated with sensory texture score, and the correlation coefficient was 0.200.

Fruit maturity was significantly positively correlated with FLC (N), WFLC (N·mm), FF (N), and W_{10} (N·mm), and the correlation coefficients were 0.200, 0.166, 0.198, and 0.201; that is, maturity was strongly correlated with all the puncture parameters of the first principal component.

Table 6. Correlation between maturity, fruit diameter, and puncture test parameters of 156 crispy pear germplasm resources.

	Sensory Texture Score	FLC/N	D/mm	WFLC/N·mm	FF/N	W_{10}/N·mm	S/N·mm^{-1}
Maturity	0.200 *	0.200 *	0.120	0.166 *	0.198 *	0.201 *	0.151
Fruit diameter	0.079	0.031	−0.127	−0.042	−0.133	−0.126	0.099

Note: * is significantly correlated (5% significant level).

The correlation between pear fruit diameter and fruit weight reached 0.948. It can be used as the main index to evaluate pear fruit size [32]. We analyzed the correlation between pear fruit diameter and texture (Table 6). The results showed that there was no correlation between fruit diameter, sensory texture score, and puncture measurement parameters. Previous studies typically controlled experimental variables by selecting uniformly sized samples for texture analysis. Based on these results, we hypothesized that when other texture-affecting variables remain constant, fruit diameter may not be a primary determinant of textural characteristics.

Based on harvest timing, pear varieties can be classified into three maturation groups: early maturity (July to mid-August), medium maturity (mid-August to mid-September), and late maturity (mid-September through October) [38,39]. The crisp pear germplasm resources in this study were divided into three categories—early, middle, and late—at a proportion of 1:48:177, and most of the tested crisp pears matured after September (Figure 6a). Late-maturity varieties accounted for a relatively high proportion of each category of resources, among which 90% of cluster E resources belonged to late-maturity varieties. Fifteen varieties with sensory scores of 10 belonged to late-maturity varieties; among them, fourteen belonged to late-maturity varieties, and the cluster with a tight flesh texture was also late-maturity (Figure 6b). Appropriate harvesting maturity has an important impact on fruit texture quality. Fully matured fruits have a sufficient accumulation of various nutrients in their bodies and have better quality and storage endurance [39,40]. Therefore, sensory texture evaluation and instrumental measurement data will be closer to each other. This property can be used as an important reference for texture character screening.

Since fruit texture undergoes dynamic changes during postharvest ripening, investigating textural changes in crisp pears during shelf life is critical for establishing reliable quality evaluation criteria and guiding pear variety selection and utilization. For the six representative crisp pear varieties, the puncture parameters were measured every 4 days after harvest (Figure 7). The six puncture parameters of 'Qiubaili' had the highest level among the six crisp pear varieties. In particular, the values of the three parameters of FLC (N), FF (N), and W_{10} (N·mm) were much higher than those of the other five varieties in the whole shelf period, and four parameters of 'Qiubaili'—FLC (N), FF (N), W_{10} (N·mm), and S (N mm^{-1})—all reached the highest values on the fourth day of shelf life and then declined. The values of the six puncture parameters of 'Shuihongxiao' and 'Chili' were the lowest among the six crisp pear varieties. The values of FLC (N), WFLC (N·mm), and D (mm) for 'Xuehua', related to rupture force, reached the highest level when the shelf life was 20 days. In contrast, the values of FF (N), W_{10} (N·mm), and S (N mm^{-1}), three hardness-related indicators, were in decline since the start of shelf life. The overall trend of five parameters of 'Korla pear', including FLC (N), WFLC (N·mm), FF (N), W_{10} (N·mm), and S (N mm^{-1}), was a decreasing one, but there was a fluctuation in the eighth day of shelf life, followed by a decline. On the whole, with the extension of shelf life, D (mm) and WFLC (N·mm) showed an upward trend, S (N mm^{-1}) decreased significantly, FF (N) and W_{10} (N. mm) showed a slight downward trend, and FLC (N) showed a relatively gentle trend (Figure 7). This phenomenon may be attributed to concurrent changes in fruit water loss and cell wall modifications during shelf life, which collectively alter textural properties, including hardness and brittleness, consequently affecting puncture measurement parameters.

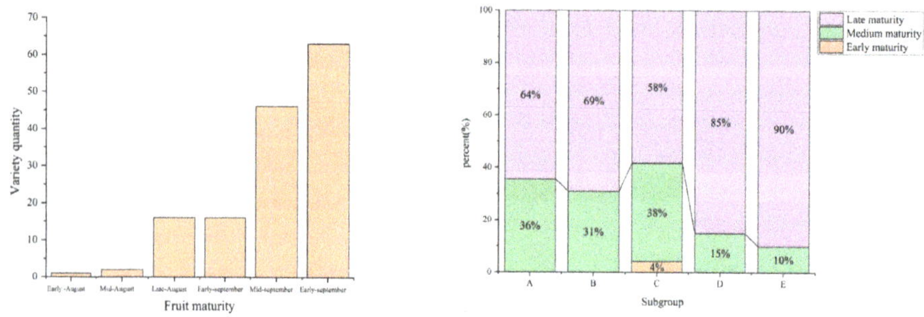

(**a**) (**b**)

Figure 6. (**a**) Fruit maturity distribution of 156 crisp pear germplasm resources; (**b**) percentage stacking histogram of the five subgroups at different maturities.

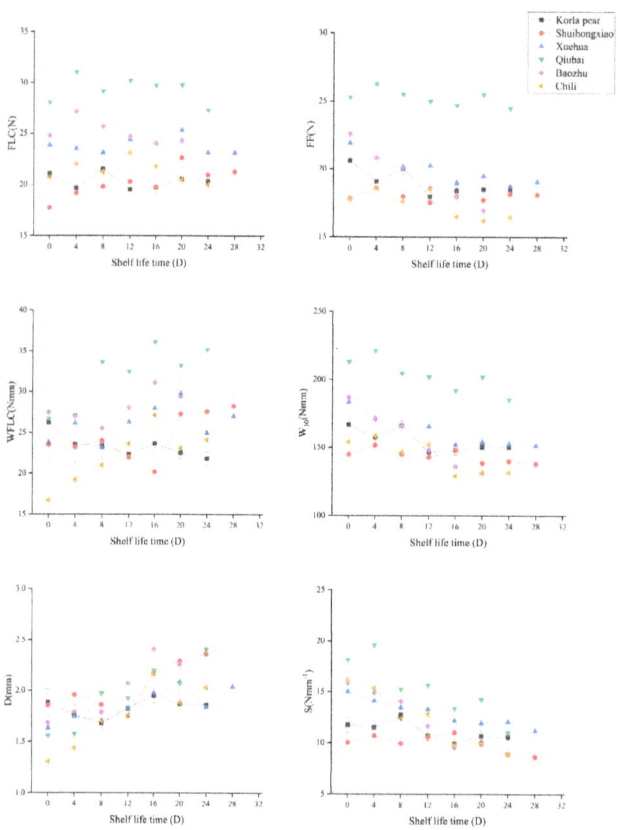

Figure 7. Changes in puncture parameters of 6 crispy pear varieties during shelf life.

4. Discussion

The puncture test serves as an instrumental simulation of the mastication process, and the textural parameters obtained through this method provide reliable quantitative indicators that closely correlate with human sensory perception during fruit consumption. In this study, the variation coefficient of most of the puncture parameters is large, which proves that the texture diversity of pear varieties is rich. These findings validate the effectiveness of puncture measurements as a reliable method for assessing and characterizing textural properties in crisp pears. Among them, the work associated with the flesh limit compression force has the largest coefficient of variation, reflecting the work required for the flesh to break, as well as the magnitude of the rupture force and the displacement of the rupture force. The flesh limit compression force, to some extent, represents the crispness of the fruit [41]. Flesh firmness and the flesh work required to attain a flesh deformation of 10 mm of pear fruit, as well as the rupture force and rupture function, were significantly and positively correlated, which was consistent with the results of puncture tests on apples [42]. The correlation analysis of puncture measurement parameters in this experiment showed significant correlations among parameters representing flesh elasticity and hardness. FF (N), FLC (N), W_{10} (N·mm), and WFLC (N·mm), which reflect flesh mastication characteristics, were significantly positively correlated. This is consistent with the results of previous studies on pears [7,14,41], apples [24,43,44], and grapes [45]. These indicators can be used to comprehensively evaluate the flesh quality properties of crispy pear fruits.

Cluster analysis is a statistical method to cluster research samples according to their inter-related properties. Because of the high correlation between the relevant indexes of texture traits, cluster analysis is usually carried out after the dimensionality of variables is reduced through principal component analysis to obtain the principal factor. Research on thin-skinned melon extracted three principal factors from eight texture parameters and six chemical indexes through principal component analysis [46]. Then, muskmelon was divided into five types formed by the main indexes obtained by systematic clustering. The assessment of flesh texture can be effectively conducted through both sensory evaluation and instrumental texture analysis. Their complementary application provides a comprehensive approach that combines subjective human perception with objective quantitative measurements, thereby yielding the most reliable and representative characterization of textural properties. Based on this, this experiment combined the two methods to cluster 156 crispy pear germplasm resources, and finally, the crisp pears were divided into five clusters, A, B, C, D, and E, corresponding to five germplasm types: loosen, crunchy, crisp, tight–crisp, and dense. The results showed that the current standard of sensory evaluation of crisp pear flesh texture was in good agreement with the results of texture instrument puncture tests. The tests could be used for the standardized evaluation of pear varieties. *Chinese Pear Genetic Resources* also includes fruit texture descriptors for soft pears in ten sensory texture categories (1. melting; 2. soft; 3. mealy; 4. fluffy; 5. sandy; 6. loose; 7. crunchy; 8. crisp; 9. tight and crisp; 10. dense), which can be used as a reference in subsequent studies to determine whether a texture evaluation system for all texture types of pear germplasm resources can be formed through this set of descriptors. In this study, a texture evaluation model was obtained through principal component analysis to describe the physical quality of crispy fruit. The puncture score model represents the value of fruit texture. The higher the puncture score, the denser the flesh texture. This is consistent with previous evaluation models representing texture quality in pear [47], fresh jujube [48,49], mulberry [50], and cucumber [34].

The textural variation among the five crisp pear germplasm clusters may be attributed to distinct physicochemical properties, with differential gene expression influencing key determinants including stone cell development, pectin metabolism, and cell wall architecture across cultivars. Using cluster analysis, we were able to classify the 14 cultivars of crisp pear into three different texture kinds, namely, soft (such as 'Housui', 'Hwangkum', 'Huangguan', etc.), crisp ('such as Mansoo'), and hard ('Butirra Rostata Morettini') [7]. Stone cells are composed of cellulose, hemicellulose, and lignin, with high physical strength, and their quantity and distribution directly affect the hardness and porosity of flesh [51,52]. The results demonstrated a significant positive correlation between stone cell content and flesh hardness, indicating that higher stone cell levels correspond to greater flesh firmness [10]. A key transcription factor (PbAGL7) which promotes stone cell content and secondary cell wall thickening was identified through transcriptome analysis of pear varieties with varying stone cell levels. Additionally, sclerenchyma cell abundance directly influences pear crispness [53]. PbrMYB4, an R2R3-MYB protein, regulates the lignification of pyritic cells by activating lignin synthesis genes. Overexpression of PbrMYB4 enhances lignin deposition and increases the cell wall thickness of ducts and wood fibers [54]. The decomposition of cellulose, hemicellulose, and pectin in the cell wall is primarily mediated by polygalacturonase (PG), pectin methylesterase (PME), pectatc lyase (PL), rhamnogalacturonase (RGase), cellulase (Cx), and β-galactosidase (TBG) [55]. These factors likely underlie the distinct textural variations observed among the five clusters of germplasm resources examined in this study. For future research on flesh texture-associated genes, these characterized resources provide a valuable foundation, enabling precise identification and evaluation of textural traits in pear cultivars.

After principal component analysis, it was found that there were overlapping distributions between sensory texture scores and puncture parameters in the principal component biplot when grouped by sensory texture scores. Using sensory texture evaluation as the dependent variable and individual texture parameters as independent variables, the authors developed a multiple linear regression model (R^2 = 0.564). This indicates that the six puncture measurement parameters could explain only 56.4% of the variation in sensory texture evaluation, a finding consistent with previous studies on apple sensory evaluation models [18,56]. Texture measurement parameters can only explain half or less of sensory juiciness and peel toughness, suggesting these textural attributes are influenced by multiple complex factors. Although sensory texture evaluation has a significant correlation with texture data, the sensory texture score is not always identical with puncture measurement data. These six puncture measurements cannot fully account for all the texture changes. Pear flesh exhibits diverse textural properties, with significant varietal differences observed among cultivars [10]. Sensory evaluation alone presents limitations for quantitative texture assessment due to the inherent subjectivity in human perception. Flesh texture variation is associated with multiple physicochemical components, including pectin [9,43], cellulose [57], lignin [58,59], and starch [60,61]. Additional texture-related components should be incorporated into future sensory texture modeling studies.

This study demonstrated that maturity significantly influences pome fruit flesh quality, while fruit size appears less critical. We hypothesize that late-ripening varieties, benefiting from extended development periods, achieve greater dry matter accumulation [62] and higher concentrations of texture-related compounds such as lignin [63]. However, regional climatic differences, particularly between northern and southern growing areas, lead to variations in pear ripening periods. Under field conditions, the maturity for a given variety is defined as the date when 70% of fruits reach full maturity [38]. Fruit maturity was additionally assessed through peel color transition, specifically from green to yellow-green or green to brown. However, the current understanding of varietal and regional differences in fruit maturity characteristics remains limited [39]. After-ripening is usually an external factor affecting fruit quality. The texture changes of crisp pear germplasm resources during shelf life showed that D (mm) and WFLC (N·mm) had an increasing trend, S (N·mm^{-1}) decreased significantly, FF (N) and W_{10} (N·mm) showed a slight decreasing trend, and FLC (N) showed a gentle trend. Optimal harvest maturity significantly influences both fruit quality and storage potential. Pears harvested at different maturity stages exhibit distinct changes in flesh texture parameters during room-temperature shelf life [7]. When harvested at optimal maturity, ripening progression shows significant correlations with both instrumental puncture measurements and sensory texture evaluations. Shelf life varied considerably among pear varieties depending on their maturation period. Early-season varieties (late July to early August) exhibited shorter shelf lives (<10 days), while late-season varieties (early September onward) demonstrated superior storage potential. Among the six tested varieties, late-maturing types showed an extended shelf life of up to 28 days, with FLC (N) displaying relatively stable trends, a characteristic potentially associated with enhanced storage capacity in late-maturing cultivars. These results significantly contribute to texture evaluation in crisp pears, enable an effective selection of premium genetic resources, and support precision breeding approaches to streamline cultivar improvement.

5. Conclusions

Using puncture tests and sensory evaluation, we assessed texture characteristics in 156 crisp pear germplasm resources. The results demonstrated a substantial diversity in texture traits. Instrumental measurements and sensory evaluations showed strong consistency, with cluster analysis dividing the pears into five clusters: loose, crunchy, crisp, tight–crisp, and dense. Principal component analysis of the six puncture measurement parameters resulted in two principal components reflecting the flesh mastication characteristics and the flesh cracking characteristics. The principal component-derived texture scores closely matched sensory evaluation trends, confirming the effectiveness of this combined approach for texture assessment. Further analysis of the influencing factors of flesh texture showed that fruit maturity and shelf life had significant effects on flesh quality, but fruit diameter did not seem to be the main factor. This study provides an important reference for the standardization, evaluation, and utilization of crisp pear variety resources.

Supplementary Materials: The following supporting information can be downloaded at https://www.mdpi.com/article/10.3390/horticulturae11040359/s1, Table S1: Comprehensive score table of fruit texture of 156 crisp-flesh pear germplasm resources; Table S2: Fruit maturity, fruit diameter, puncture data, and cluster group statistics of crisp-flesh pear varieties.

Author Contributions: Y.M. and X.D., writing—original draft, visualization, data curation, software; N.L., C.Y. and X.Y., formal analysis, software; L.T., Y.Z., H.H., D.Q., J.X. and C.L., resources; Y.M. and X.D., conceptualization, writing—review and editing; X.D. supervision. All authors have read and agreed to the published version of the manuscript.

Funding: This study was supported by The Science and Technology Innovation Project of the Chinese Academy of Agricultural Sciences (CAAS-ASTIP-RIP), the Earmarked Fund for the China Agriculture Research System (CARS-28-01), and National Natural Science Foundation of China project (32272676).

Data Availability Statement: The original contributions presented in the study are included in the article/Supplementary Materials; further inquiries can be directed to the corresponding author.

Conflicts of Interest: The authors declare no conflicts of interest.

References

1. Cao, Y.F.; Li, S.L.; Huang, L.S.; Sun, J.Z.; Tan, X.W. Survey of germplasm resources of pear in China and comprehensive evaluation of excellent germplasm. *China Fruits* **2000**, *4*, 42–44. [CrossRef]
2. Wang, W.H.; Wang, G.P.; Tian, L.M.; Li, X.G.; Lv, X.L.; Zhang, Y.X.; Zhang, J.H.; Cao, Y.F. New China fruit science research 70 years—Pear. *J. Fruit Sci.* **2019**, *36*, 1273–1282. [CrossRef]

3. Harker, F.R.; Redgwell, R.J.; Hallett, I.C.; Murray, S.H.; Carter, G. Texture of Fresh Fruit. In *Horticultural Reviews*; John Wiley & Sons, Ltd.: Hoboken, NJ, USA, 1997; pp. 121–224. ISBN 978-0-470-65064-6.
4. Cano-Salazar, J.; López, M.L.; Echeverría, G. Relationships between the instrumental and sensory characteristics of four peach and nectarine cultivars stored under air and CA atmospheres. *Postharvest Biol. Technol.* **2013**, *75*, 58–67. [CrossRef]
5. Jaeger, S.R.; Andani, Z.; Wakeling, I.N.; MacFie, H.J.H. Consumer preferences for fresh and aged apples: A cross-cultural comparison. *Food Qual. Prefer.* **1998**, *9*, 355–366. [CrossRef]
6. Corrigan, V.K.; Hurst, P.L.; Boulton, G. Sensory characteristics and consumer acceptability of 'Pink Lady' and other late-season apple cultivars. *N. Z. J. Crop Hortic. Sci.* **1997**, *25*, 375–383. [CrossRef]
7. Wang, Y.X.; Wang, X.M.; Guan, J.F. Flesh Texture Characteristic Analysis of Pear. *Agric. Sci. China* **2014**, *47*, 4056–4066. [CrossRef]
8. Atkinson, R.G.; Gunaseelan, K.; Wang, M.Y.; Luo, L.; Wang, T.; Norling, C.L.; Johnston, S.L.; Maddumage, R.; Schröder, R.; Schaffer, R.J. Dissecting the role of climacteric ethylene in kiwifruit (*Actinidia chinensis*) ripening using a 1-aminocyclopropane-1-carboxylic acid oxidase knockdown line. *J. Exp. Bot.* **2011**, *62*, 3821–3835. [CrossRef]
9. Lunn, J.E.; MacRae, E. New complexities in the synthesis of sucrose. *Curr. Opin. Plant Biol.* **2003**, *6*, 208–214. [CrossRef]
10. Dong, X.G.; Tian, L.M.; Cao, Y.F.; Zhang, Y.; Qi, D. Factor analysis and comprehensive evaluation of fruit quality in cultivars of *Pyrus pyrifolia* (*Burm. f.*) Nakai from south China. *J. Fruit Sci.* **2014**, *31*, 815–822. [CrossRef]
11. Cao, Y.F.; Zhang, S.L. *Chinese Pear Genetic Resources*; China Agriculture Press: Beijing, China, 2020; ISBN 978-7-109-26808-1.
12. Harker, F.R.; Marsh, K.B.; Young, H.; Murray, S.H.; Gunson, F.A.; Walker, S.B. Sensory interpretation of instrumental measurements 2: Sweet and acid taste of apple fruit. *Postharvest Biol. Technol.* **2002**, *24*, 241–250. [CrossRef]
13. Infante, R.; Meneses, C.; Byrne, D.H. Present Situation of Peach Breeding Programs: Post Harvest and Fruit Quality Assessment. *Acta Hortic.* **2006**, *713*, 121–124. [CrossRef]
14. Lozano, L.; Iglesias, I.; Puy, J.; Echeverria, G. Performance of an Expert Sensory Panel and Instrumental Measures for Assessing Eating Fruit Quality Attributes in a Pear Breeding Programme. *Foods* **2023**, *12*, 1426. [CrossRef] [PubMed]
15. Hampson, C.R.; Quamme, H.A.; Hall, J.W.; MacDonald, R.A.; King, M.C.; Cliff, M.A. Sensory evaluation as a selection tool in apple breeding. *Euphytica* **2000**, *111*, 79–90. [CrossRef]
16. Mitchell, J. Food Texture and Viscosity: Concept and Measurement. *Int. J. Food Sci. Technol.* **2003**, *38*, 839–840. [CrossRef]
17. Ross, C.F. Sensory science at the human–machine interface. *Trends Food Sci. Technol.* **2009**, *20*, 63–72. [CrossRef]
18. Bejaei, M.; Stanich, K.; Cliff, M.A. Modelling and Classification of Apple Textural Attributes Using Sensory, Instrumental and Compositional Analyses. *Foods* **2021**, *10*, 384. [CrossRef] [PubMed]
19. Kilcast, D.; Fillion, L. Understanding consumer requirements for fruit and vegetable texture. *Nutr. Food Sci.* **2001**, *31*, 221–225. [CrossRef]
20. Río Segade, S.; Orriols, I.; Giacosa, S.; Rolle, L. Instrumental Texture Analysis Parameters as Winegrapes Varietal Markers and Ripeness Predictors. *Int. J. Food Prop.* **2011**, *14*, 1318–1329. [CrossRef]
21. Li, Y.H.; Chang, R.F.; Zhang, L.S.; Wang, Z.Y.; Chen, H.; Han, J.C.; Liu, G.J. The Optimization of Texture Determination of Fresh Peach by Using Texture Analyzer TPA. *J. Hebei Agric. Sci.* **2016**, *20*, 95–100. [CrossRef]
22. He, G.Q.; Huang, M.H.; Zhang, E.Z.; Xin, M.; Huang, M.K.; Tan, R.Y.; Huang, Z.Y. Optimization for mango texture profile analysis and characterization of texture to different maturaity of mango. *Sci. Technol. Food Ind.* **2016**, *37*, 122–126. [CrossRef]
23. Ma, Q.H.; Wang, G.X.; Liang, L.S. Establishment of the Detecting Method on the Fruit Texture of Dongzao by Puncture Test. *Agric. Sci. China* **2011**, *44*, 1210–1217. [CrossRef]
24. Yang, L.; Xiang, L.; Wang, Q.; Zhang, C.X.; Cong, P.H.; Tian, Y. Study on texture properties of apple flesh by using texture profile analysis. *J. Fruit Sci.* **2014**, *31*, 977–985. [CrossRef]
25. Li, J.K.; Lin, Y.; Zhang, P.; Qin, G.Z.; Li, B.Q.; Tian, S.P. Effect of 1-Methylcyclopropene Treatment at Different Times Postharvest on the Texture of Apple Fruits. *Food Sci.* **2013**, *34*, 277–281. [CrossRef]
26. Wang, X.M.; Guan, J.F.; Wang, Y.X.; Liu, Y. Effect of 1-MCP on flesh texture of "Whangkeumbae" pear during ambient temperature storage. *J. Hebei Agric. Univ.* **2013**, *36*, 46–49. [CrossRef]
27. Wang, F.; Jiang, S.L.; Chen, Q.J.; Ou, C.Q.; Zhang, W.J.; Hao, N.N.; Ma, L.; Li, L.W. Changes in fruit texture of crisp-flesh pear during fruit ripening. *J. Fruit Sci.* **2016**, *33*, 950–958. [CrossRef]
28. Li, Y.H.; Zhang, L.S.; Chang, R.F.; Wang, S.Y.; Chen, H.; Liu, G.J. Change of Texture Properties of Three Peach Varieties During Postharvest Storage by Texture Profile Analysis. *N. Hortic.* **2016**, *4*, 133–137. [CrossRef]
29. Yuan, C.L.; Dong, X.Y.; Li, P.H.; Li, D.L.; Duan, Y.X. Changes in Texture Properties of Crisp Peach during Postharvest Storage by Texture Profile Analysis. *Food Sci.* **2013**, *34*, 273–276. [CrossRef]
30. Zhang, Y.; Liang, Y.L.; Pan, Q.W.; Zhang, W. Correlation between the Sensory Evaluation and Texture Profile Analysis of Kiwifruit. *Sci. Technol. Food Ind.* **2018**, *39*, 243–247+252. [CrossRef]
31. Reng, Z.H.; Zhang, K.M.; Li, Z.W.; Nong, S.Z.; Zhang, P. Study on the evaluation of texture parameters of grape berry during storage by using texture profile analysis. *Sci. Technol. Food Ind.* **2011**, *32*, 375–378. [CrossRef]

32. Zhang, Y.; Cao, Y.; Huo, H.; Xu, J.; Tian, L.; Dong, X.; Qi, D.; Liu, C. An assessment of the genetic diversity of pear (*Pyrus* L.) germplasm resources based on the fruit phenotypic traits. *J. Integr. Agric.* **2022**, *21*, 2275–2290. [CrossRef]
33. Shen, S.Y.; Wang, Z.Q.; Zhang, Q.; Yang, J.; Han, F.; Zhong, C.H.; Wang, C.H.; Huang, W.J. Analysis of fruit quality and sensory evaluation of 36 kiwifruit (*Actinidia*) germplasm accessions. *J. Integr. Plant Biol.* **2023**, *41*, 540–551. [CrossRef]
34. Yang, Y.H.; Song, X.F.; Zhao, Y.H.; Li, X.L.; Cui, H.N.; Jia, J.H.; Yan, L.Y. Evaluation of fruit texture traits of cucumber germplasm resources. *Jiangsu Agric. Sci.* **2023**, *51*, 145–152. [CrossRef]
35. Tang, R. Comprehensive Evaluation and Genetic Trend Analysis of Skin Texture of 50 Watermelon Germplasm Resource. Master's Thesis, Northeast Agricultural University, Harbin, China, 2024.
36. Blanckenberg, A.; Muller, M.; Theron, K.I.; Crouch, E.M.; Steyn, W.J. Harvest maturity and ripeness differentially affects consumer preference of 'Forelle', 'Packham's Triumph' and 'Abate Fetel' pears (*Pyrus communis* L.). *Sci. Hortic.* **2016**, *207*, 131–139. [CrossRef]
37. Zhang, M.-Y.; Xue, C.; Xu, L.; Sun, H.; Qin, M.-F.; Zhang, S.; Wu, J. Distinct transcriptome profiles reveal gene expression patterns during fruit development and maturation in five main cultivated species of pear (*Pyrus* L.). *Sci. Rep.* **2016**, *6*, 28130. [CrossRef] [PubMed]
38. Wan, C.Y.; Mi, L.; Guo, D.; Qiao, Y.S.; Huo, H.Z.; Chen, B.Y.; Li, J.F.; Chen, X.P. Preliminary screening of *Pyrus pyrifolia* Nakai combination with different mature periods based on fuzzy synthetic evaluation of fruit quality. *J. Northwest AF Univ. (Nat. Sci. Ed.)* **2018**, *46*, 99–107. [CrossRef]
39. Lee, B.-R.; Cho, J.-H.; Wi, S.G.; Yang, U.; Jung, W.-J.; Lee, S.-H. The Sucrose-to-Hexose Ratio is a Significant Determinant for Fruit Maturity and is Modulated by Invertase and Sucrose Re-Synthesis During Fruit Development and Ripening in Asian Pear (*Pyrus pyrifolia* Nakai) Cultivars. *Hortic. Sci. Technol.* **2021**, *39*, 141–151. [CrossRef]
40. Byun, J.; Kim, D.H.; Lee, D.; Kang, I.; Chang, K.; Shin, S.L. Changes of Pectic Substances and Polygalacturonase Activity during Fruit Development of Various Peach Cultivars with Degrees of Fruit Softening. *J. Korean Soc. Hortic. Sci.* **2003**, *44*, 503–507.
41. Gao, H.S.; Jia, Y.R.; Wei, J.M.; Ran, X.T.; Le, W.Q. Studies on the Post-harvested Fruit Texture Changes of 'Yali' and 'Jingbaili' Pears by Using Texture Analyzer. *Hortic. Plant J.* **2012**, *39*, 1359–1364. [CrossRef]
42. Brookfield, P.L.; Nicoll, S.; Gunson, F.A.; Harker, F.R.; Wohlers, M. Sensory evaluation by small postharvest teams and the relationship with instrumental measurements of apple texture. *Postharvest Biol. Technol.* **2011**, *59*, 179–186. [CrossRef]
43. Gálvez-López, D.; Laurens, F.; Devaux, M.F.; Lahaye, M. Texture analysis in an apple progeny through instrumental, sensory and histological phenotyping. *Euphytica* **2012**, *185*, 171–183. [CrossRef]
44. Du, X.M.; Zhao, Q.C.; Lv, K.; Liu, J.Y.; Cheng, S.F.; Ma, Y.S. Comparison of Texture Determination Method and Correlation Analysis with Sensory Evaluation of 5 Kinds of Apple. *Sci. Technol. Food Ind.* **2020**, *41*, 240–246. [CrossRef]
45. Zhang, W.; Mayinur, J.M.L.; Wang, M.; Han, S.A.; Xie, H.; Pan, Q.M. Analysis on the Flesh Texture, Cell Architecture andIts Physiological Characteristics of Different Grape Varieties. *Acta Bot. Boreali-Occident. Sin.* **2022**, *42*, 1870–1879. [CrossRef]
46. Pan, H.B.; Liu, D.; Shao, Q.X.; Gao, G.; Qi, H.Y. Analysis and Comprehensive Evaluation of Textural Quality of Ripe Fruits from Different Varieties of Oriental Melon (*Cucumis melo* var. makuwa Makino). *Food Sci.* **2019**, *40*, 35–42. [CrossRef]
47. Xu, Y.Q.; Tian, L.M.; Cao, Y.F.; Dong, X.G.; Qi, D.; Huo, H.L. Evaluation and analysis of flesh texture of six pear varieties with different shelf life after cold storage. *China Fruits* **2024**, *9*, 14–23. [CrossRef]
48. Wu, S.; Jia, Y.L.; Zhi, F.J.; Wei, W. Multivariate Statistical Analysis of 19 Characters of 31 Jujube Resources. *J. Hebei Agric. Sci.* **2020**, *24*, 56–62+70. [CrossRef]
49. Yang, Z.; Wang, Z.L. Evaluation and Cluster Analysis of Jujube Fruit Texture Based on TPA Method. *Xinjiang Acad. Agric. Sci.* **2019**, *56*, 1860–1868. [CrossRef]
50. Fan, Z.P. Analysis and Comprehensive Evaluation of Fruit Quality of Different Mulberry Varieties. Master's Thesis, Hebei Agriculture University, Baoding, China, 2020.
51. Yan, C.; Yin, M.; Zhang, N.; Jin, Q.; Fang, Z.; Lin, Y.; Cai, Y. Stone cell distribution and lignin structure in various pear varieties. *Sci. Hortic.* **2014**, *174*, 142–150. [CrossRef]
52. Wang, Y.; Cao, X.H.; Nian, R.; You, K.Y.; Zhu, D.S. Research Progress on Factors Controlling the Formation of Stone Cells in Pear Fruits and Their Effects on Fruit Texture. *Food Sci.* **2024**, *45*, 340. [CrossRef]
53. Gong, X.; Qi, K.; Zhao, L.; Xie, Z.; Pan, J.; Yan, X.; Shiratake, K.; Zhang, S.; Tao, S. PbAGL7–PbNAC47–PbMYB73 complex coordinately regulates PbC3H1 and PbHCT17 to promote the lignin biosynthesis in stone cells of pear fruit. *Plant J.* **2024**, *120*, 1933–1953. [CrossRef]
54. Liu, D.; Xue, Y.; Wang, R.; Song, B.; Xue, C.; Shan, Y.; Xue, Z.; Wu, J. PbrMYB4, a R2R3-MYB protein, regulates pear stone cell lignification through activation of lignin biosynthesis genes. *Hortic. Plant J.* **2025**, *11*, 105–122. [CrossRef]
55. Payasi, A.; Mishra, N.N.; Chaves, A.L.S.; Singh, R. Biochemistry of fruit softening: An overview. *Physiol. Mol. Biol. Plants Int. J. Funct. Plant Biol.* **2009**, *15*, 103–113. [CrossRef]
56. Cliff, M.A.; Bejaei, M. Inter-correlation of apple firmness determinations and development of cross-validated regression models for prediction of sensory attributes from instrumental and compositional analyses. *Food Res. Int.* **2018**, *106*, 752–762. [CrossRef] [PubMed]

57. Hiwasa, K. European, Chinese and Japanese pear fruits exhibit differential softening characteristics during ripening. *J. Exp. Bot.* **2004**, *55*, 2281–2290. [CrossRef] [PubMed]
58. Asrey, R.; Patel, V.B.; Singh, S.K.; Sagar, V.R. Factors affecting fruit maturity and maturity standards—A review. *J. Food Sci. Technol.* **2008**, *45*, 381–390.
59. Prasad, K.; Jacob, S.; Siddiqui, M.W. Chapter 2—Fruit Maturity, Harvesting, and Quality Standards. In *Preharvest Modulation of Postharvest Fruit and Vegetable Quality*; Siddiqui, M.W., Ed.; Academic Press: Cambridge, MA, USA, 2018; pp. 41–69. ISBN 978-0-12-809807-3.
60. Lin, H.T.; Xi, Y.F.; Chen, S.J. Postharvest Softening Physiological Mechanism of Huang hua Pear Fruit. *Agric. Sci. China* **2003**, *36*, 349–352.
61. Shiga, T.M.; Soares, C.A.; Nascimento, J.R.; Purgatto, E.; Lajolo, F.M.; Cordenunsi, B.R. Ripening-associated changes in the amounts of starch and non-starch polysaccharides and their contributions to fruit softening in three banana cultivars. *J. Sci. Food Agric.* **2011**, *91*, 1511–1516. [CrossRef] [PubMed]
62. Lopez, G.; Behboudian, M.H.; Echeverria, G.; Girona, J.; Marsal, J. Instrumental and Sensory Evaluation of Fruit Quality for 'Ryan's Sun' Peach Grown under Deficit Irrigation. *HortTechnology* **2011**, *21*, 712–719. [CrossRef]
63. Li, X.; Xu, C.; Korban, S.S.; Chen, K. Regulatory Mechanisms of Textural Changes in Ripening Fruits. *Crit. Rev. Plant Sci.* **2010**, *29*, 222–243. [CrossRef]

horticulturae

MDPI

Article

Efficient Identification and Classification of Pear Varieties Based on Leaf Appearance with YOLOv10 Model

Niman Li [1,2,3], Yongqing Wu [1,*], Zhengyu Jiang [1], Yulu Mou [2,3], Xiaohao Ji [2,3], Hongliang Huo [2,3] and Xingguang Dong [2,3,*]

1 School of Software, Liaoning Technical University, Huludao 125105, China; liniman0518@163.com (N.L.); jzy6203@163.com (Z.J.)
2 Institute of Pomology, Chinese Academy of Agricultural Sciences, Huludao 125105, China; mouyulude@163.com (Y.M.); jixiaohao2006@163.com (X.J.); huohongliang@caas.cn (H.H.)
3 Key Laboratory of Germplasm Resources Utilization of Horticultural Crops, Ministry of Agriculture and Rural Affairs, Huludao 125105, China
* Correspondence: yqwuyywu@163.com (Y.W.); dongxingguangde@126.com (X.D.)

Abstract: The accurate and efficient identification of pear varieties is paramount to the intelligent advancement of the pear industry. This study introduces a novel approach to classifying pear varieties by recognizing their leaves. We collected leaf images of 33 pear varieties against natural backgrounds, including 5 main cultivation species and inter-species selection varieties. Images were collected at different times of the day to cover changes in natural lighting and ensure model robustness. From these, a representative dataset containing 17,656 pear leaf images was self-made. YOLOv10 based on the PyTorch framework was applied to train the leaf dataset, and construct a pear leaf identification and classification model. The efficacy of the YOLOv10 method was validated by assessing important metrics such as precision, recall, F1-score, and mAP value, which yielded results of 99.6%, 99.4%, 0.99, and 99.5%, respectively. Among them, the precision rate of nine varieties reached 100%. Compared with existing recognition networks and target detection algorithms such as YOLOv7, ResNet50, VGG16, and Swin Transformer, YOLOv10 performs the best in pear leaf recognition in natural scenes. To address the issue of low recognition precision in Yuluxiang, the Spatial and Channel reconstruction Convolution (SCConv) module is introduced on the basis of YOLOv10 to improve the model. The result shows that the model precision can reach 99.71%, and Yuluxiang's recognition and classification precision increased from 96.4% to 98.3%. Consequently, the model established in this study can realize automatic recognition and detection of pear varieties, and has room for improvement, providing a reference for the conservation, utilization, and classification research of pear resources, as well as for the identification of other varietal identification of other crops.

Keywords: identification; classification; pear; leaves; YOLOv10

1. Introduction

Pear (*Pyrus*) is the world's major fruit tree species. China is the world's largest pear-producing country, with pear production and area accounting for about 70% of the world [1]. China is also one of the three major centers of pear diversity in the world, so the varietal resources are extremely rich, with more than 3000 preserved resource types. The varieties in China cover 5 major species, including *P. bretschneideri*, *P. pyrifolia*, *P. ussuriensis*, *P. sinkiangensis*, and *P. communis*, and more than 130 varieties were widely used in production practice. Given that pear plants hold significant economic value as a cash crop, extensive varietal selection and breeding programs have been implemented by research institutions across China. And over 180 representative and excellent pear varieties were developed, around the fruit quality, disease resistance, stress resistance, different ripening periods, and other objectives of genetic improvement, which further enrich the diversity of pear varieties [2]. As the exchange of pear germplasm resources increases, the diversity of pear cultivars is constantly enriched. However, this has led to challenges in standardized variety identification, classification, and naming, which now constrain the development of the industry. At the same time, it is also a major issue that needs to be solved in the collection, preservation, and innovative utilization of pear germplasm resources. At present, the identification of varieties in production still relies mainly on the analysis of phenotypic characteristics, but phenotypes are susceptible to environmental factors. For example, the fruits of Akiziki and Aikansui are both flattened and round in shape, with a brown outer skin. Their fruits cannot be completely distinguished solely by the naked eye. In addition, both have elliptical-shaped leaves with sharp serrated margins. Even experienced professionals cannot guarantee the correct classification of pear varieties by observing their leaves. The identification error rate is still very high. With the rapid development of high-throughput sequencing and molecular biology techniques, pear varieties can be accurately identified using molecular markers [3]. However, the time-consuming and costly limitations of molecular marker technology have greatly restricted its application. In recent years, the fields of machine vision and artificial intelligence have rapidly emerged. Combining modern information technology with agriculture can gradually make the planting, production, and development of agricultural products more intelligent [4]. Machine vision technology has the advantages of high efficiency, high precision, and low cost in recognizing and classifying agricultural products, so it has gradually become a popular research direction for agricultural recognition and classification [5].

In plant variety identification studies, leaves are easier to collect than other organs. The leaves of any plant have unique characteristics that vary in size and shape, such as shape, tip, veins, and margins. But as far as a plant is concerned, the morphological characteristics of the leaves are relatively stable, so they can be used as a basis for plant identification and classification [6]. The method of classifying plant varieties by recognizing leaf images is feasible. In traditional agricultural practices, the classification of pear leaves is based on empirical knowledge and subjective evaluation, but it is difficult to accurately recognize small differences in pear leaf characteristics with the naked eye. This manual recognition method based on leaf shape, size, and color is not only human resource-intensive, but inefficient, with a high error rate, and the efficiency is greatly affected by human factors [7]. For example, the leaves of *P. communis* are ovate, suborbicular, or elliptic; the edges of the leaves are mostly crenate or obtuse serrate. The leaves of *P. ussuriensis* are nearly ovoid, and the edges of the leaves are mostly sharply serrated. When the two are categorized, observers with a wealth of experience and knowledge may be able to make a general distinction between the two when they look carefully. Both Huagai and Nanguoli belong to *P. ussuriensis* varieties, and the appearances of the leaves are highly similar, which are almost difficult to recognize by the naked eye. Therefore, how to efficiently and accurately recognize and classify pear varieties through the leaves is a difficult task, although it is of great significance to the development of the pear industry and the protection of pear resources.

With the rise of computer vision technology and the maturity of deep learning techniques, Convolutional Neural Networks (CNNs) have made significant progress in image recognition, target detection, image generation, and other fields [8], becoming one of the most important networks in the field of deep learning [9]. The CNN model automatically extracts leaf feature information from multi-species image data for variety classification, overcoming the limitations of manual methods [10]. For the field of plant leaf recognition and classification in complex natural scenes, some progress has been made in recent years. Liu et al. [11] used a Deep Convolutional Neural Network (DCNN) and leaf image input to classify 14 apple cultivars. Siravenha et al. [12] proposed a method to classify plants by analyzing leaf textures. Grinblat et al. [13] focused on leaf vein patterns and used DCNN to classify three different legume species. Zhang et al. [14] constructed a seven-layer CNN to automatically classify 32 species of plants by leaf images. Although deep learning has achieved significant accuracy in leaf classification, challenges still exist. The inference speed is acceptable for individual images, but the complex architecture of standard CNNs (such as ResNet, that focus on deep design) demands extensive computational resources for training. Although laboratory-grade GPUs can handle inference well, agricultural edge devices such as orchard robots often require lighter models. In addition, model compression techniques such as quantization often yield accuracy degradation. These factors necessitate trade-offs between performance and deployability in agricultural edge devices.

The "You Only Look Once" (YOLO) target detection algorithm was first proposed by Redmon et al. [15] in 2016. The YOLO series is the leading single-stage algorithm in deep learning object detection methods. Its basic principle is to divide the input image into multiple grids and predict the category and location of objects in each grid. YOLO-series algorithms have the advantage of faster speed and higher real-time performance, and have made significant progress in target recognition, with better performance and application prospects in quickly recognizing plant leaves. From YOLOv1 to YOLOv10, its performance has been continuously improved, gradually reaching a balance between recognition speed and accuracy [16–21]. Many recent studies have demonstrated that the YOLO family of target detection algorithms can effectively achieve the task of leaf species identification and classification. Yang et al. [22] used the YOLOX algorithm combined with the homemade tea bud dataset to establish a tea bud classification model, which recognized and classified four types of tea buds, in which the recognition accuracy for the yellow mountain species could reach 90.54%. Susa et al. [23] used the YOLOv3 algorithm to build a cotton plant classification system, which recognized and detected cotton leaves, and could classify damaged and healthy cotton leaves. Soeb et al. [24] proposed a solution to the problem of tea leaf disease detection based on the training of the YOLOv7 model, using datasets of diseased tea leaves from four famous tea gardens in Bangladesh. This helps in the rapid identification and classification of tea leaf diseases, minimizing economic losses to the greatest extent.

YOLOv10 embodies a breakthrough in real-time object detection, achieving unprecedented performance. It selects YOLOv8 as the baseline model and proposes a new model design based on it [25]. YOLOv10 uses a dual label strategy to improve accuracy without excessive impact on detection speed. Eliminating the need for non-maximum suppression (NMS) in earlier models reduces latency. YOLOv10 integrates two core modules in the model design: the compact inverted block (CIB) and partial self-attention (PSA) module. This design reduces the computational complexity and improves the model's ability to extract sample features. Tests on performance show that YOLOv10 outperforms previous state-of-the-art models across various model scales. Compared with the baseline model YOLOv8-N, YOLOv10-N has reduced its parameter by 28% and FLOPs by 23%, reducing model complexity. And the latency has been reduced by 70%, which is a great improvement. When considering the detection accuracy and computational efficiency, YOLOv10-N is able to achieve good results for the detection of samples.

Plant recognition and classification using neural networks and leaf image inputs has become the focus and challenge of precision agriculture and has attracted extensive attention from many scholars. However, the application of the YOLO series of algorithms to the task of recognizing and classifying pear leaf varieties, a challenging class of leaf image recognition problems, has not been reported.

Based on the above background, we combine the YOLOv10 algorithm with pear leaf images to construct a pear leaf recognition and classification model as a way to classify pear varieties. The proposed method strikes an appropriate balance between accuracy and efficiency, and is able to quickly and accurately recognize pear leaves and classify the variety. A dataset of images of pear leaves consisting of 33 varieties with 500 to 600 sample images in each category was crafted to meet multi-task classification requirements. YOLOv10 was trained using the dataset of images of pear leaves, and in this way, a recognition and classification model was constructed to classify pear varieties by recognizing the features of pear leaves. The trained model was evaluated using the self-made dataset, and satisfactory results were obtained. In order to verify the effectiveness of the method, we trained YOLOv7, ResNet50, VGG16, and Swin Transformer using the same dataset and under the same experimental environment. Upon comparing the evaluation results of five models, it was found that all metrics of YOLOv10 outperformed those of the other four models. Moreover, while ensuring high recognition accuracy, it improved the speed of recognition. It performed well in both inter-species and intra-species recognition of pears. This proves that the model is suitable for the pear leaf recognition and classification task. On the basis of YOLOv10, Spatial and Channel reconstruction Convolution (SCConv) is introduced to optimize the model. The results indicate that the performance of the optimized model in recognition and classification, as well as the complexity of the model, have been improved. The study of the pear leaf classification model is of great significance for improving its pear leaf identification ability, promoting scientific research and innovation, and promoting the intelligent development of pear industry. And it provides certain scientific references for the classification, protection, and production research of pear varieties and variety identification of other crops.

2. Materials and Methods

2.1. Image Acquisition and Dataset Building

Training a pear leaf recognition and classification model requires feeding a large number of pear leaf images as sample images into the network architecture. The quality of the dataset directly affects the accuracy and efficiency of the model. High quality image data should not only be visually clear and accurate, but also representative, diverse, and balanced in content. The varieties in China cover 5 main cultivation species. Therefore, we selected 6 representative varieties each from *P. bretschneideri*, *P. pyrifolia*, *P. ussuriensis*, and *P. communis*, and selected 3 varieties from *P. sinkiangensis*. And in order to make the dataset more abundant, we also selected 6 varieties from *P.*hybrid. The leaf photos of pear varieties were collected from the "National Pear and Apple Germplasm Resource Repository (Xingcheng)". Images of pear leaves were captured under natural conditions using a mobile phone with a rear camera, the iPhone 13 model, with a resolution of 12 MP. All camera optimizations such as HDR are disabled to avoid affecting model generalization. Placing the phone at a distance of 30–40 cm from the leaves for shooting, images were collected at different times of the day (8:00–18:00) to cover changes in natural lighting and ensure model robustness. During the shooting, there were cloudy and sunny days. The leaves selected were adult leaves, fully expanded, with no obvious signs of nutrient deficiencies, pathogen infection, or insect damage. In the images of dense foliage, there was leaf shading and leaf overlap. And when photographed under well-lit conditions, some leaves cast light and shadows from other leaves. Frontal images of different leaves of 33 pear varieties were taken, and 500 to 600 images were collected for each variety. The images were captured in JPG format with an image size of 3024 × 4032 pixels. A total of 18,316 images of pear leaves were captured during the entire shooting process, due to the fact that the task of capturing leaf images is carried out outdoors, and many factors in the natural environment are inevitable. For example, during filming, excessive wind can cause leaves to move, resulting in blurred leaf features, such as shape and margin, in the captured image. In addition, due to the outdoor cultivation of pear trees, some leaves are affected by pests and diseases, resulting in severe damage to the leaves. Therefore, in order to ensure the quality of the dataset, we manually filtered the images, and finally, a total of 17,656 pear leaf images were used to compose the pear leaf dataset. Some representative samples of the dataset are shown in Figure 1. Table 1 shows the information about the 33 pear leaf samples.

Figure 1. Some representative samples of the leaf-datasets; (**1**) Anli, (**2**) Bartlett, (**3**) Cangxi Xueli, (**4**) Chili, (**5**) Cuiguan, (**6**) Dangshan Suli, (**7**) Dayexue, (**8**) Fojianxi, (**9**) Beurré Hardy, (**10**) Harrow Sweet, (**11**) Hongshaobang, (**12**) Hongxiangsu, (**13**) Huachangba, (**14**) Huagai, (**15**) Huangguan, (**16**) Whangkeumbae, (**17**) Jianbali, (**18**) Jinfeng, (**19**) Jingbaili, (**20**) Conference, (**21**) Korla Pear, (**22**) Mandingxue, (**23**) Nanguoli, (**24**) Akiziki, (**25**) Dr. Jules Guyot, (**26**) Seerkefu, (**27**) Shuihongxiao, (**28**) Xiaoxiangshui, (**29**) Xinli 7, (**30**) Xuehuali, (**31**) Yali, (**32**) Yuluxiang, (**33**) Early Red Comice.

Table 1. Leaf sample information of 33 species of pears.

Species	Name	Origin	Number of Leaf Photos in the Dataset
P. communis	Bartlett	Britain	520
	Beurré Hardy	France	524
	Conference	Britain	514
	Dr. Jules Guyot	France	500
	Early Red Comice	Britain	553
	Harrow Sweet	Canada	552
P. pyrifolia	Akiziki	Japan	513
	Cangxi Xueli	China	500
	Dayexue	China	553
	Hongshaobang	China	504
	Mandingxue	Korea	566
	Whangkeumbae	China	564
P. bretschneideri	Chili	China	510
	Dangshan Suli	China	515
	Fojianxi	China	515
	Shuihongxiao	China	500
	Xuehuali	China	500
	Yali	China	533
P. ussuriensis	Anli	China	567
	Huagai	China	566
	Jianbali	China	542
	Jingbaili	China	575
	Nanguoli	China	544
	Xiaoxiangshui	China	573
P. sinkiangensis	Huachangba	China	554
	Korla Pear	China	572
	Seerkefu	China	562
*P.*hybrid	Cuiguan	China	509
	Hongxiangsu	China	546
	Huangguan	China	547
	Jinfeng	China	506
	Xinli 7	China	543
	Yuluxiang	China	514

The pear leaf dataset was randomly divided into a training set, validation set, and testing set in the ratio of 6:3:1 for training, validation, and prediction processes, respectively. The image data were manually labeled using the LabelImg tool (can be in https://github.com/HumanSignal/labelImg get, accessed on 19 February 2025). Target leaves in the image were marked with rectangular bounding boxes and assigned category labels, and then an annotation file in the .txt format was generated that contained the category of the leaf, the coordinates x and y of the rectangular centroid, and the width w and height h relative to the image.

2.2. *YOLOv10*

YOLOv10 is a real-time object detection method proposed by Wang et al. in 2024. YOLOv10 (https://github.com/THU-MIG/yolov10, accessed on 30 April 2025) has improved detection accuracy and speed compared to previous versions. The dual label assignment strategy of this model eliminates NMS requirements while maintaining detection performance. To achieve high performance and efficiency in modeling, YOLOv10 proposes a CIB structure. Replacing some basic blocks of the model with CIB reduces the redundancy and improves the efficiency of the model [26]. The structure of CIB is shown in (1), where DW (Depthwise Convolution) represents a depthwise separable convolution layer. In addition, a PSA module design is proposed. The channel features are uniformly divided into two parts, one of which is processed by the multi-head self-attention module (MHSA) and a feed-forward network (FFN) before being fused with the other part [27]. This design enhances the model's ability to capture diverse contextual information and boosts detection accuracy while keeping computational overhead low. The structure of the PSA is shown in (2).

$$CIB = [3 \times 3DW \rightarrow 1 \times 1 \rightarrow 3 \times 3DW \rightarrow 1 \times 1 \rightarrow 3 \times 3DW] \quad (1)$$

$$PSA = [MHSA \rightarrow FFN] \times N\text{psa} \quad (2)$$

A series of real-time end-to-end detectors YOLOv10-N/S/M/B/L/X with different model scales was successfully implemented. YOLOv10-N/S/M/B/L/X is trained from scratch on COCO [28]. During the training process, we used the SGD optimizer (momentum of 0.937) for 500 epochs. The initial learning rate was 0.01. The experimental results demonstrate their performance and efficiency over previous state-of-the-art models. Among them, YOLOv10-N achieves a balance between accuracy and efficiency in edge devices, making it suitable for plant phenotype applications. Therefore, for the purpose of this study, we chose YOLOv10-N for our experiment. The network architecture of YOLOv10 is shown in Figure 2.

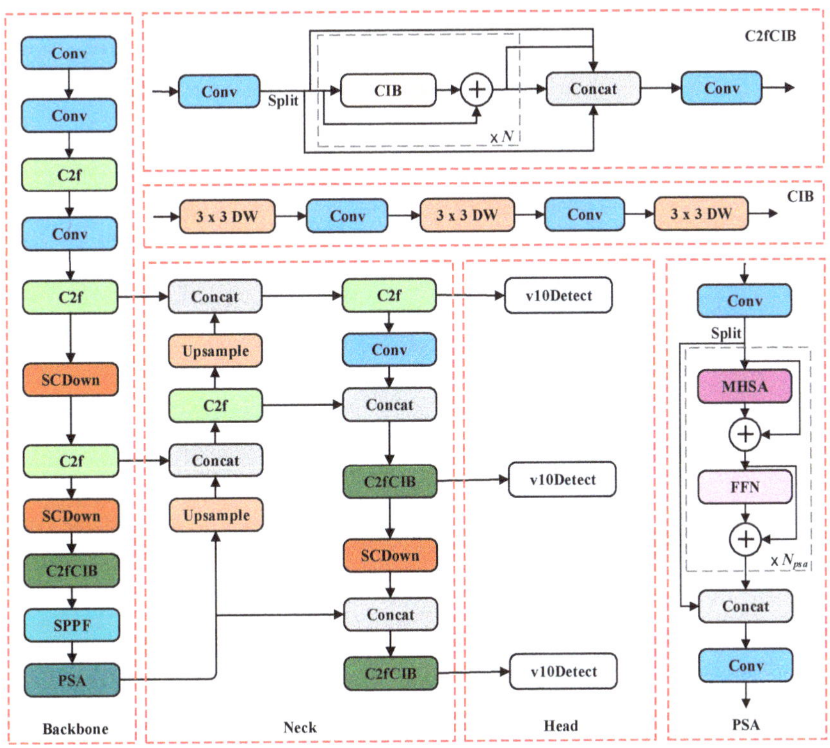

Figure 2. Network architecture diagram of YOLOv10.

2.3. Selection of Other Network Models

YOLOv7 is a real-time object detector proposed by Wang et al. in early 2023. It uses a trainable free package that provides a significant improvement in object detection accuracy without incurring additional inference cost. YOLOv7 focuses on optimizing the training process, which includes optimization of modules to improve the accuracy of object detection. In addition, YOLOv7 proposes "extend" and "compound scaling" methods, which effectively reduces the number of parameters and computation and improves the detection rate. The model is trained from scratch on the MS COCO dataset without using any other datasets or pre-trained weights. The results show that YOLOv7 outperforms other object detectors such as YOLOR, YOLOX, YOLOv5, and DETR in terms of accuracy and speed.

ResNet (Residual Neural Network), also known as traditional neural network, is a neural network proposed by He et al. [29]. It won the first place in ImageNet detection, ImageNet localization, COCO detection, and COCO segmentation in the ILSVRC & COCO 2015 competitions, and profoundly influenced the design of subsequent deep neural networks. ResNet incorporates a batch normalization (BN) layer, combined with normalized initialization and intermediate normalization, to accelerate training and mitigate gradient vanishing or exploding issues. Additionally, residual structures are introduced to address the degradation problem, facilitating the development of ultra-deep network architectures. Among them, ResNet50 is a CNN with 50 layers. In this work, we used ResNet50 for pre-training on ImageNet and trained our dataset with pre-trained weights. The input images were uniformly scaled to 224 $\times$ 224 pixels. We used the SGD optimizer (initial learning rate of 0.001, momentum of 0.9) to optimize the model. The batch size was 16, with continuous iterations of 100.

VGG is a deeper convolutional network proposed by the Visual Geometry Group [30] at the University of Oxford. It builds on AlexNet [31] to further explore the effects of depth and network performance. In the 2014 ImageNet Challenge, training from scratch on ImageNet, VGG19 achieved first place in the localization task, and VGG16 came in second place in classification. Among them, VGG16 is a convolutional network with 16 network layers, which uses several consecutive 3 $\times$ 3 convolutional kernels instead of the larger convolutional kernels (11 $\times$ 11, 7 $\times$ 7, 5 $\times$ 5) in AlexNet. Under the condition of the same receptive field, the depth of the network is improved, and the performance of the network model is thus improved. In the experiment, we trained our dataset using pre-trained weights of VGG16 on ImageNet. The model used SGD as the optimizer (initial learning rate of 0.001, momentum of 0.9) for 100 iterations, with a batch size set to 16. And its image input size remains at 224 $\times$ 224.

Swin Transformer is a hierarchical vision transformer proposed by Liu et al. [32] in 2021. It is based on the idea of the ViT model [33] and uses a hierarchical structure similar to CNN to process images, allowing the model to flexibly handle images of different scales and reducing computational complexity. The model introduces a shifted windowing mechanism, which facilitates cross-window information learning and enhances the capture of local contextual information. This approach improves the model's expressive power and global perception while effectively lowering computational complexity, achieving an optimal balance between performance and efficiency. Swin Transformer demonstrates state-of-the-art performance in COCO object detection, outperforming previous leading methods. It is widely applied in various visual tasks, including image classification, object detection, and segmentation.

2.4. SCConv Module

SCConv is an efficient convolution module proposed by Li et al. [34] in 2023. It reduces spatial and channel redundancy in convolutional neural networks through a Spatial Reconstruction Unit (SRU) and Channel Reconstruction Unit (CRU). Two units are combined in order, first reducing redundancy in the spatial dimension through the SRU, and then reducing redundancy in the channel dimension through the CRU. Specifically, the SRU processes feature maps by dividing them into features that contain rich spatial information and redundant features that do not contain too much information. Subsequently, these two features are recombined to obtain spatially refined features. The CRU divides the channel of spatially refined features into two parts, and then extracts representative high-level features and supplementary shallow features separately. Finally, through global average pooling and an attention mechanism, high-level features and shallow features are fused to obtain the final channel refinement features. The SCConv module enhances the feature extraction capability of the model, significantly reducing the parameters and computational costs while ensuring accuracy.

2.5. Evaluation Metrics

Commonly used evaluation metrics include precision (P), recall (R), average precision (AP), and mean average precision (mAP), which are defined [35] as shown below.

The precision rate denotes the probability that all samples predicted to be positive are actually positive. The formula for precision is shown in Equation (3), where true positive (TP) denotes the number of samples that are predicted to be correct and are actually correct; false positive (FP) denotes the number of samples that are predicted to be correct but are actually incorrect.

$$P = \frac{TP}{TP + FP} \tag{3}$$

The recall rate denotes the probability that a sample that is actually positive is predicted to be positive and it measures the ability of the model to correctly identify the sample. The recall rate is calculated as shown in Equation (4), where true positive (TP) denotes the number of samples that are predicted to be correct and are actually correct; false negative (FN) denotes the number of samples that are predicted to be wrong but are actually correct.

$$R = \frac{TP}{TP + FN} \tag{4}$$

The average precision is the area enclosed by the axes below the precision–ecall curve at different detection thresholds, which is plotted with precision as the y-axis and recall as the x-axis. As recall increases, a robust model can maintain high accuracy. In general, AP indicates the detection accuracy of the model for each category, and the higher the AP value, the better the detection performance of the model. mAP denotes the average of the accuracy of each category, which is used to measure the detection performance of the model for all the categories and is used as the main metric to evaluate the overall performance of the model. YOLO prioritizes the precision and speed, and mAP reflects this by considering both [36]. Among them, mAP50 indicates the value of mAP when the IOU threshold is 0.5. mAP50-95 represents mAP at multiple IOU thresholds, with larger values indicating more accurate prediction boxes. The formulas for AP and mAP are shown in (5) and (6), where C denotes the total number of categories.

$$AP = \int_0^1 P(r)dr \tag{5}$$

$$mAP = \frac{\sum_{i=1}^{C} APi}{C} \tag{6}$$

2.6. Experimental Environment

This experiment uses the Python programming language and is based on the Windows 11 operating system with an Intel(R) Xeon(R) Gold 6330 CPU processor and 80 GB of RAM. GPU training and inference are performed on the AutoDL experiment platform. The details of the experimental environment are shown in Table 2.

Table 2. The experimental conditions.

Experimental Environment	Details
Programming language	Python 3.9
Operating system	Windows 11
Deep learning framework	PyTorch 1.11.0
Development environment	CUDA 11.3
GPU	NVIDIA GeForce RTX 3090

2.7. Statistical Analysis

Data analysis was conducted using BM SPSS 20, with $p < 0.05$ indicating statistical significance in one-way ANOVA (Duncan's test). The error value (%) was used to indicate the size of the error in the actual value, and was calculated using Microsoft Excel 2007.

3. Results and Discussion

3.1. Image and Label Datasets

After labeling the leaf samples in the images using the labeling tool, the labeled dataset was generated. The number and distribution of labels were counted, and the result is shown in Figure 3. The horizontal axis in Figure 3a represents the category of the label in the training set, and the vertical axis represents the quantity of labels. Figure 3b represents the boxes labeled around each leaf instance in the pear leaf dataset, reflecting the statistical information on the size as well as the number of all the boxes. The YOLO algorithm usually uses these bounding boxes to learn how to detect objects with similar categories in a new image and to localize the recognized objects using anchor boxes. Figure 3c displays the distribution of labels, from which one can roughly see the position of the centroid of the leaf in the pear leaf dataset relative to the whole image, and the height-to-width ratio of the leaf samples in the figure relative to the whole image.

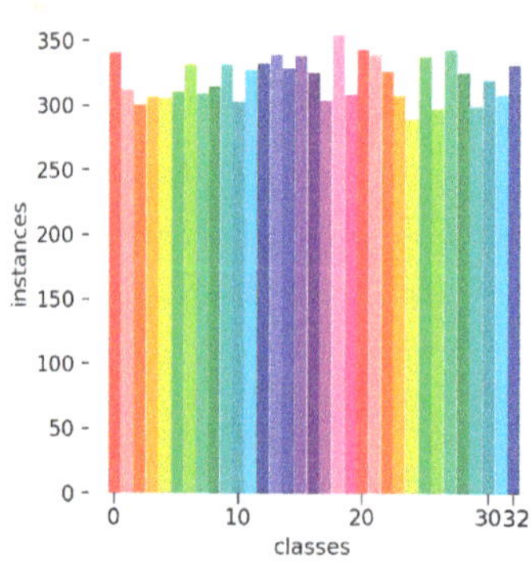

(a)

(b)

Figure 3. *Cont.*

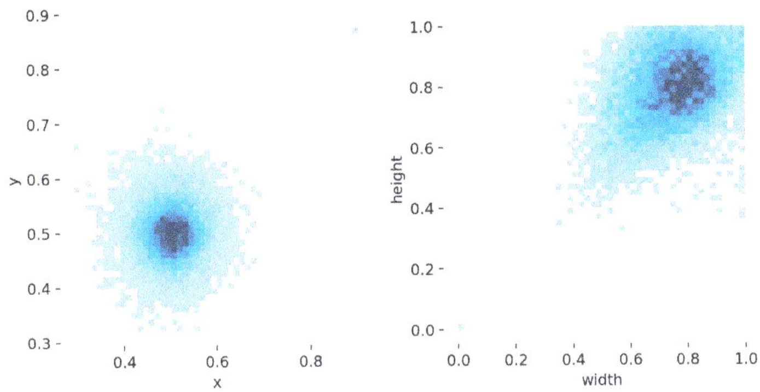

(c)

Figure 3. Labels and label distribution: (**a**) Class and number of training set; the horizontal axis 0–32 represents pear variety: (0) Anli, (1) Bartlett, (2) Cangxi Xueli, (3) Chili, (4) Cuiguan, (5) Dangshan Suli, (6) Dayexue, (7) Fojianxi, (8) Beurré Hardy, (9) Harrow Sweet, (10) Hongshaobang, (11) Hongxiangsu, (12) Huachangba, (13) Huagai, (14) Huangguan, (15) Whangkeumbae, (16) Jianbali, (17) Jinfeng, (18) Jingbaili, (19) Conference, (20) Korla Pear, (21) Mandingxue, (22) Nanguoli, (23) Akiziki, (24) Dr. Jules Guyot, (25) Seerkefu, (26) Shuihongxiao, (27) Xiaoxiangshui, (28) Xinli 7, (29) Xuehuali, (30) Yali, (31) Yuluxiang, (32) Early Red Comice. (**b**) Size and quantity of bounding boxes, (**c**) Location and size of the labels in the images.

3.2. Model Training

The YOLOv10 source code documentation was obtained via GitHub, and the generalization ability of the training model was enhanced using the officially provided pre-training weights file. The model went through 100 epochs using SGD as the gradient optimization function, and the batch size was set to 16. During the training model phase, the input images were uniformly scaled to 640 × 640 pixels, and 8 threads were used for processing. Figure 4 shows the YOLOv10 training results. Figure 5 shows the loss curves generated by YOLOv10 during training as well as validation. Combining the two figures, the value of the loss function shows a decreasing trend, indicating that the SGD optimizer continuously optimizes the model by updating parameters such as the network weights. YOLOv10 has a convergence process in both phases, where the value of the loss function decreases rapidly until the epoch reaches 10, accompanied by a rapid increase in recall, mAP50, mAP50-95, and precision. After the epoch reaches 10, the loss curve decreases slowly until it almost no longer decreases, and at the same time, the values of recall, mAP50, mAP50-95, and precision level off, indicating that the model basically converges. In addition, YOLOv7, ResNet50, VGG16, and Swin Transformer were trained using the same pear leaf dataset, and the training environment and training parameter settings were kept the same as for YOLOv10.

3.3. Comparative Analysis of Different Models on the Self-Made Dataset

In order to evaluate whether YOLOv10 is more suitable for the pear leaves recognition and classification task compared with other algorithms, we chose five models, YOLOv10, YOLOv7, ResNet50, VGG16 and Swin Transformer, to conduct experiments and analyze the results. In the above setting, the pear leaf dataset was used to train each of the five models, with 100 iterations for each model. Subsequently, the models were evaluated on the validation set to provide a comprehensive analysis of their ability to recognize classification. The details of precision, recall, F1-score, and mAP of the recognition and classification results of the five models are shown in Table 3. From Table 3, it can be seen that YOLOv10 shows superior performance in terms of detection precision. Its precision (99.6%), recall

(99.4%), and mAP50 (99.5%) are significantly higher than those of the other four models. Loss function is an important index to quantify the gap between the prediction result and the real result, and it is also the basic standard to evaluate the quality of the model. The smaller the value of the loss function, the closer the prediction result of the model is to the real result. The variation in the loss values for pear leaf recognition and classification for the five models is shown in Figure 6. As can be seen from Figure 6, all five models have a convergence process, and YOLOv10 and YOLOv7 converge faster than ResNet50, VGG16, and Swin Transformer. The loss value of the YOLOv10 model is always in a low state, which indicates that the prediction result of this model for the pear leaf types is closer to the actual results. Figure 7 shows the variation in precision and recall for the five classification models. As can be seen from Figure 7, the precision and recall values of YOLOv10 stabilize faster and have higher values compared to other models.

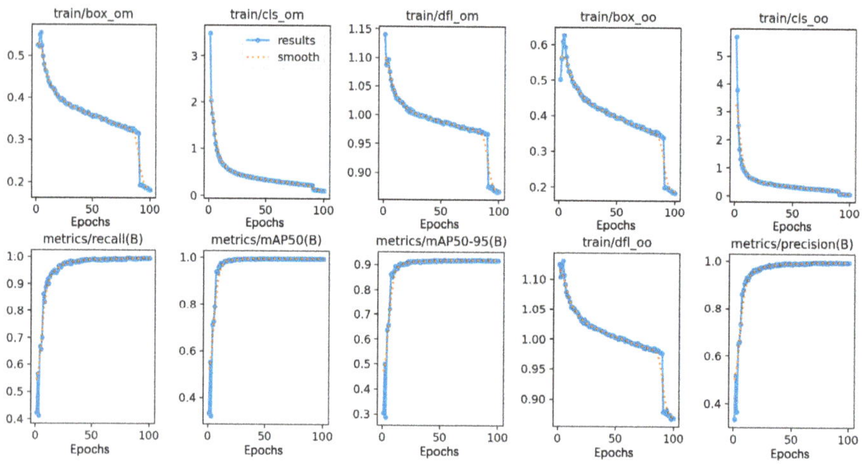

Figure 4. YOLOv10 training result. Loss function indicators during the training phase: train/box_om—bounding box regression loss (one-to-many head); train/box_oo—bounding box regression loss (one-to-one head); train/cls_om—classification loss (one-to-many head); train/cls_oo—classification loss (one-to-one head); train/dfl_om—box position distribution learning (one-to-many head); train/dfl_oo—lightweight distribution learning (one-to-one head).

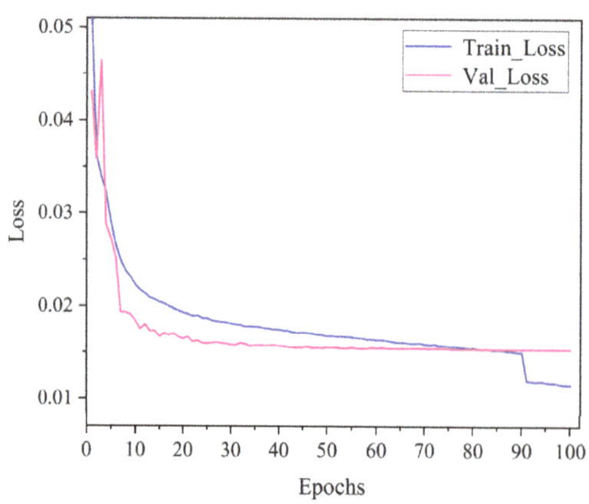

Figure 5. Changes in loss values of YOLOv10 recognition and classification model.

Table 3. Precision, recall, F1-scores, and mAP for recognition and classification results of different models.

Model	Epochs	Precision (%)	Recall (%)	mAP50 (%)	F1-Score
YOLOv10	100	99.60 ± 0.72 a	99.40 ± 0.93 a	99.50 ± 0.17 a	0.99
YOLOv7	100	97.20 ± 3.0 b	96.00 ± 4.06 bc	98.90 ± 1.37 b	0.97
ResNet50	100	96.79 ± 2.62 b	96.76 ± 2.07 b	98.53 ± 0.38 b	0.96
VGG16	100	95.27 ± 2.82 c	95.24 ± 2.73 c	97.87 ± 0.54 c	0.95
Swin Transformer	100	97.92 ± 0.87 b	94.85 ± 1.37 c	98.64 ± 0.75 b	0.97

Note: Each value of Precision, recall, and mAP50% was expressed as mean ± standard deviation. Different letters indicate significant differences at the 0.05 level in Duncan's test of one-way ANOVA.

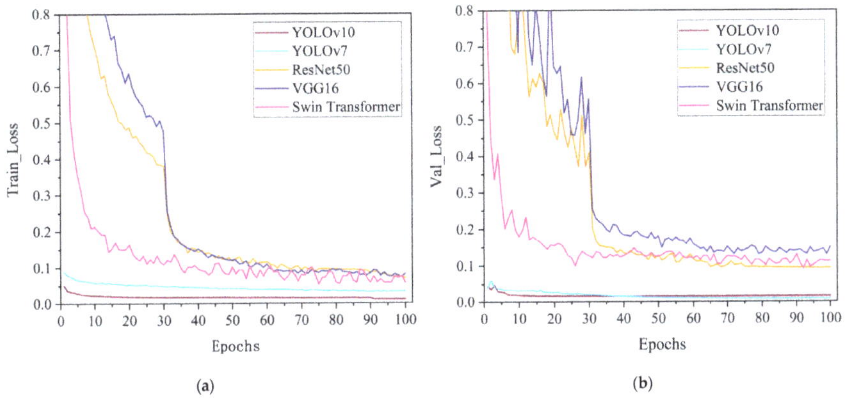

Figure 6. Comparison of loss function curves: (**a**) Loss function curve during the training phase, (**b**) Loss function curve during the verification phase.

In addition, we calculated the parameters, floating-point operations (FLOPs), and the training time for the five models, and the specific information is shown in Table 4. From the table, we can see that VGG16 has the most parameters and YOLOv10 has the least number of parameters. And compared with YOLOv7 of the same series, YOLOv10 has 33.05 M fewer parameters and 97.2 G fewer FLOPs, which shortens the training time. Among the five models, ResNet50 has the fewest FLOPs. But in terms of recognition precision, it is far inferior to YOLOv10. This indicates that YOLOv10 has improved the precision of object detection by optimizing the network structure and training strategy. And this model design effectively reduces the complexity of the model, and the feature extraction ability of the model is improved. While maintaining high performance, it reduces the parameters and FLOPs, and lowers resource consumption. More suitable for deployment on edge devices. Therefore, when considering the overall evaluation of recognition accuracy and efficiency, YOLOv10 becomes a more suitable choice for the pear leaf blade recognition and classification task. The pear leaf recognition and classification model will continue to be constructed using YOLOv10.

Table 4. Comparison of the training parameters, floating point operations, and training time of each network.

Model	Parameter/M	FLOPs/G	Training-Time
YOLOv10	2.59	8.5	11 h 49 m 37 s
YOLOv7	35.64	105.7	32 h 31 m 30 s
ResNet50	23.58	4.12	43 h 31 m 7 s
VGG16	134.4	15.5	27 h 10 m 9 s
Swin Transformer	28.3	4.5	13 h 48 m 28 s

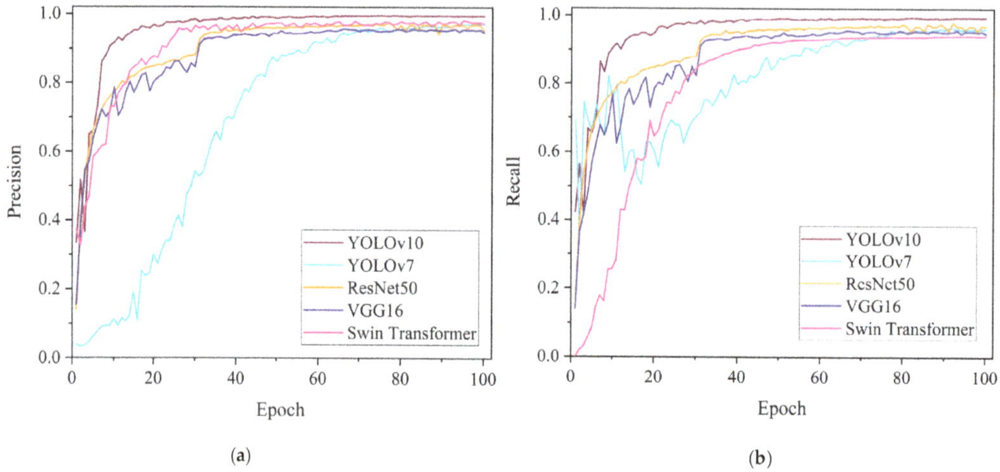

Figure 7. Result curves of five models: (**a**) Comparison of precision, (**b**) Comparison of recall.

3.4. Comparative Experiment on Leaf Recognition and Classification Performance of Different Pear Varieties

The pear leaf recognition and classification model constructed using YOLOv10 in this article can accurately identify pear leaves of different varieties. At the same time, in order to verify the reliability of the results and evaluate the ability of the model to recognize and distinguish leaf related features of different varieties, while maintaining consistency in the training platform configuration, we compared the recognition results of YOLOv10 and the other four models for each variety. The specific information is shown in Figure 8. From Figure 8a, it can be seen that YOLOv7 has the highest recognition accuracy when identifying Bartlett, Dangshan Suli, Hongxiangsu, and Huangguan varieties. In addition, YOLOv10 can achieve the highest recognition accuracy on 29 other varieties. Overall, the model constructed using YOLOv10 has higher detection accuracy for 33 pear leaf samples, with an average accuracy of 99.5%. The higher the recall rate, the lower the probability of missed detections. From Figure 8b, it can be seen that YOLOv7 has the highest recall rate when identifying Cangxi Xueli and Conference varieties. When identifying Hongxiangsu and Mandingxue, ResNet50 can achieve a higher recall rate. And in the recognition of Jinfeng, YOLOv10, YOLOv7, and ResNet50 can all achieve a 100% recall rate. The remaining 28 varieties have the highest recall rate on YOLOv10. Compared with the other four models, YOLOv10 has a more stable recall rate among different varieties.

We used the same testing set to perform pear leaf recognition, classification, and prediction on five models, and randomly selected 70 pear leaf images in the test set for prediction. Figure 9 shows the prediction of several images in a natural environment. From Figure 9, it can be seen that there are differences in the prediction results of the five models for the same image. Among them, YOLOv10 has the highest accuracy in identifying and classifying leaves. From Figure 9i,n, it can be seen that YOLOv7 and ResNet50 mistakenly identified Anli as Jingbaili. This is because both Anli and Jingbaili are varieties of *P. ussuriensis*, with extremely high similarity in leaf characteristics, making it difficult to identify and classify them. From this, it can be seen that, compared with other models, YOLOv10 has fewer misclassification phenomena of different types due to high similarity of leaf surface features.

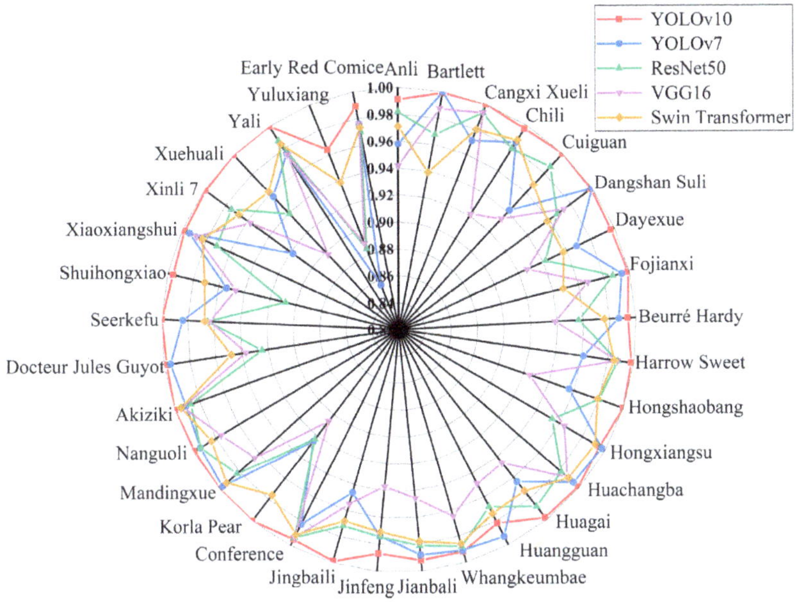

(a)

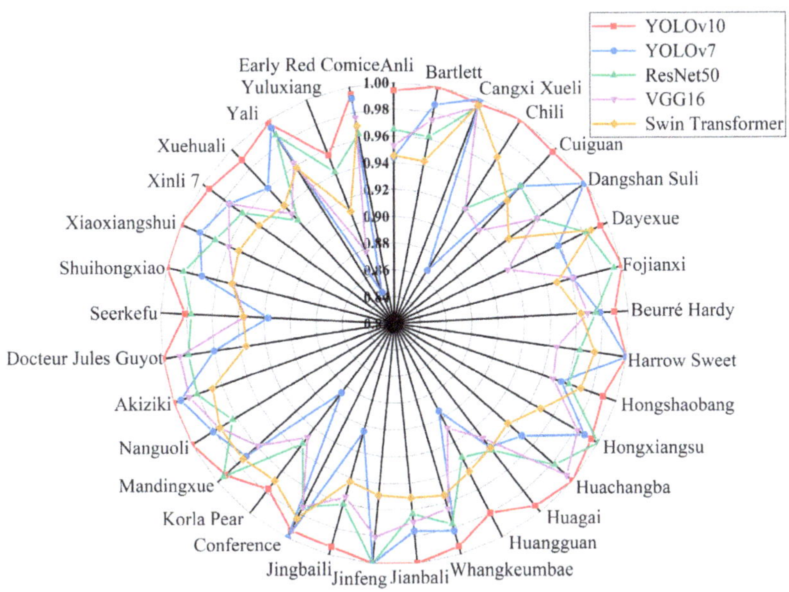

(b)

Figure 8. Recognition accuracy and recall of YOLOv10, YOLOv7, ResNet50, VGG16, and Swin Transformer models: (**a**) Precision of 33 varieties in different models, (**b**) Recall rates of 33 varieties in different models.

Figure 9. Pear leaf recognition and classification results. Among them, (**a–e**) are the recognition results of YOLOv10; (**f–j**) are the recognition results of YOLOv7; (**k–o**) are the recognition results of ResNet50; (**p–t**) are the recognition results of VGG16; (**u–y**) are the recognition results of Swin Transformer.

The confusion matrices of the five recognition and classification models are shown in Figure 10. It describes the accuracy of the model in identifying and classifying 33 pear leaf varieties in the dataset. In the confusion matrix, columns represent the predicted labels of the model for each class of instances, rows represent the true labels of each class of instance labels, and the values on the diagonal represent the correct detection rate. Dark colors indicate high rates, while light colors indicate low rates [37]. This visualization helps to observe the differences between instances belonging to different categories. The diagonals of the five confusion matrices are all dark, indicating that these five models have relatively high accuracy for pear leaf classification. However, as shown in Figure 8, the five models have poor recognition and classification performance for Yuluxiang. Analyzing Figure 10, it was found that YOLOv10 incorrectly identified some samples as Hongxiangsu and Huangguan for Yuluxiang's recognition and classification. YOLOv7 recognized them as Cuiguan, Huangguan, Korla Pear. These were identified as Cuiguan, Huangguan,

Korla Pear, Whangkeumbae in ResNet50. VGG16 recognized them as Huangguan, Jingbaili, Korla Pear. And Swin Transformer recognized them as Hongshaobang, Seerkefu, Xiaoxiangshui. Among them, Cuiguan, Hongxiangsu, Huangguan, and Yuluxiang are all *P.*hybrid. Yuluxiang's parents are Korla Pear and Xuehuali. Huangguan's parents include Xuehuali, and Hongxiangsu's parents include Korla Pear. There is a certain correlation between the three varieties, so there are some interferences when identifying them. However, Cuiguan's parents are Kosui and Hangqing × Shinseiki, and there is not much correlation with Yuluxiang. Except for YOLOv10, which incorrectly identified it as a variety with the same parent, the other four models also identified it as a variety in other populations. From this, it can be seen that YOLOv10 has a stronger ability to extract leaf feature information and perform recognition and classification, and has better recognition performance for some varieties with extremely high leaf feature similarity. The pear leaf recognition and classification model constructed using YOLOv10 can accurately identify and classify pear leaves of various varieties.

Table 5 shows the results of identifying and classifying 33 pear varieties using the YOLOv10 recognition and classification model. From Table 5, it can be seen that the YOLOv10 recognition and classification model has the highest recognition accuracy for Cangxi Xueli, Cuiguan, Hongshaobang, Korla Pear, Mandingxue, Akiziki, Seerkefu, Xuehuali, and Yali, reaching 100% (Error value = 0.44).

Table 5. Test results of the YOLOv10n model on 33 pear varieties.

Leaf-Variety	Precision (%)	Error Value (%)	Recall (%)	Error Value (%)	mAP50 (%)	Error Value (%)	mAP50-95 (%)	Error Value (%)
Anli	99.1	−0.47	99.4	0.04	99.5	0.04	79.6	−13.42
Bartlett	99.9	0.34	100	0.65	99.5	0.04	88.4	−3.85
Cangxi Xueli	100	0.44	99.6	0.24	99.5	0.04	81.8	−11.03
Chili	99.7	0.14	100	0.65	99.5	0.04	92.5	0.61
Cuiguan	100	0.44	99.6	0.24	99.5	0.04	94.7	3.00
Dangshan Suli	99.9	0.34	100	0.65	99.5	0.04	91.6	−0.37
Dayexue	99.7	0.14	99.4	0.04	99.5	0.04	94.6	2.89
Fojianxi	99.9	0.34	100	0.65	99.5	0.04	89.9	−2.22
Beurré Hardy	99.4	−0.16	98.9	−0.46	99.5	0.04	86.4	−6.03
Harrow Sweet	99.8	0.24	100	0.65	99.5	0.04	82.2	−10.59
Hongshaobang	100	0.44	98.9	−0.46	99.5	0.04	95.8	4.20
Hongxiangsu	99.5	−0.06	99.4	0.04	99.5	0.04	90.9	−1.13
Huachangba	99.9	0.34	100	0.65	99.5	0.04	92.5	0.61
Huagai	99.8	0.24	99.4	0.04	99.5	0.04	94	2.24
Huangguan	98.2	−1.37	98	−1.37	99.1	−0.36	91.5	−0.48
Whangkeumbae	99.2	−0.37	99.4	0.04	99.5	0.04	91.6	−0.37
Jianbali	99.2	−0.37	100	0.65	99.5	0.04	96.6	5.07
Jinfeng	98.7	−0.87	100	0.65	99.5	0.04	95	3.33
Jingbaili	99.9	0.34	99.4	0.04	99.5	0.04	89.9	−2.22
Conference	99.5	−0.06	99.4	0.04	99.5	0.04	88.1	−4.18
Korla Pear	100	0.44	97.7	−1.67	99.5	0.04	94.5	2.79
Mandingxue	100	0.44	99.2	−0.16	99.5	0.04	95.1	3.44
Nanguoli	99.9	0.34	100	0.65	99.5	0.04	93.7	1.91
Akiziki	100	0.44	100	0.65	99.5	0.04	99.1	7.79
Dr. Jules Guyot	99.8	0.24	100	0.65	99.5	0.04	93	1.15
Seerkefu	100	0.44	98.1	−1.27	99.5	0.04	94.7	3.00
Shuihongxiao	99.6	0.04	100	0.65	99.5	0.04	91.7	−0.26
Xiaoxiangshui	99.8	0.24	100	0.65	99.5	0.04	92.7	0.83
Xinli 7	99.9	0.34	99.4	0.04	99.5	0.04	97.2	5.72
Xuehuali	100	0.44	98.9	−0.46	99.5	0.04	92.4	0.50
Yali	100	0.44	99.8	0.45	99.5	0.04	95.4	3.76
Yuluxiang	96.4	−3.18	95.5	−3.88	98.6	−0.87	93.4	1.59
Early Red Comice	98.9	−0.67	99.4	0.04	99.5	0.04	93.5	1.70

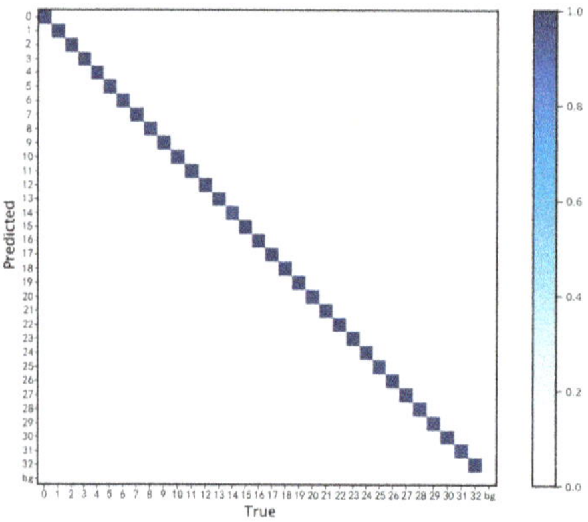

(**a**) YOLOv10

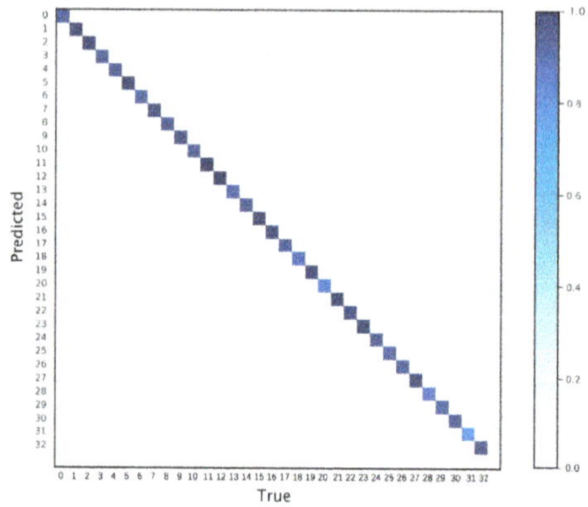

(**b**) YOLOv7

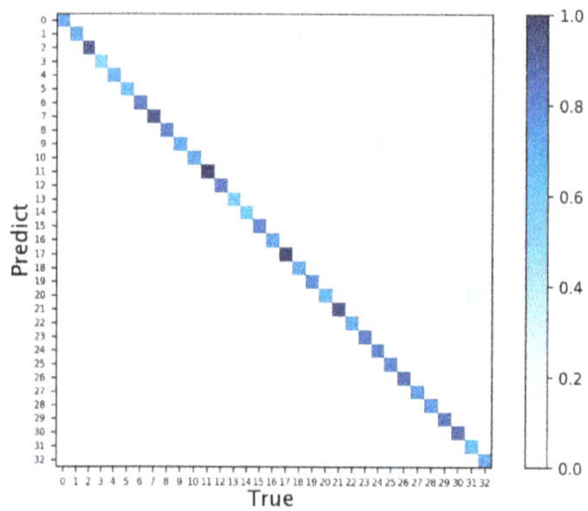

(**c**) ResNet50

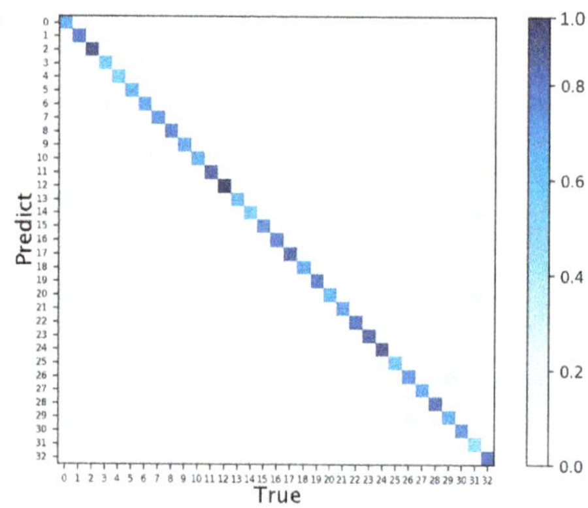

(**d**) VGG16

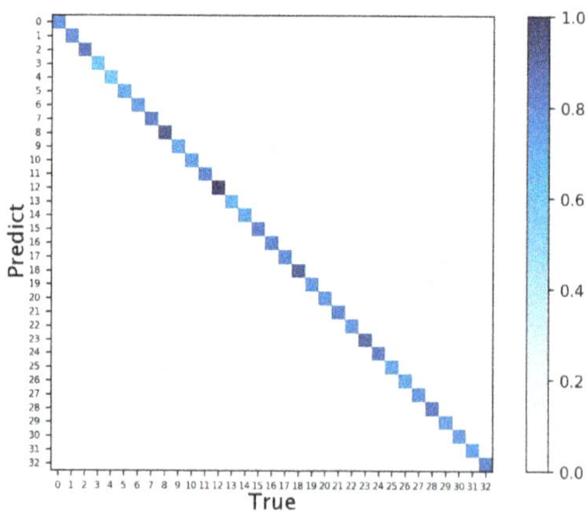

(**e**) Swin Transformer

Figure 10. Confusion matrix diagram for all five models; (0) Anli, (1) Bartlett, (2) Cangxi Xueli, (3) Chili, (4) Cuiguan, (5) Dangshan Suli, (6) Dayexue, (7) Fojianxi, (8) Beurré Hardy, (9) Harrow Sweet, (10) Hongshaobang, (11) Hongxiangsu, (12) Huachangba, (13) Huagai, (14) Huangguan, (15) Whangkeumbae, (16) Jianbali, (17) Jinfeng, (18) Jingbaili, (19) Conference, (20) Korla Pear, (21) Mandingxue, (22) Nanguoli, (23) Akiziki, (24) Dr. Jules Guyot, (25) Seerkefu, (26) Shuihongxiao, (27) Xiaoxiangshui, (28) Xinli 7, (29) Xuehuali, (30) Yali, (31) Yuluxiang, (32) Early Red Comice.

3.5. Improved YOLOv10 Model

The pear leaf classification model based on YOLOv10 performs well in pear leaf recognition and classification, with better indicators than other models, showing great prospects in accurately identifying pear leaf types. However, through the analysis of the recognition and classification results of 33 pear leaves, it is found that the precision of the model for Yuluxiang's recognition and classification is 96.4% (Error value = −3.18), which is lower compared to other varieties. Therefore, we used YOLOv10 as the baseline model to improve the model. SCConv is an efficient convolution module that compresses CNN models and improves their performance by reducing spatial and channel redundancy between features in the convolutional layer. We introduced this convolution module in the Backbone section of YOLOv10 and used SCConv instead of Bottleneck in C2f. The improved model was used to identify and classify pear leaves, and the result shows that the precision, recall, and mAP of the model are 99.71%, 99.46%, and 99.6%, respectively. Compared with YOLOv10, they have increased by 0.11%, 0.06%, and 0.1% respectively. The model has 2.52 M parameters and 7.9 G FLOPs. Compared with YOLOv10, these have decreased by 0.07 M and 0.6 G, respectively. And the precision of Yuluxiang's recognition and classification can reach 98.3%. By introducing the SCConv module, the redundant features of the model have been reduced, and the parameters and computational costs have been reduced. Model complexity is reduced, and detection performance is improved.

4. Discussion

The object detection algorithm based on deep learning is practical for the variety classification problem in agricultural production and research. This is also the mainstream method commonly used in current intelligent agricultural production. Intelligent recognition and classification of plant varieties can not only reduce labor and time costs, but also improve the accuracy of recognition and classification. In the research of plant variety recognition, plant leaves have the advantages of easy collection and stable features. Therefore, in order to classify pear varieties, we used pear leaves as experimental samples and used YOLOv10 based on the PyTorch framework to perform variety recognition and classification tasks on pear leaves, considering recognition accuracy and efficiency. We constructed a pear leaf recognition and classification model that can recognize 33 pear varieties. These 33 varieties cover 27 representative cultivation species and 6 *P.*hybrid varieties. The pear leaf image data we collected were captured using a cell phone, which can capture as many images as possible from different angles. Taking photos of leaves in natural environments may result in occlusions, cluttered backgrounds, and uneven lighting in the captured images. Not only did this enrich the pear leaf dataset, but it also improved the practicality of the model. YOLOv10, as the latest algorithm released in the YOLO series, has improved in both efficiency and accuracy. It proposes an efficiency driven and accuracy driven model design, which solves the problems of computational redundancy and performance limitations. Generally speaking, YOLO has more FLOPs and parameter quantities in the classification head than in the regression head. However, compared to the classification head, the regression head has a greater impact on the performance of YOLO. Therefore, simplifying the classification head architecture can reduce computational overhead. And by using point-by-pointwise convolution to modulate dimensions and performing spatial downsampling using deep convolution, computational costs are reduced and higher efficiency is achieved.

The application of YOLOv10 in pear leaf recognition and classification tasks has shown overall good results. However, through analysis of the recognition results of individual varieties, it was found that the model had lower precision in identifying Yuluxiang. Therefore, the SCConv convolution module was introduced in YOLOv10 to optimize the model.

The results showed that the precision, recall, and mAP of the optimized model were all improved, for more accurate identification of Yuluxiang. At the same time, the parameters and FLOPs are reduced, and the complexity of the model is reduced. This indicates that there is still space for improvement in the model. Therefore, we need to further study the model and conduct more testing and performance optimization.

Future work will collect more leaf images of different pear varieties to enrich the dataset, in order to train better models and improve their generalization performance and efficiency on more pear varieties, and deploy the model on mobile devices, truly contributing to pear variety classification tasks and the intelligent pear industry. This method can also be used for the recognition, classification, and detection of other fruit tree varieties and other plants. Our model was only tested on the pear leaf image dataset. The performance in other professional fields has not been tested yet. We can try to apply this model to real-world research and production to achieve its practicality. YOLOv10 performs well in real-time detection and excels in accuracy and latency, but still has some issues. For example, balancing the weight of the model and the accuracy of the model for specific problems still requires appropriate design [38]. Therefore, in future work, specific modular design of models can be carried out for different application scenarios.

5. Conclusions

The recognition and classification of pear varieties play a crucial role in improving the efficiency of pear production and professionalism. In this study, we propose an efficient and lightweight YOLOv10 pear leaf recognition and classification model, taking pear leaves as an example, specifically designed for the task of identifying and classifying pear leaves. In order to provide sufficient pear variety leaf images for the training model to achieve high generalization ability, we created a pear leaf dataset containing 17,656 images, including a total of 33 pear variety leaves. In order to ensure the universality and representativeness of the dataset, we selected 6 varieties each from *P. bretschneideri*, *P. pyrifolia*, *P. ussuriensis*, *P. communis*, *P.*hybrid, and selected 3 varieties from *P. sinkiangensis*. We used the LabelImg tool to annotate pear leaf images, trained YOLOv10 using the training set, and validated the weights of the trained model using the validation set. To investigate whether YOLOv10 is the most suitable for building pear leaf recognition and classification model, we trained YOLOv7, ResNet50, VGG16, and Swin Transformer on the same dataset in the same experimental environment. Using pre-trained models to train the network accelerates the training process, improves network performance, accelerates convergence speed, and enhances the model's generalization ability. The results indicate that the model constructed using YOLOv10 has high precision, recall, and F1-score in pear leaf recognition and classification tasks. Under the consideration of recognition precision and speed, it outperforms the other four models. And compared with other models, YOLOv10 is more accurate in identifying five cultivated varieties and *P.* hybrid. When there is a close genetic relationship between varieties, YOLOv10 also has an advantage. Overall, the classification model constructed using YOLOv10 has better performance in identifying pear varieties.

Due to the poor performance of the model in recognizing and classifying Yuluxiang, we attempted to improve YOLOv10. On the basis of YOLOv10, the SCConv module was introduced, and pear leaves were tested again. The result shows that the improved model improved detection performance while reducing model complexity. Yuluxiang's precision has increased from 96.4% to 98.3%. This indicates that there is still room for improvement in the model. Therefore, further improvements can be made to the model in the future. The model established in this study can achieve the recognition and classification of pear varieties. In the pear germplasm resource bank, this model can assist in quickly identifying and recording the morphological characteristics of different varieties and assist researchers in efficiently managing pear germplasm resources. In addition, the successful application of YOLOv10 in pear variety recognition can also provide relevant technical references for the classification and detection of other plants.

Author Contributions: Conceptualization, X.J. and H.H.; methodology, Z.J. and Y.M.; writing—original draft preparation, N.L.; writing—review and editing, X.D. and Y.W. All authors have read and agreed to the published version of the manuscript.

Funding: This study was supported by The Science and Technology Innovation Project of the Chinese Academy of Agricultural Sciences (CAAS-ASTIP-RIP), the Earmarked Fund for the China Agriculture Research System (CARS-28-01), and National Natural Science Foundation of China project (32272676).

Data Availability Statement: The original data presented in the study are openly available in GitHub at https://github.com/gss-pear/leaf-dataset (accessed on 27 April 2025).

Acknowledgments: The authors would like to acknowledge the contributions of the participants in this study and the support provided by the Earmarked Fund for the China Agriculture Research System.

Conflicts of Interest: The authors declare no conflicts of interest.

Abbreviations

The following abbreviations are used in this manuscript:

CIB	Compact Inverted Block
Conv	Convolutional Layer
C2f	Cross Stage Partial Fusion with 2 convolutions
DW	Depthwise Convolution
FFN	Feed-Forward Network
PSA	Partial Self-Attention
SPPF	Spatial Pyramid Pooling Fast
SCDown	Spatial Channel Downsampling

References

1. Hussain, S.Z.; Naseer, B.; Qadri, T.; Fatima, T.; Bhat, T.A. Pear (*Pyrus Communis*)—Morphology, Taxonomy, Composition and Health Benefits. In *Fruits Grow in Highland Regions of the Himalayas: Nutritional and Health Benefits*; Springer International Publishing: Cham, Switzerland, 2021; pp. 35–48.
2. *Fruits and Nuts*; Springer Science & Business Media: Berlin/Heidelberg, Germany, 2007.
3. Liu, J.; Wang, H.; Fan, X.; Zhang, Y.; Sun, L.; Liu, C.; Fang, Z.; Zhou, J.; Peng, H.; Jiang, J. Establishment and application of a Multiple nucleotide polymorphism molecular identification system for grape cultivars. *Sci. Hortic.* **2024**, *325*, 112642. [CrossRef]
4. Sun, H.; Zhang, S.; Ren, R.; Su, L. Surface defect detection of "Yuluxiang" pear using convolutional neural network with class-balance loss. *Agronomy* **2022**, *12*, 2076. [CrossRef]
5. Zhang, L.; Yang, Q.; Sun, Q.; Feng, D.; Zhao, Y. Research on the size of mechanical parts based on image recognition. *J. Vis. Commun. Image Represent.* **2019**, *59*, 425–432. [CrossRef]
6. Wang, Z.; Cui, J.; Zhu, Y. Review of plant leaf recognition. *Artif. Intell. Rev.* **2023**, *56*, 4217–4253. [CrossRef]
7. Aslam, T.; Qadri, S.; Qadri, S.F.; Nawaz, S.A.; Razzaq, A.; Zarren, S.S.; Ahmad, M.; Rehman, M.U.; Hussain, A.; Hussain, I.; et al. Machine learning approach for classification of mangifera indica leaves using digital image analysis. *Int. J. Food Prop.* **2022**, *25*, 1987–1999. [CrossRef]
8. Gu, J.; Wang, Z.; Kuen, J.; Ma, L.; Shahroudy, A.; Shuai, B.; Liu, T.; Wang, X.; Wang, G.; Cai, J.; et al. Recent advances in convolutional neural networks. *Pattern Recognit.* **2018**, *77*, 354–377. [CrossRef]
9. Li, Z.; Liu, F.; Yang, W.; Peng, S.; Zhou, J. A survey of convolutional neural networks: Analysis, applications, and prospects. *IEEE Trans. Neural Netw. Learn. Syst.* **2021**, *33*, 6999–7019. [CrossRef]
10. Zhang, Y.; Peng, J.; Yuan, X.; Zhang, L.; Zhu, D.; Hong, P.; Wang, J.; Liu, Q.; Liu, W. MFCIS: An automatic leaf-based identification pipeline for plant cultivars using deep learning and persistent homology. *Hortic. Res.* **2021**, *8*, 172. [CrossRef]
11. Liu, C.; Han, J.; Chen, B.; Mao, J.; Xue, Z.; Li, S. A novel identification method for apple (*Malus domestica* Borkh.) cultivars based on a deep convolutional neural network with leaf image input. *Symmetry* **2020**, *12*, 217. [CrossRef]

12. Siravenha, A.C.; Carvalho, S.R. Plant classification from leaf textures. In Proceedings of the 2016 International Conference on Digital Image Computing: Techniques and Applications (DICTA), Gold Coast, QLD, Australia, 30 November–2 December 2016; IEEE: Piscataway, NJ, USA, 2016; pp. 1–8.
13. Grinblat, G.L.; Uzal, L.C.; Larese, M.G.; Granitto, P.M. Deep learning for plant identification using vein morphological patterns. *Comput. Electron. Agric.* **2016**, *127*, 418–424. [CrossRef]
14. Zhang, C.; Zhou, P.; Li, C.; Liu, L. A convolutional neural network for leaves recognition using data augmentation. In Proceedings of the 2015 IEEE International Conference on Computer and Information Technology; Ubiquitous Computing and Communications; Dependable, Autonomic and Secure Computing; Pervasive Intelligence and Computing, Liverpool, UK, 26–28 October 2015; IEEE: Piscataway, NJ, USA, 2015; pp. 2143–2150.
15. Redmon, J.; Divvala, S.; Girshick, R.; Farhadi, A. You only look once: Unified, real-time object detection. In Proceedings of the IEEE Conference on Computer Vision and Pattern Recognition, Las Vegas, NV, USA, 27–30 June 2016; pp. 779–788.
16. Redmon, J.; Farhadi, A. Yolov3: An incremental improvement. *arXiv* **2018**, arXiv:1804.02767.
17. Bochkovskiy, A.; Wang, C.Y.; Liao, H.Y.M. Yolov4: Optimal speed and accuracy of object detection. *arXiv* **2020**, arXiv:2004.10934.
18. Ultralytics yolov5. Available online: https://github.com/ultralytics/yolov5 (accessed on 5 March 2025).
19. Li, C.; Li, L.; Jiang, H.; Weng, K.; Geng, Y.; Li, L.; Ke, Z.; Li, Q.; Cheng, M.; Nie, W.; et al. YOLOv6: A single-stage object detection framework for industrial applications. *arXiv* **2022**, arXiv:2209.02976.
20. Wang, C.Y.; Bochkovskiy, A.; Liao, H.Y.M. YOLOv7: Trainable bag-of-freebies sets new state-of-the-art for real-time object detectors. In Proceedings of the IEEE/CVF Conference on Computer Vision and Pattern Recognition, Vancouver, BC, Canada, 17–24 June 2023; pp. 7464–7475.
21. Lu, J.; Yu, M.; Liu, J. Lightweight strip steel defect detection algorithm based on improved YOLOv7. *Sci. Rep.* **2024**, *14*, 13267. [CrossRef]
22. Yang, M.; Yuan, W.; Xu, G. Yolox target detection model can identify and classify several types of tea buds with similar characteristics. *Sci. Rep.* **2024**, *14*, 2855.
23. Susa, J.A.B.; Nombrefia, W.C.; Abustan, A.S.; Macalisang, J.; Maaliw, R.R. Deep learning technique detection for cotton and leaf classification using the YOLO algorithm. In Proceedings of the 2022 International Conference on Smart Information Systems and Technologies (SIST), Nur-Sultan, Kazakhstan, 28–30 April 2022; IEEE: Piscataway, NJ, USA, 2022; pp. 1–6.
24. Alam Soeb, J.; Jubayer, F.; Tarin, T.A.; Al Mamun, M.R.; Ruhad, F.M.; Parven, A.; Mubarak, N.M.; Karri, S.L.; Meftaul, I.M. Tea leaf disease detection and identification based on YOLOv7 (YOLO-T). *Sci. Rep.* **2023**, *13*, 6078.
25. Wang, A.; Chen, H.; Liu, L.; Chen, K.; Lin, Z.; Han, J. Yolov10: Real-time end-to-end object detection. *Adv. Neural Inf. Process. Syst.* **2024**, *37*, 107984–108011.
26. Qiu, X.; Chen, Y.; Cai, W.; Niu, M.; Li, J. LD-YOLOv10: A lightweight target detection algorithm for drone scenarios based on YOLOv10. *Electronics* **2024**, *13*, 3269. [CrossRef]
27. Aktouf, L.; Shivanna, Y.; Dhimish, M. High-Precision Defect Detection in Solar Cells Using YOLOv10 Deep Learning Model. *Solar* **2024**, *4*, 639–659. [CrossRef]
28. Lin, T.Y.; Maire, M.; Belongie, S.; Hays, J.; Perona, P.; Ramanan, D.; Dollár, P.; Zitnick, C.L. Microsoft coco: Common objects in context. In Proceedings of the Computer Vision–ECCV 2014: 13th European Conference, Zurich, Switzerland, 6–12 September 2014; Proceedings, part v 13. Springer International Publishing: Berlin/Heidelberg, Germany, 2014; pp. 740–755.
29. He, K.; Zhang, X.; Ren, S.; Sun, J. Deep residual learning for image recognition. In Proceedings of the IEEE Conference on Computer Vision and Pattern Recognition, Las Vegas, NV, USA, 27–30 June 2016; pp. 770–778.
30. Simonyan, K.; Zisserman, A. Very deep convolutional networks for large-scale image recognition. *arXiv* **2014**, arXiv:1409.1556.
31. Krizhevsky, A.; Sutskever, I.; Hinton, G.E. ImageNet classification with deep convolutional neural networks. In *Communications of the ACM*; AcM: New York, NY, USA, 2017; pp. 84–90.
32. Liu, Z.; Lin, Y.; Cao, Y.; Hu, H.; Wei, Y.; Zhang, Z.; Lin, S.; Guo, B. Swin transformer: Hierarchical vision transformer using shifted windows. In Proceedings of the IEEE/CVF International Conference on Computer Vision, Montreal, QC, Canada, 11–17 October 2021; pp. 10012–10022.
33. Dosovitskiy, A.; Beyer, L.; Kolesnikov, A.; Weissenborn, D.; Zhai, X.; Unterthiner, T.; Dehghani, M.; Minderer, M.; Heigold, G.; Gelly, S.; et al. An image is worth 16x16 words: Transformers for image recognition at scale. *arXiv* **2020**, arXiv:2010.11929.
34. Li, J.; Wen, Y.; He, L. Scconv: Spatial and channel reconstruction convolution for feature redundancy. In Proceedings of the IEEE/CVF Conference on Computer Vision and Pattern Recognition, Vancouver, BC, Canada, 17–24 June 2023; pp. 6153–6162.
35. Wang, Y.; Yang, C.; Yang, Q.; Zhong, R.; Wang, K.; Shen, H. Diagnosis of cervical lymphoma using a YOLO-v7-based model with transfer learning. *Sci. Rep.* **2024**, *14*, 11073. [CrossRef] [PubMed]
36. Vijayakumar, A.; Vairavasundaram, S. Yolo-based object detection models: A review and its applications. *Multimed. Tools Appl.* **2024**, *83*, 83535–83574. [CrossRef]

37. Chen, G.; Hou, Y.; Cui, T.; Li, H.; Shangguan, F.; Cao, L. YOLOv8-CML: A lightweight target detection method for Color-changing melon ripening in intelligent agriculture. *Sci. Rep.* **2024**, *14*, 14400. [CrossRef]
38. Nguyen, P.T.; Nguyen, G.L.; Bui, D.D. LW-UAV–YOLOv10: A lightweight model for small UAV detection on infrared data based on YOLOv10. *Geomatica* **2025**, *77*, 100049. [CrossRef]

horticulturae

MDPI

Article

De Novo Assembly and Comparative Analysis of the Mitochondrial Genomes for Six *Rubus* Species

Yujie Shi [1,2,3], Zhen Chen [1,2], Jingyong Jiang [4], Qianfan Li [5] and Wei Zeng [1,2,3,*]

1 Zhejiang Key Laboratory for Restoration of Damaged Coastal Ecosystems, School of Life Sciences, Taizhou University, Taizhou 318000, China; shiyujie@tzc.edu.cn (Y.S.); chenzh@tzc.edu.cn (Z.C.)
2 Zhejiang Provincial Key Laboratory of Plant Evolutionary Ecology and Conservation, School of Life Sciences, Taizhou University, Taizhou 318000, China
3 Zhejiang International Science and Technology Cooperation Base for Biomass Resources Development and Utilization, School of Life Sciences, Taizhou University, Taizhou 318000, China
4 Institute of Horticulture, Taizhou Academy of Agricultural Sciences, Taizhou 318000, China; jjy5971@163.com
5 State Key Laboratory of Subtropical Silviculture, College of Forestry and Biotechnology, Zhejiang A&F University, Hangzhou 311300, China; qianfanli@stu.zafu.edu.cn
* Correspondence: zengw@tzc.edu.cn

Abstract: *Rubus* is a genus of small berry-producing shrubs, valued for their medicinal properties and as a food source. This genus is a large, globally distributed group that includes over 700 species. Despite numerous plastid and nuclear genomes having been reported for *Rubus*, there is a notable lack of research on its mitogenomes. We utilized PMAT to assemble the mitogenomes of six *Rubus* species according to long-read HiFi reads and annotated them through homologous alignment. Subsequently, we compared their characteristic differences within *Rubus* mitogenomes. The complete mitogenomes of *R. parviflorus*, *R. spectabilis*, *R. idaeus*, *R. armeniacus*, and *R. caesius* all exhibit master circle structures, with lengths ranging from 360,869 bp to 447,754 bp. However, *R. chamaemorus* displays a double-circle structure composed of two small circular molecules, spanning 392,134 bp. These mitogenomes encode a total of 54–61 genes, including 33–34 PCGs, 17–24 tRNAs, and 3 rRNA genes. Compared to the other five *Rubus* species, *R. chamaemorus* has fewer sequence repeats. These six species exhibit similar codon usage patterns. A large number of gene transfers were detected between organellar genomes of six *Rubus* species. Additionally, two phylogenetic trees were constructed using 31 mitogenomes and 94 chloroplast genomes, revealing a minor conflict within *Rubus*. Overall, this study clarifies the mitogenome characteristics of *Rubus* and provides valuable insights into the evolution of the genus.

Keywords: mitogenome; double-circle structure; phylogeny; rearrangement; sequence collinearity

1. Introduction

Rubus L. is a genus in the family Rosaceae, comprising approximately 700 species, with 208 of those being found in China. Of these 208, 67% are endemic, making China, along with Western Europe and Eastern North America, geographic hotspots [1,2]. Fruits of *Rubus* species are edible when fresh [3] and abundant in phenols, terpenoids, minerals, vitamin C, and anthocyanins [4,5]. The most widely grown cultivars are selections of red raspberries (*R. idaeus*) and blackberries from the subgenus *Rubus*. There are also medicinal uses for certain *Rubus* species. For instance, the dried fruit of *R. chingii* serves as a traditional Chinese medicine [6], while the roots, stems, and leaves of some *Rubus* species are reported to be used to treat diarrhea and aid in wound healing [7,8]. Furthermore, phenomena such as hybridization, apomixis, polyploidization, and introgression are common in *Rubus* [9,10]. Ploidy levels vary significantly among different subgenera and species, with chromosome numbers ranging from diploid to fourteen-ploid (2n = 14–98) [11]. Focke categorized the genus *Rubus* into twelve subgenera [12] (*Chamaemorus, Dalibarda, Chamaebatus, Comaropsis, Cylactis, Orobatus, Dalibardastrum, Malachobatus, Anoplobatus, Idaeobatus, Lampobatus*, and *Rubus*), whereas the Flora of China recognized eight groups (*Chamaemorus, Chamaebatus, Cylactis, Dalibardastrum, Malachobatus, Idaeobatus, Lampobatus*, and *Rubus*) within the genus [13]. Consequently, a worldwide, unified classification system for *Rubus* is still lacking. Different classification systems exhibit considerable discrepancies in grouping and the delineation of subordinate taxa. However, current classification systems also face issues such as incorrect classification of species and varieties, questionable assignment of certain taxa, and ambiguous descriptions of species morphology. Additionally, a multitude of existing research outcomes, based on palynology [14], cytology, and molecular taxonomy [15,16], often conflict with the subordinate divisions of the classification systems, indicating a need for further refinement. This, in turn, hampers the development and utilization of the genus.

Mitochondria are semi-autonomous organelles found within eukaryotic cells. They play a crucial role in cellular respiration and energy conversion [17]. Mitochondria are primarily responsible for regulating cell growth, division, apoptosis, and the synthesis and breakdown of certain compounds [18]. These key metabolic processes exhibit semi-autonomous characteristics and possess their own genetic material and regulatory mechanisms. Their genetic variation can be valuable for studying plant phylogeny and patterns of diversity and can serve as a key marker for variant identification and development [19,20]. Plant mitogenomes consistently exhibit complex and dynamic structures. For instance, they display extreme variation in genome structure and size, a profusion of repetitive sequences, and the incorporation of foreign DNA sequences [21,22]. These factors have hindered the study of mitogenomes in relation to that of other plastid genomes. Thus, the majority of current plant systematics research focuses on nuclear and chloroplast genomes, with the complete assembly of plant mitochondria persisting as a bottleneck in evolutionary biology. With the advancement of high-throughput sequencing technology and the emergence of next-generation genomics, software programs suitable for sequencing and assembling the mitogenome have been developed, such as GetOrganelle v1.7.5 [23], GSAT v1.12 [24], and PMAT v2.1.0 [25]. These advances have made the sequencing and assembly of mitogenomes more accurate and efficient, providing valuable tools to deepen our understanding of the genetic characteristics and phylogenies of plants.

In the family Rosaceae, 34 mitogenomes have been reported and comparatively analyzed [26]. The results indicate that repetitive sequences primarily influence dynamic changes in the mitogenome structure through homologous recombination and genomic rearrangement. Currently, genomic research on *Rubus* species primarily concentrates on the chloroplast and nuclear genomes [1,2,5,27–32], with relatively few studies on its mitogenomes [3,26,33]. This paucity of mitogenome data has impeded phylogenetic studies within this genus. Consequently, there is an urgent need to analyze the mitogenomes of additional *Rubus* species. This will facilitate a comprehensive understanding of diversity within the genus, aid in advancing future phylogenetic research, and may support more effective protection and utilization of these genetic resources.

We assembled and annotated the mitogenomes of six *Rubus* species, conducted comparative analysis, and further analyzed their phylogeny. Together with 31 other mitogenomes and 94 plastid genomes, the evolutionary relationships and genetic backgrounds of *Rubus* were clarified. For examining the topological differences in the phylogenetic trees derived from plastids and mitochondria, we conducted multiple collinearity analyses within

Rubus and identified a substantial number of gene rearrangements. The diversity of the mitogenomes within *Rubus* species and the complexity of their phylogeny may be closely related to frequent recombination and introgression. The purpose of this study is to analyze the mitogenome and provide a baseline for further exploration of the genetic variation, phylogeny, and molecular breeding of *Rubus*.

2. Materials and Methods

2.1. Data Acquisition, Genome Assembly, and Annotation

Based on published genome data, we downloaded PacBio sequencing data for nine *Rubus* species from the NCBI database, including *R. caesius* (ERR12875186), *R. armeniacus* (SRR30502861), *R. parviflorus* (SRR30533059), *R. spectabilis* (SRR30534156), *R. argutus* (SRR2161389), *R. idaeus* (SRR18212518), *R. leucanthus* (SRR29481259), *R. occidentalis* (SRR6675257), and *R. chamaemorus* (SRR30534586).

According to HiFi reads, we used PMAT v2.1.0 software to assemble plant mitogenomes with the "autoMito" model [25,34]. The initial assembly sequence was visualized using Bandage v0.8.1 software [35], and the contigs of chloroplast and nuclear genes were manually deleted. The HiFi reads were used to map the mitogenomes of *Rubus* to identify repetitive sequences through the Minimap v2.24 tool [36]. After uncovering these repeat sequences, six complete mitogenomes were generated (*R. caesius, R. armeniacus, R. parviflorus, R. spectabilis, R. idaeus*, and *R. chamaemorus*), while the remaining three species (*R. argutus, R. leucanthus*, and *R. occidentalis*) did not produce complete mitogenomes due to insufficient HiFi sequence data. Furthermore, the chloroplast genomes of four *Rubus* species (*R. spectabilis, R. armeniacus, R. caesius*, and *R. chamaemorus*) were also assembled using PMAT v2.1.0 software with the default parameters.

The complete mitogenomes of six *Rubus* species were annotated through the PMGA (http://47.96.249.172:16084/annotate.html, accessed on 10 January 2025) and GeSeq (https://chlorobox.mpimp-golm.mpg.de/geseq.html, accessed on 10 January 2025) tools [37,38], and two existing *Rubus* mitogenomes (*R. chingii* and *R. suavissimus*) were downloaded from the NCBI database as BLAST v2.9.0 reference sequences in GeSeq. The annotation of chloroplast genomes used CPGAVAS2 (http://47.96.249.172:16019/analyzer/home, accessed on 10 January 2025) and GeSeq online tools (https://chlorobox.mpimp-golm.mpg.de/geseq.html, accessed on 10 January 2025) [38,39]. Furthermore, tRNA and rRNA in the mitochondrial and chloroplast genomes were annotated with the tRNAscan-SE v2.0 tool and BLASTN v2.9.0 software [40,41], respectively. Since there may be errors in the annotation information, we used Geneious v21.0.7 software [42] to visually compare the information annotated using different tools and manually corrected them one by one. Finally, six complete mitogenomes were visualized via the PMGmap tool (http://47.96.249.172:16086/home/, accessed on 10 January 2025) [43], and all assembled sequences were uploaded to the GenBank database to obtain accession numbers. Details for all species accession numbers are recorded in Table S1.

2.2. Repeat Sequence Identification

We identified three types of repeat sequences in six complete mitogenomes, including simple sequence repeats (SSRs), dispersed sequence repeats, and tandem repeats (TRs). SSRs were detected using MISA-web (https://webblast.ipk-gatersleben.de/misa/index.php?action=1, accessed on 10 January 2025) [44], and the minimum number of repeats for SSR motifs with lengths of 1–6 were set to 10, 5, 4, 3, 3, and 3, respectively. TRs were identified via the Tandem Repeat Finder (https://tandem.bu.edu/trf/trf.html, accessed on 10 January 2025) [45] using the default parameters. And the REPuter tool (https://bibiserv.cebitec.uni-bielefeld.de/reputer, accessed on 10 January 2025) [46] was used to identify dispersed sequence repeats (including forward repeats, reverse repeats, palindromic repeats, and complementary repeats) with a Hamming distance of 3, and the minimum repeat size and maximum repeat time were set to 30 and 5000, respectively.

2.3. RNA Editing Predictions

All protein coding genes (PCGs) from six *Rubus* mitogenomes were extracted with the PhyloSuite v1.2.3 tool [47]. Then, we utilized the Deepred-Mt tool (http://47.96.249.172:16084/deepredmt.html, accessed on 10 January 2025) [48] to predict all potential editing sites for C to U, employing the default parameters, and ultimately set the threshold for the results to be greater than 0.9.

2.4. Ka/Ks Value Analysis

To evaluate selection pressures on PCGs in the *Rubus* mitogenomes, we used the PhyloSuite tool to extract shared PCGs in the mitogenomes of *R. armeniacus*, *R. caesius*, *R. chamaemorus*, *R. idaeus*, *R. parviflorus*, *R. spectabilis*, *R. chingii*, and *R. suavissimus*, with *Arabidopsis thaliana* as an outgroup. Then, the MAFFT v7.149b tool [49] was used to align sequences, and the KaKs_Calculator v2.0 tool [50] was used to determine the Ka and Ks ratio of each PCG.

2.5. Codon Usage Bias Analysis

The relative synonymous codon usage (RSCU) value was evaluated using the Genepionee cloud platform (http://cloud.genepioneer.com:9929/#/, accessed on 10 January 2025), which was also used to calculate the number of codon usages corresponding to each amino acid. The RSCU is the ratio of the frequency of codon observations to the frequency of codon usage. When the value of RSCU is less than 1, then codon usage is less frequent than expected, and the converse is also true. Finally, a circular heat map for the total number of codons used was plotted with the Chiplot online tool (https://www.chiplot.online/, accessed on 10 January 2025).

2.6. Mitochondrial and Plastid Sequence Migration Analysis

Six homologous sequences between *Rubus* mitochondria and plastids (MTPTs) were examined via BLAST, with the parameters set as identity > 70%, e-value < 1×10^{-5}, and length greater than 30 bp. We analyzed the homologous sequence regions to determine the length, number, and annotation information of the migrated sequences. Sequence migration maps of the mitochondrial and plastid genomes were drawn using the "Advanced Circos" program in TBtools v2.210 [51].

2.7. Phylogenetic Analysis

To explore the evolutionary relationships of *Rubus* taxa in a broader context, we downloaded 25 plant mitogenomes from the NCBI database along with our six newly assembled mitogenomes for analysis, including four Magnoliids (*Cinnamomum camphora*, *Magnolia liiflora*, *Machilus yunnanensis*, and *Liriodendron tulipifera*), four Monocots (*Oryza sativa*, *Zea mays*, *Allium cepa*, and *Cocos nucifera*), and 23 Eudicots. The PCGs common to all species were extracted through PhyloSuite v1.2.3 software, the MAFFT v7.149b tool was used for multiple sequence alignment, and the trimAI tool was used for sequence trimming. Subsequently, we used IQ-TREE v2.2.0.3 to build a maximum likelihood (ML) tree with 5000 bootstrap replicates, using Magnoliids as the outgroup. Moreover, due to the limited number of mitogenomes of *Rubus*, we obtained all chloroplast genomes for this genus from the NCBI database, comprising 90 *Rubus* species. An ML phylogenetic tree based on chloroplast PCGs was constructed using the same method as above. Finally, the

ML phylogenetic trees of both mitochondria and plastids were visualized via the Chiplot online tool (https://www.chiplot.online/, accessed on 10 January 2025) [52].

2.8. Collinearity Analysis of Mitochondrial Genome

In order to compare collinearity between *Rubus* mitogenomes, we detected homologous fragments between *R. caesius* and six other *Rubus* taxa by using MUMmer v4 with the default parameters [53]; only homologous fragments with alignment lengths greater than 500 bp were retained for analysis. The collinearity results were visualized using NGenomeSyn v1.39 [54].

3. Results

3.1. Characteristics of Rubus Mitogenomes

Based on HiFi sequencing data, we successfully assembled six complete mitogenomes of *Rubus*. Interestingly, five mitogenomes conformed to a single master circle (Figures 1A, S1 and S2), while the mitogenome of *R. chamaemorus* had a double-circle conformation (Figure 1B). There were significant differences in their sizes, with *R. armeniacus* having the longest mitogenome, at 447,754 bp, whereas *R. spectabilis* had the shortest, spanning only 360,869 bp (Table S2). The GC content of the six mitogenomes was relatively stable, ranging from 43.8% to 44.6%, with a mean of 44.3%. Furthermore, we compared the annotated information for these six mitogenomes with two previously reported mitogenomes (*R. chingii* and *R. suavissimus*) [33], revealing that *R. armeniacus* had the most genes (61), including 34 PCGs, 24 tRNAs, and 3 rRNAs; *R. spectabilis* and *R. chingii* had fewer annotated genes (54). The mitogenomes of the eight *Rubus* species all contained 24 core PCGs, including ATP synthesis genes (*atp1/4/6/8/9*), cytochrome c biogenesis genes (*ccmB/C/FC/FN*), a ubiquinol cytochrome c reductase gene (*cob*), cytochrome c oxidase genes (*cox1/2/3*), maturases (*matR*), and transport membrane protein (*mttB*) and NADH dehydrogenase genes (*nad1/2/3/4/4L/5/6/7/9*), but there were certain differences in the copy numbers (Table S3). Among them, *atp1*, *atp4*, *cox3*, and *nad4L* genes had two copies in some species (Figure 1C). Among the variable genes, there were large differences among species, e.g., *R. caesius* lacked the *rps12* gene, *R. parviflorus* lacked *rps14*, *R. suavissimus* lacked *rps3*, and *R. chingii* lacked both *rpl16* and *rps7*.

3.2. Repeat Sequence Analysis and RNA Editing Prediction

Repeat sequences included SSRs, dispersed repeats, and tandem repeats. A total of 1037 SSRs were detected in the mitogenomes of eight *Rubus* species (Figure 2A). Among them, *R. chingii* had the most SSRs, with 168 loci, while *R. chamaemorus* had the least, with 97. Although the SSR numbers among the eight *Rubus* species varied significantly, they all contained six types of nucleotide repeats. Among seven species, mononucleotide repeats dominated, ranging from 33 to 50 loci (accounting for 0.30% to 0.40% of the total), whereas *R. caesius* had the largest number of tetranucleotide repeats (accounting for 0.33% of the total). Furthermore, we detected tandem repeats and dispersed repeats in the mitogenomes of all eight species (Figure 2B). The number of tandem repeats ranged from 29 (*R. chamaemorus*) to 47 (*R. chingii*), with large differences among species, with their numbers trending in a pattern consistent with those of SSRs. For dispersed repeats, we detected a total of 4089 sequences, ranging from 254 (*R. chamaemorus*) to 646 (*R. parviflorus*), and most species had four dispersed repeat types. However, *R. armeniacus* and *R. caesius* lacked complementary repeats, and *R. chamaemorus* lacked reverse repeats. Within this general class, forward repeats (169–346 sequences) predominated in seven species, but palindromic repeats were the most common *R. caesius* sequences (at 281, accounting for 51% of all). Although the dispersed sequence repeats varied in number and type among the eight species, trends in the distribution of their lengths were similar, mainly distributed between 30 and 49 bp, accounting for 64.26 to 77.56% of the total (Figure 2C). There were one and two dispersed repeats with spans greater than 5000 bp in *R. armeniacus* and *R. caesius* mitogenomes, respectively, with sizes of 5341 bp, 7700 bp, and 10,422 bp (Figures 1A and S1). Furthermore, *R. chamaemorus* and *R. spectabilis* had fewer repeats than other *Rubus* species, which may be the main reason for their smaller mitogenomes.

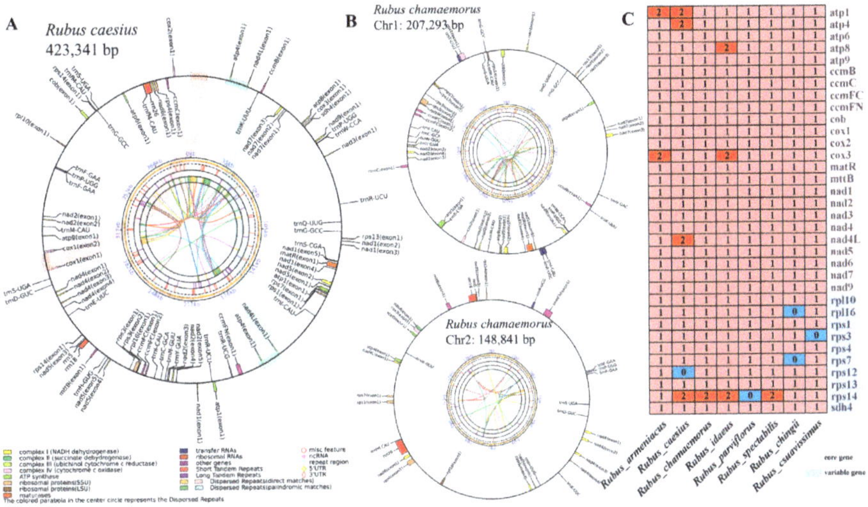

Figure 1. Mitochondrial genome assembly and annotation. (**A**) The mitogenome map of *R. caesius*. (**B**) The mitogenome map of *R. chamaemorus*. Genes with different functions are given different colors. The colored curves in the circles represent dispersed sequence repeats. (**C**) Comparison of the number of protein-coding genes in eight *Rubus* mitogenomes.

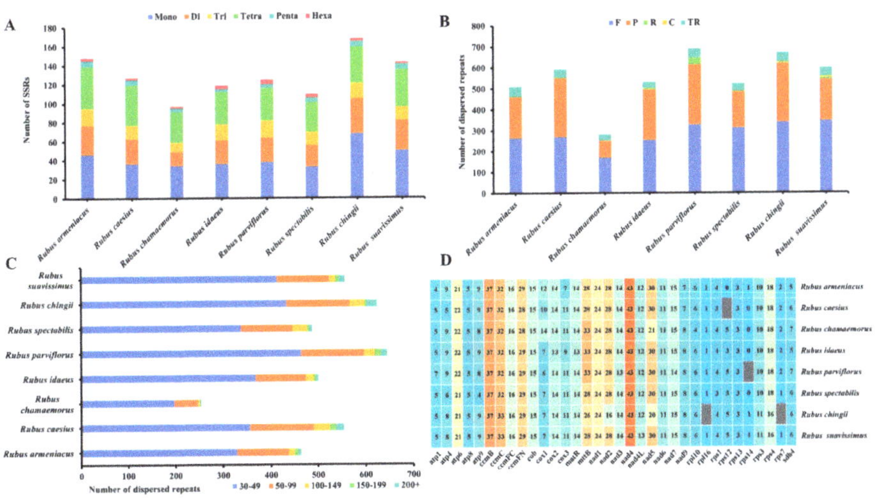

Figure 2. Repeated sequence analysis and prediction of RNA editing sites in the mitogenomes. (**A**) Identification of SSR loci in eight *Rubus* mitogenomes; (**B**) identification of dispersed sequence repeats and tandem repeats in eight *Rubus* mitogenomes; (**C**) distribution of dispersed sequence repeats in eight *Rubus* mitogenomes; (**D**) prediction of RNA editing sites for PCGs of eight *Rubus* mitogenomes. From blue to red represents the number of RNA editing sites from fewer to more.

In this study, we predicted 34 PCGs to further understand gene expression in eight *Rubus* mitogenomes (Figure 2D) and identified 3890 RNA editing sites, 465 to 497 within each genome. Although there were some differences in overall number of editing sites, trends for the editing sites of specific genes were similar, with *nad4* having the most RNA editing sites at 43, followed by *ccmB* with 37, while most ribosomal proteins usually had fewer editing sites. Additionally, *rpl16* and *rps7* had only one editing site in some species, and *rps12* and *rps14* sometimes even had none. It is worth noting that the number of editing sites for the second codon was dominant in all eight species, ranging from 298 to 308 (61.57 to 62.60% of the total). However, the number of editing sites for the third codon was much lower than that for the first and second codons (Figures S3 and S4). These RNA editing events often lead to amino acid changes that contribute to improved protein stability [55].

3.3. Analysis of Selection Pressure on Shared PCGs Among Rubus Species

To investigate the impact of selection pressure on the evolution of mitogenomes, we used 26 PCGs shared in eight *Rubus* mitogenomes to calculate the ratio of non-synonymous substitutions (Ka) to synonymous substitutions (Ks) (Figure 3). Our results showed that the Ka/Ks ratio distribution trends of 26 PCGs were similar across these taxa, with a majority of the genes undergoing purifying selection (Ka/Ks ratios < 1). Notably, the Ka/Ks values for *ccmB* and *sdh4* were >1 in all *Rubus* species, while *atp4* and *atp6* exceeded 1 in two species, meaning that they had likely undergone positive selection in response to environmental changes or other selective pressures.

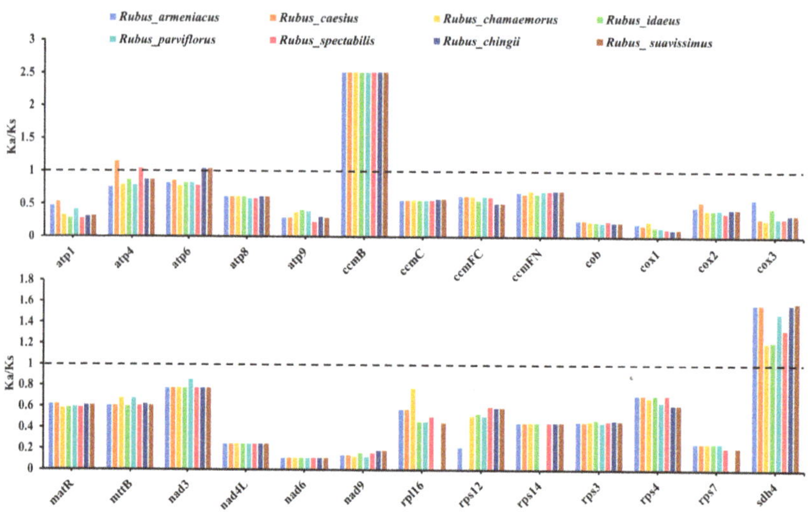

Figure 3. Ka/Ks ratios for 26 PCGs in the mitogenomes of eight *Rubus* species. Different species have different colors.

3.4. Codon Usage Bias Analysis of Rubus mitogenomes

To compare the codon usage bias of PCGs in the *Rubus* mitogenomes, we calculated the total number of codons in each *Rubus* mitogenome; they varied between 9369 and 10,020 (Figure 4B). There were 64 different codon types, encoding 20 amino acids and 1 stop codon, of which UUU was most commonly used at 342 to 373 copies. Among the 20 amino acids, leucine was the most common, ranging from 998 to 1078 copies (10.60 to 10.77% of the total), followed by serine (840 to 911 copies, 8.97 to 9.15% of the total), while cysteine had the fewest codons in *R. armeniacus*, *R. caesius*, *R. parviflorus*, and *R. spectabilis* (144–156), and tryptophan had the fewest (140–153) in the other species. RSCU values reflect the results of genetic drift in natural selection, mutation, and codon-use preferences [18]. We estimated RSCU values for the 64 codon types by species (Figure 4A), and the results showed that 31 codons were used more frequently than expected, i.e., RCSU values > 1, with methionine (AUG) bias being the strongest. Usage frequencies of the remaining 32 codons were lower than expected (RSCU < 1), but codon types with the lowest values varied by species. Notably, there was no codon-usage preference for tryptophan (UGG) (RSCU = 1). Therefore, except tryptophan, the codons of all amino acids showed preference for usage, and methionine, like tryptophan, had only one codon, while most amino acids had at least two. Arginine, serine, and leucine each had six different codons.

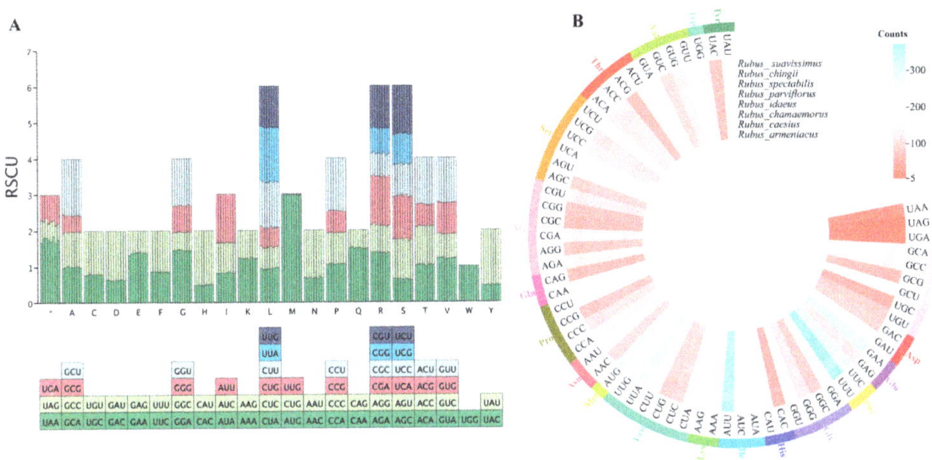

Figure 4. Codon usage bias analysis of mitogenomes from eight *Rubus* species. (**A**) Comparison of relative synonymous codon usage (RSCU) values for different species. Each thin bar from left to right represents a species, namely *R. armeniacus*, *R. caesius*, *R. chamaemorus*, *R. idaeus*, *R. parviflorus*, *R. spectabilis*, *R. chingii*, and *R. suavissimus*. A to Y on the X-axis represent alanine, cysteine, aspartic acid, glutamic acid, phenylalanine, glycine, histidine, isoleucine, lysine, leucine, methionine, asparagine, proline, glutamine, arginine, serine, threonine, valine, tryptophan, and tyrosine. (**B**) Total numbers of amino acid codons in protein-coding genes of different species.

3.5. Sequence Migration Analysis

We analyzed homologous fragments of the mitochondrial and chloroplast genomes from six *Rubus* species, and fragments with a similarity of >70% were identified as organelle migrating sequences (MTPTs), thus identifying 55 MTPTs (*R. armeniacus*), 43 MTPTs (*R. idaeus*), 36 MTPTs (*R. caesius*), 30 MTPTs (*R. spectabilis*), 42 (*R. parviflorus*), and 35 MTPTs (*R. chamaemorus*). These MTPTs varied greatly in total length, ranging from 23,446 (*R. spectabilis*) to 67,012 bp (*R. armeniacus*), accounting for 6.50 to 14.97% of the total mitogenome (Figure 5A,B). Among them, the longest MTPT in *R. armeniacus* spanned 13,207 bp, which was much larger than the longest MTPTs in the other five species (between 5878 and 8958 bp) (Figure S5). Further genetic annotation of these MTPTs showed that most contained homologous sequences of chloroplast genes, but most PCGs lost their integrity during the transfer from chloroplasts to mitochondria, retaining only partial sequences, and the numbers of complete, intact genes were fairly small. For example, 47 complete genes were detected in the MTPTs of *R. armeniacus*, including 27 PCGs (*accD*, *atpB/E/F/H/I*, *ccsA*, *ndhC/J/K*, *petG*, *psaA/I/J/C*, *psbZ*, *rbcL*, *rpl20/32/33*, *rps2/7/14/18*, and *ycf3/4*), 3 rRNA genes, and 17 tRNA genes. The fewest complete genes (17) were detected in the MTPTs of *R. spectabilis*, containing only 8 PCGs and 9 tRNA genes (Table S4). In addition, we compared the distribution of these complete genes in six species. Interestingly, we found that 10 complete migration genes coexisted in six *Rubus* mitogenomes, namely *trnW-CCA*, *trnP-UGG*, *trnN-GUU*, *trnM-CAU*, *trnI-CAU*, *trnH-GUG*, *trnD-GUC*, *rps14*, *psbZ*, and *petG* (Figure 5C). Furthermore, some genes existed only in one taxon. For example, the *atpA* gene existed only in *R. spectabilis*; the *rpl23* gene existed only in *R. idaeus*; and the *psbE/F/J/L* genes existed only in *R. parviflorus*. These genes are closely related to plant photosynthesis and self-repair [56] and may be preserved by organellar genomes of *Rubus* species during the adaptive evolution process [33].

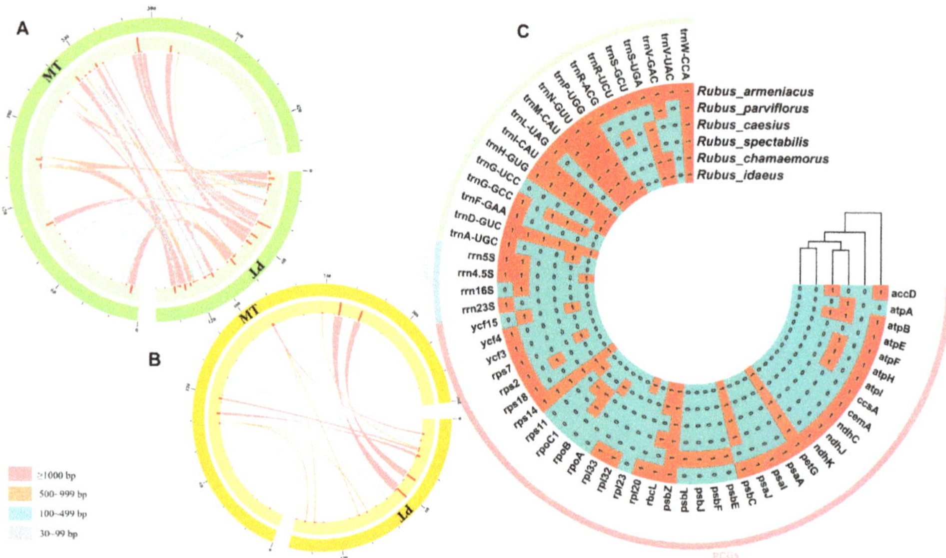

Figure 5. Migration sequence analysis between mitochondrial and plastid genomes. (**A**) Homogeneous fragments between the mitochondrial and chloroplast genomes of *R. armeniacus*. (**B**) Homogeneous fragments between the mitochondrial and chloroplast genomes of *R. spectabilis*. Two arcs of different lengths represent the mitochondrial and plastid genome, the histogram in the inner circle represents the length of MTPTs, and the lines between the two arcs represents MTPTs. Detailed information is recorded in Table S4. (**C**) Comparative analysis of complete genes in MTPT sequences from six *Rubus* species.

3.6. Multidimensional Systemic Generation Analysis

To further explore the maternal phylogeny of *Rubus* species, an ML evolutionary tree was constructed based on 20 conserved mitochondrial PCGs from 31 plant species (Figure 6A). Phylogenetic analysis showed that eight *Rubus* species formed an independent branch and had a close genetic relationship to *Glycine max*. Among the eight *Rubus* taxa, *R. idaeus* and *R. spectabilis* formed a clade, and *R. chingii* and *R. suavissimus* formed a clade, which together formed a branch with *R. armeniacus* and then formed a large clade with *R. caesius* and *R. parviflorus*. *Rubus chamaemorus* was located basally with a relatively distant relationship to the other *Rubus* species, which is also reflected in the different conformation of its mitogenome (Figure 1B). The reconstructed ML phylogenetic tree helps clarify evolutionary relationships among *Rubus* species, and its overall topology was reasonably consistent with the APG IV classification system [57].

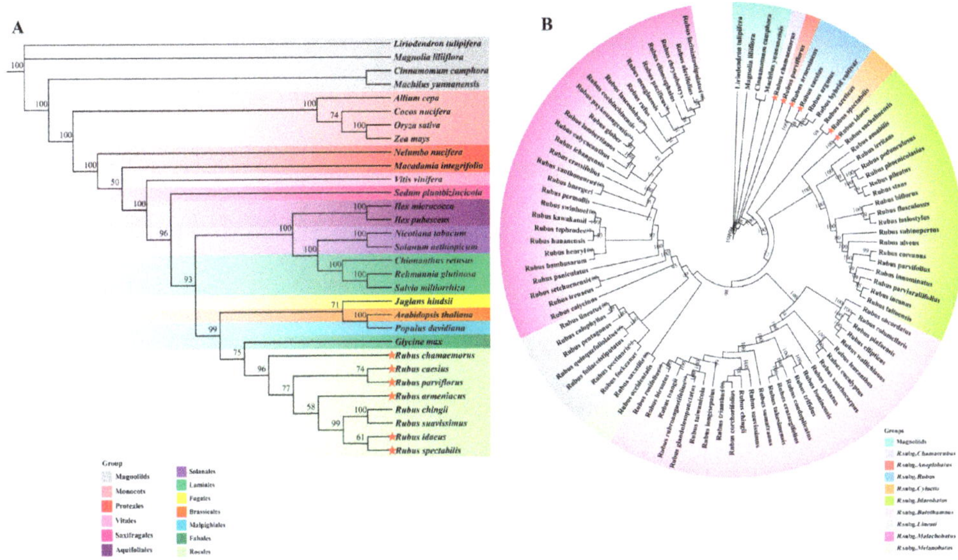

Figure 6. Phylogenetic analysis. (**A**) The ML phylogenetic tree constructed based on PCGs shared in the mitogenomes of 31 species; (**B**) the ML phylogenetic tree constructed based on PCGs shared in the plastid genomes of 94 species. Support rates for different nodes are displayed on branches. The red stars represent the newly assembled genomes for this study.

To explore phylogenetic relationships within *Rubus* in greater detail, we reconstructed an ML evolutionary tree based on 65 conserved PCGs from the chloroplast genomes of 90 *Rubus* species (Figure 6B). Based on the subgenera system proposed by predecessors [1,2], we could assign these 90 species to 9 subgenera, of which *R.* subg. *Malachobatus* contained the most species (28 species), and *R.* subg. *Batothamnus* contained 26 species, while *R.* subg. *Anoplobatus* and *R.* subg. *Chamaerubus* both contained only 1 species. Among the six newly assembled genomes, *R. chamaemorus* was located basally within the genus, consistent with our mitochondrial ML phylogeny, indicating that *R. chamaemorus* may be an ancient taxon. Notably, the maternal phylogenetic relationships involving *R. caesius* were inconsistent between the two trees. In the mitochondrial phylogeny, *R. caesius* and *R. parviflorus* were sister groups, while in the chloroplast-based phylogeny, *R. caesius* and *R. armeniacus* had a close genetic relationship, consistent with their membership in subg. *Rubus*. This inconsistency could arise through differential evolutionary rates between the mitogenome and the chloroplast genome or through the complexities of reticulate evolution, especially since *R. caesius* is known to hybridize with members of other subgenera [9].

3.7. Collinearity Analysis of Mitochondrial Genome

In this study, we used the mitogenome of *R. caesius* as a reference and conducted a collinearity analysis with the mitogenomes of six other *Rubus* taxa to elucidate conflicts in the relationship between *R. caesius* and these other *Rubus* species (Figure 7). The results showed that 44, 75, 60, 54, 54, and 74 fragments homologous to *R. caesius* were identified in *R. armeniacus*, *R. parviflorus*, *R. spectababilis*, *R. idaeus*, *R. chamaemorus*, and *R. chingii*, respectively. *Rubus caesius* and *R. chamaemorus* had the shortest homologous fragments, spanning 259,071 bp, accounting for 61.20% of the *R. caesius* mitogenome, which was consistent with the phylogenetic trees. Noteworthily, *R. caesius* and *R. armeniacus* had the longest homologous fragments, spanning 343,826 bp, accounting for 81.22% of the *R. caesius* mitogenome, while the homologous fragments with *R. parviflorus* had a length of 276,356 bp, which was consistent with the phylogenetic tree constructed from the chloroplast genomes. Additionally, we discovered a substantial number of sequence rearrangements within the mitogenomes of *Rubus* species, contributing to the diversity and complexity of the genomes. Consequently, to further refine the phylogenetic relationships of this genus, it is essential to expand the mitogenome database for *Rubus* species.

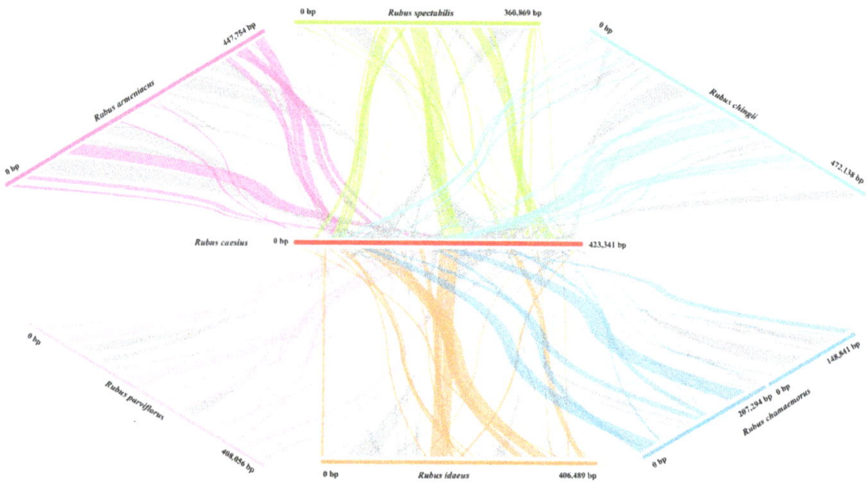

Figure 7. Collinearity analysis among the mitogenomes of *Rubus caesius* and six other *Rubus* species. Different colored bars represent the mitogenomes of different taxa. Among them, the colored lines between the mitogenome of *R. caesius* and the other species indicate good similarity, and the gray lines indicate reversal. All collinear fragments were greater than 500 bp.

4. Discussion

Mitochondria are essential organelles within eukaryotic cells, fulfilling the energy demands of physiological processes. Their genomes exhibit complex structural features, with notable variations, including linear, circular, branched, and network structures [3]. The widespread occurrence of recombination mediated by repetitive sequences enables multiple structures to coexist within the same species [58,59]. Typically, the structure of plant mitogenomes consists of master circular DNA molecules. However, numerous studies in recent years have clarified the diversity of plant mitogenome conformations. For instance, the mitogenome of *Spodiopogon sagittifolius* exhibits a typical master circular structure spanning 500,699 bp [60], whereas the mitogenome of *Echinacanthus longipes* comprises five chromosomes, including three linear and two circular chromosomes, with significant size differences [61]. The mitogenome of *Cymbidium ensifolium* is extremely complex, with a total length of 560,647 bp, comprising 19 circular chromosomes ranging in size from 21,995 to 48,212 bp [18]. Remarkably, even closely related plants of the same genus may evolve mitogenomes with different conformations. In *Mikania*, the mitogenome of *M. cordata* is a branched linear DNA molecule, but *M. micrantha* possesses a typical circular structure [59]. The mitogenomes of *R. armeniacus*, *R. caesius*, *R. spectrabilis*, *R. idaeus*, and *R. parviflorus* all presented typical master circular structures, which were the same as most species of Rosaceae, whereas *R. chamaemorus* displayed a double-ring structure. This implied that the length and structure of the small genome may have undergone convergent evolution to some extent. Additionally, the mitogenomes of angiosperms are in a state of dynamic change [62], influenced by recombination, the accumulation of repetitive sequences, and gene transfer, resulting in genomes with varying conformations and sizes [63]. Past studies have described very broad variations in the size of plant mitogenomes, typically ranging from 66 kb to 12 Mb [64,65]. To date, in *Rubus*, the mitogenome size may vary much less, ranging only from 360,869 bp in *R. spectabilis* to 472,138 bp in *R. chingii*, a phenomenon similar to that observed in *Silene* [21]. As far as *Rubus* mitogenomes are currently described, they exhibit a moderate length within the family Rosaceae (ranging 277.76 to 535.73 kb) [3]. Size differences in the mitogenomes of these congeneric species may be attributed to the accumulation of repetitive sequences or the integration of foreign DNA [66,67]. Notably, the GC content is a key indicator to assess species evolution and amino acid composition [68]. In the mitogenomes of *Rubus* species, the GC content ranged from 43.8 to 44.6%, indicating that despite the diversity in the structure and size of the *Rubus* mitogenomes throughout evolution, the GC content remained relatively stable. Similar results have been observed in other species of the family Rosaceae [3].

Generally, the *rps10* gene is absent from the mitogenome of most plants, with its function being supplanted by other nuclear-encoded genes. This phenomenon is also observed in model plants such as *Arabidopsis* and *Populus* [34,69]. In this study, we determined that our six studied *Rubus* mitogenomes also lack it, along with the *rps2* and *rps11* genes, consistent with the mitogenomes of *M. micrantha* [59] and *Mentha spicata* [70]. It is speculated that these genes may have been lost early in the evolution of vascular plants [71]. Additionally, the mitogenomes of *R. caesius*, *R. parviflorus*, and *R. suavissimus* lacked the *rps12*, *rps14*, and *rps3* genes, respectively, and *R. chingii* was missing both the *rpl16* and *rps7* genes. The number of tRNAs varied more significantly among these species, ranging from 17 to 24 genes. This indicates that during the evolution of *Rubus* species, there were varying degrees of gene loss or retention among individual taxa.

Codons are the fundamental units responsible for accurately identifying and transmitting genetic information [72]. There are significant differences in the usage of the 64 codons among different species, which is believed to be the result of long-term evolutionary selection [73]. By studying codon-usage patterns, we can uncover the evolutionary patterns of genes and predict the regulatory mechanisms involved in gene expression [74]. In general, ATG serves as a common start codon in all plants, although a few PCGs in the mitochondrial genome utilize ACG or CTG as the initiation codon [75]. In our study, the *nad1*, *nad4L*, and *rps4* genes in the mitogenomes of six *Rubus* taxa begin with ACG, as well as the *cox1* genes in *R. idaeus*, *R. armeniacus*, *R. caesius*, and *R. chamaemorus*. Comparable findings have been reported in other plants, including *Echinacanthus longipes* [61], *Ilex rotunda* [76], *Lycopodium japonica* [77], *Prunus pedunculata* [78], and *P. davidiana* [79]. This phenomenon may be attributed to an RNA editing event. For *Rubus* species, the distribution of the amino acid composition is comparable to that of other angiosperms [34,59,80]. Our analysis revealed that within the mitogenomes of eight *Rubus* taxa, 31 codons had an RSCU value greater than 1, suggesting a significant enrichment of A/T bases at the third codon position of these gene sequences. A strong bias towards A/T bases at the third codon position is a common phenomenon in most plant mitogenomes [81,82].

It is well established that the ratio of the synonymous substitution rate (Ka) to the non-synonymous substitution rate (Ks) of a gene can reflect its relative molecular evolution rate. A Ka/Ks ratio greater than 1, equal to 1, and less than 1 indicates that the gene has undergone positive selection, neutral selection, and negative selection, respectively [83]. In our study, most genes within the mitogenomes of *Rubus* species appeared to be under negative selection, indicating that these PCGs and key functional genes are highly conserved. A few genes exhibited signs of positive selection, with slight differences among different species. Notably, the Ka/Ks values of the *ccmB* and *sdh4* genes in all *Rubus* species were greater than 1, while the Ka/Ks value of the *ccmB* gene was the highest, suggesting that it is an important functional gene involved in adaptation. This selection model has similar findings in other plant species [61,84]. For instance, *sdh4* exhibits the highest Ka/Ks value in the mitogenomes of both *P. davidiana* and *Taihangia rupestris* var. *ciliata*, whereas *ccmFN* has the highest value in *Rosa chinensis* [78,79,85].

The mitogenomes of higher plants contain numerous repetitive sequences, comprising three types: simple sequence repeats, tandem repeats, and dispersed sequence repeats. These can be utilized to develop molecular markers for studying phylogeny [59,84]. These sequences are crucial for plant evolution, influencing mitogenome structure, resulting in size variations, intermolecular recombination, heterodimerization, and isomerization [59,86,87]. Among the eight *Rubus* taxa, the longest repeat sequence ranged widely from 237 (*R. chamaemorus*) to 10,422 bp (*R. caesius*). Furthermore, *R. caesius* exhibited the highest proportion of such sequences in its mitogenome (at 13.51%), and *R. chamaemorus* had the lowest proportion at only 3.39%. The proportion of repeat sequences in four other taxa was relatively consistent, ranging from 7.19 to 9.73%. There was a positive correlation between the proportion of these repetitive sequences and the size of the mitogenome, except that the mitogenome of *R. caesius* was smaller than that of *R. armeniacus* or *R. chingii*, yet its proportion of repetitive sequences was much larger, containing a substantial number of longer repetitive sequences. Consequently, the amplification of mitogenome size was not solely due to the accumulation of repeat sequences but was also influenced by factors such as gene sequence migration and gene loss [88,89].

Sequence migration between organellar genomes is frequent, encompassing the transfer of sequence fragments from plastids to mitochondria and the movement of gene fragments from nuclei to mitochondria [66,89]. Understanding transfer relationships is crucial for elucidating the evolution of plant mitogenomes. In this study, we identified homologous genes shared between the plastids and mitochondrial genomes (MTPTs) of six *Rubus* species. These MTPTs constituted 6.50% (*R. spectabilis*), 8.09% (*R. idaeus*), 8.28% (*R. chamaemorus*), 8.50% (*R. parviflorus*), and 8.62% (*R. caesius*) of five mitogenomes, respectively, well within the range observed in most plants (1–12%) [90]. But the MTPT sequences of *R. armeniacus* constituted 14.97% of the mitogenome, a proportion higher than that observed in most plants, yet comparable to the MTPT proportion reported in *R. suavissimus* (13.16%) [33] and lower than that in *R. chingii* (16.34%). Generally, a substantial number of tRNA genes in angiosperms have been transferred from the plastid genome to the mitogenome. We annotated MTPTs in *Rubus*, and as observed in other plants such as *M. micrantha* [59], *S. sagittifolius* [60], *T. rupestris* var. *ciliata*, and *P. davidiana* [79,85], we also identified plastid-derived tRNA genes in six *Rubus* mitogenomes. To preserve functionality within the mitogenome, tRNA genes typically exhibit a high level of integrity, signifying their strong conservation and potentially indispensable roles in the mitogenome. In contrast, transferred PCGs and rRNA genes display less integrity than tRNA genes. The occurrence of plastid-derived tRNAs in plant mitogenomes is a well-documented phenomenon that may offer insights into early gene transfer events [91]. For example, *trnM-CAU* and *trnD-GUC* were identified in gymnosperms [92] and dicotyledonous plants [59], respectively. These genes were also present in the mitogenomes of six *Rubus* species. However, some plastid-derived genes were found only in one species, such as *trnV-UAC*, which was unique to *R. armeniacus*.

RNA editing events encompass post-transcriptional modifications of RNA sequences, including substitutions, deletions, or insertions [66]. They are prevalent in the mitochondrial and chloroplast genomes and play a crucial role in plant gene expression. Current research indicates that substitution is the most common type of editing, primarily involving the conversion of C to U [66,77]. In this study, we examined RNA editing sites in eight *Rubus* mitogenomes, identifying C-to-U editing sites ranging in number from 465 (*R. chingii*) to 497 (*R. parviflorus*), similar to other angiosperms such as *M. micrantha* (473) and *cordata* (492) [59], *Oryza sativa* (491) [93], *A. thaliana* (441) [33,94], *P. davidiana* (502) [79], and *T. ru-*

pestris var. *ciliata* (470) [85]. Among all the RNA editing events in *Rubus* mitogenomes, the *nad4* gene exhibited the highest editing efficiency. This is consistent with past findings for *R. suavissimus* and *C. camphora* [33,66]. This editing may affect the stability of enzymes within the mitochondrial respiratory chain complex [33,66]. Additionally, the second position in the codon triplet has been identified as the most prevalent site for RNA editing, which we observed consistently across all *Rubus* mitogenomes.

Mitogenomes help reveal the unique evolutionary paths of angiosperms [95]. The content, structure, and genetic arrangement of plastids are crucial for exploring the evolutionary relationships among plants [96]. In most terrestrial plants, the chloroplast genome size remains relatively constant, and gene losses are rare during biological evolution [97]. On the other hand, the mitogenomes are relatively complex and vary greatly in size, but their PCGs exhibit low substitution rates and high homogeneity [98]. These characteristics make mitogenome analysis better at uncovering older and more fundamental phylogenetic relationships within the tree of life [99]. For some taxa, there are clear topological differences between phylogenies based on the distinct evolutionary paths of plastid and mitochondrial genomes [76,100]. Several studies have investigated the intricate phylogeny within *Rubus*. For instance, Howarth et al. [101] conducted a phylogenetic analysis of the *ndhF* gene and concluded that Hawaiian Islands endemics (*R. hawaiensis* and *R. macraei*) originated from distinct ancestors, a finding that challenges the results of morphological studies. Wang et al. [15] employed multiple chloroplast and nuclear genes to examine the phylogenetic relationships among 142 *Rubus* taxa, revealing that evolutionary events among different subgenera and species exhibited reticulate patterns. Carter et al. [1] constructed phylogenetic trees for 87 wild *Rubus* taxa based on nearly 1000 target genes. The findings suggested that the low resolution and topological conflicts observed among different subgenera were not due to insufficient molecular signals but rather to hybridization and incomplete lineage sorting (ILS).

Our study investigated the phylogeny of *Rubus* by examining both mitochondria and plastids. The phylogenetic tree constructed from mitochondrial PCGs largely aligns with the APG IV classification system [57]. However, there was a certain conflict regarding the phylogenetic placement of *R. caesius* within the genus, as reflected in discrepancies between mitochondrial and plastid data, which may be attributed to ILS. In the phylogenetic tree constructed using mitochondrial data, *R. caesius* and *R. parviflorus* were depicted as sister groups, whereas the plastid tree suggested a closer genetic relationship between *R. caesius* and *R. armeniacus*. To further investigate the reasons behind this conflict, we performed a collinearity analysis involving *R. caesius* and six other *Rubus* species. Collinearity analysis is essential for clarifying evolutionary relationships among species [102]. The results indicated that *R. caesius* and *R. parviflorus* possessed a greater number of homologous fragments, whereas the collinear sequences with *R. armeniacus* were more extensive.

Thus, the entire mitogenome of *R. caesius* exhibited greater homology with *R. armeniacus*, whereas *R. caesius* shared a more similar collinear relationship with *R. parviflorus* concerning the PCGs of the mitogenome. This observation led to the construction of a phylogenetic tree based on PCGs from the mitogenome, which indicated a closer genetic affinity between *R. caesius* and *R. parviflorus*. Generally, the mitogenome holds significant value for the phylogeny of species. However, to obtain more accurate phylogenetic relationships, comprehensive analysis is necessary, which should be conducted using multiple datasets and algorithms. Moreover, the mitogenome exhibits a high degree of variability, indicating that it has undergone numerous instances of genetic recombination over the course of its extensive evolutionary history. Previous research has identified gene rearrangement as a key factor in the reconfiguration of the mitogenome within the family Rosaceae [26]. Similar results have been observed in *Rubus* mitogenomes, which may be a potential driving force for their adaptation and evolution.

5. Conclusions

For the first time, this study analyzed the structures and annotations of the mitogenomes of six *Rubus* species. The complete mitogenomes of *R. parviflorus*, *R. spectabilis*, *R. idaeus*, *R. armeniacus*, and *R. caesius* were all master circle structures with lengths ranging from 360,869 to 447,754 bp. In contrast, the mitogenome of *R. chamaemorus* had a double-ring structure, consisting of two small circular structures with a span of 392,134 bp. The mitogenomes of the six *Rubus* taxa encoded 33–34 PCGs, 17–24 tRNAs, and 3 rRNA genes. The difference in the number of PCGs among these species was mainly due to a lack of *rps12* and *rps14* genes in *R. caesius* and *R. parviflorus*, respectively. Further comparative analysis showed that *R. chamaemorus* had a lower number of SSRs, tandem repeats, and dispersed repeats than other studied *Rubus* taxa. However, all of these *Rubus* species showed similar codon-usage patterns. The Ka/Ks ratios for most coding genes were less than 1, indicating that these genes had been subjected to negative selection, highlighting the conservation of mitochondrial genes in these species. We also conduct comparative analysis of RNA editing, genome collinearity, and sequence migration among *Rubus* taxa, providing a broader understanding of their mitochondrial genetic characteristics. Maternal phylogeny of *Rubus* species was clarified based on 31 mitogenomes and 94 plastid genomes. This study not only elucidated the general characteristics of *Rubus* mitogenomes but also contributed to the phylogeny of the genus. However, due to frequent hybridization, asexual fusion, polyploidy, and introgression within this genus, further clarification of its classification requires more extensive sampling and analyses of additional taxa.

Supplementary Materials: The following are available online at https://www.mdpi.com/article/10.3390/horticulturae11050559/s1, Figure S1. Mitochondrial genome annotation circle maps of *Rubus armeniacus* and *Rubus idaeus*. Figure S2. Mitochondrial genome annotation circle maps of *Rubus parviflorus* and *Rubus spectabilis*. Figure S3. Prediction of RNA editing sites for 34 protein coding genes in *Rubus*. Figure S4. Prediction of RNA editing sites for 34 protein coding genes in *Rubus*. Figure S5. Migration sequence analysis between mitochondrial and plastid genomes. Table S1. Accession numbers for plant mitochondrial and chloroplast genomes. Table S2. Characteristic information on the mitochondrial genomes of 8 *Rubus* plants. Table S3. Functional annotation of mitochondrial genome of 8 *Rubus* species. Table S4. The homologous DNA fragments between mitochondrial genome and chloroplast genome of 6 *Rubus* plants. Table S5. Collinear fragments of the mitogenome of *Rubus caesius* and the mitogenomes of six remaining *Rubus* plants.

Author Contributions: Y.S., Z.C., J.J., Q.L. and W.Z. conceived and performed the original research project. Y.S., Z.C., Q.L. and J.J. designed the experiments and analyzed the data. Y.S. refined the project and wrote the manuscript with contributions from all authors. Z.C. and W.Z. supervised the experiments and revised the writing. Z.C. and W.Z. obtained the funding for the research project. All authors have read and agreed to the published version of the manuscript.

Funding: This work was funded by Taizhou 500 talent program (W.Z., grant number: Z2024136), the Basic Public Welfare Research Project of Zhejiang Province (Z.C., grant number: LGN22C020001), and Startup Funding of Taizhou University for the Biomass Polysaccharide Metabolism Institute (W.Z., grant number: T20231801002).

Data Availability Statement: The mitochondrial genomes of *R. parviflorus* (PV329689), *R. spectabilis* (PV353732), *R. idaeus* (PV339482), *R. armeniacus* (PV340594), *R. caesius* (PV296168), and *R. chamaemorus* (PV242989; PV242990) are accessioned in GeneBank. The chloroplast genomes of *R. spectabilis* (PV330320), *R. armeniacus* (PV282406), *R. caesius* (PV282407), and *R. chamaemorus* (PV282408) are also accessioned in GeneBank.

Conflicts of Interest: The authors declare that they have no conflicts of interest.

References

1. Carter, K.A.; Liston, A.; Bassil, N.V.; Alice, L.A.; Bushakra, J.M.; Sutherland, B.L.; Mockler, T.C.; Bryant, D.W.; Hummer, K.E. Target Capture Sequencing Unravels *Rubus* Evolution. *Front. Plant Sci.* **2019**, *10*, 1615. [CrossRef] [PubMed]
2. Huang, T.; Chen, J.; Hummer, K.E.; Alice, L.A.; Wang, W.; He, Y.; Yu, S.; Yang, M.; Chai, T.; Zhu, X.; et al. Phylogeny of *Rubus* (Rosaceae): Integrating Molecular and Morphological Evidence into an Infrageneric Revision. *Taxon* **2023**, *72*, 278–306. [CrossRef]
3. Zhang, H.; Yan, M.; Li, L.; Jiang, Z.; Xiong, Y.; Wang, Y.; Akogwu, C.O.; Tolulope, O.M.; Zhou, H.; Sun, Y.; et al. Assembly and Comparative Analysis of the Complete Mitochondrial Genome of Red Raspberry (*Rubus idaeus* L.) Revealing Repeat-Mediated Recombination and Gene Transfer. *BMC Plant Biol.* **2025**, *25*, 85. [CrossRef] [PubMed]
4. Huang, X.; Wu, Y.; Zhang, S.; Yang, H.; Wu, W.; Lyu, L.; Li, W. Overexpression of *RuFLS2* Enhances Flavonol-Related Substance Contents and Gene Expression Levels. *Int. J. Mol. Sci.* **2022**, *23*, 14230. [CrossRef] [PubMed]
5. Yang, Y.; Zhang, K.; Xiao, Y.; Zhang, L.; Huang, Y.; Li, X.; Chen, S.; Peng, Y.; Yang, S.; Liu, Y.; et al. Genome Assembly and Population Resequencing Reveal the Geographical Divergence of Shanmei (*Rubus corchorifolius*). *Genom. Proteom. Bioinform.* **2022**, *20*, 1106–1118. [CrossRef]
6. Hua, Y.; Dai, B.; Luo, Y.; Ding, Y. Integrated Analysis of Multiple Metabolome and Transcriptome Revealed the Accumulation of Flavonoids and Associated Molecular Regulation Mechanisms in *Rubus chingii* Hu at Different Developmental Stages. *Plant Physiol. Biochem.* **2023**, *204*, 108085. [CrossRef]
7. Kumar, V.; Sharma, A.; Sharma, N.; Saini, R.; Dev, K.; El-Shazly, M.; Bari, A.B.A. A Review of Botany, Traditional Applications, Phytochemistry, Pharmacological Applications, and Toxicology of *Rubus ellipticus* Smith Fruits. *Naunyn-Schmiedeberg's Arch. Pharmacol.* **2024**, *397*, 4483–4497. [CrossRef]
8. Tao, Y.; Bao, J.; Zhu, F.; Pan, M.; Liu, Q.; Wang, P. Ethnopharmacology of *Rubus idaeus* Linnaeus: A Critical Review on Ethnobotany, Processing Methods, Phytochemicals, Pharmacology and Quality Control. *J. Ethnopharmacol.* **2023**, *302*, 115870. [CrossRef]
9. Alice, L.; Campbell, C. Hybridization and Gene Flow Between Distantly Related Species of *Rubus* (Rosaceae): Evidence from Nuclear Ribosomal DNA Internal Transcribed Spacer Region Sequences. *Syst. Bot.* **2001**, *26*, 769–778.
10. Sochor, M.; Vašut, R.J.; Sharbel, T.F.; Trávníček, B. How Just a Few Makes a Lot: Speciation via Reticulation and Apomixis on Example of European Brambles (*Rubus* subgen. *Rubus*, Rosaceae). *Mol. Phylogenetics Evol.* **2015**, *89*, 13–27. [CrossRef]
11. Thompson, M.M.; Zhao, C.M. Chromosome Numbers of *Rubus* species in southwest China. *Acta Hortic.* **1993**, *352*, 493–502. [CrossRef]
12. Focke, W.O. *Species Ruborum: Monographiae Generis Rubi Prodromus*; Botanica, B., Ed.; Schweizerbart: Stuttgart, Germany, 1914.
13. Lu, L.; David, E.B. *Flora of China*; *Rubus* Linnaeus; 1: 492. 1753; Science Press: Beijing, China, 2003; Volume 9, pp. 195–285.
14. Xiong, X.-H.; Zhou, X.-M.; Li, M.; Xu, B.; Deng, H.-N.; Yu, Q.; Gao, X.-F. Pollen Morphology in *Rubus* (Rosaceae) and Its Taxonomic Implications. *Plant Syst. Evol.* **2019**, *305*, 705–716. [CrossRef]
15. Wang, Y.; Chen, Q.; Chen, T.; Tang, H.; Liu, L.; Wang, X. Phylogenetic Insights into Chinese *Rubus* (Rosaceae) from Multiple Chloroplast and Nuclear DNAs. *Front. Plant Sci.* **2016**, *7*, 968. [CrossRef] [PubMed]
16. Wang, Y.; Wang, X.; Chen, Q.; Zhang, L.; Tang, H.; Luo, Y.; Liu, Z. Phylogenetic Insight into Subgenera *Idaeobatus* and *Malachobatus* (*Rubus*, Rosaceae) Inferring from ISH Analysis. *Mol. Cytogenet.* **2015**, *8*, 11. [CrossRef]
17. Birky, C.W. The Inheritance of Genes in Mitochondria and Chloroplasts: Laws, Mechanisms, and Models. *Annu. Rev. Genet.* **2001**, *35*, 125–148. [CrossRef]
18. Shen, B.; Shen, A.; Liu, L.; Tan, Y.; Li, S.; Tan, Z. Assembly and Comparative Analysis of the Complete Multichromosomal Mitochondrial Genome of *Cymbidium ensifolium*, an Orchid of High Economic and Ornamental Value. *BMC Plant Biol.* **2024**, *24*, 255. [CrossRef]
19. Rydin, C.; Wikström, N.; Bremer, B. Conflicting Results from Mitochondrial Genomic Data Challenge Current Views of Rubiaceae Phylogeny. *Am. J. Bot.* **2017**, *104*, 1522–1532. [CrossRef]
20. Gray, M.W.; Cedergren, R.; Abel, Y.; Sankoff, D. On the Evolutionary Origin of the Plant Mitochondrion and Its Genome. *Proc. Natl. Acad. Sci. USA* **1989**, *86*, 2267–2271. [CrossRef]
21. Wu, Z.; Liao, X.; Zhang, X.; Tembrock, L.R.; Broz, A. Genomic Architectural Variation of Plant Mitochondria—A Review of Multichromosomal Structuring. *J. Syst. Evol.* **2022**, *60*, 160–168. [CrossRef]
22. Kozik, A.; Rowan, B.A.; Lavelle, D.; Berke, L.; Schranz, M.E.; Michelmore, R.W.; Christensen, A.C. The Alternative Reality of Plant Mitochondrial DNA: One Ring Does Not Rule Them All. *PLoS Genet.* **2019**, *15*, e1008373. [CrossRef]
23. Jin, J.-J.; Yu, W.-B.; Yang, J.-B.; Song, Y.; dePamphilis, C.W.; Yi, T.-S.; Li, D.-Z. GetOrganelle: A Fast and Versatile Toolkit for Accurate de Novo Assembly of Organelle Genomes. *Genome Biol.* **2020**, *21*, 241. [CrossRef]
24. He, W.; Xiang, K.; Chen, C.; Wang, J.; Wu, Z. Master Graph: An Essential Integrated Assembly Model for the Plant Mitogenome Based on a Graph-Based Framework. *Brief. Bioinform.* **2023**, *24*, bbac522. [CrossRef] [PubMed]

25. Bi, C.; Shen, F.; Han, F.; Qu, Y.; Hou, J.; Xu, K.; Xu, L.; He, W.; Wu, Z.; Yin, T. PMAT: An Efficient Plant Mitogenome Assembly Toolkit Using Low-Coverage HiFi Sequencing Data. *Hortic. Res.* **2024**, *11*, uhae023. [CrossRef]
26. Sun, M.; Zhang, M.; Chen, X.; Liu, Y.; Liu, B.; Li, J.; Wang, R.; Zhao, K.; Wu, J. Rearrangement and Domestication as Drivers of Rosaceae Mitogenome Plasticity. *BMC Biol.* **2022**, *20*, 181. [CrossRef] [PubMed]
27. VanBuren, R.; Bryant, D.; Bushakra, J.M.; Vining, K.J.; Edger, P.P.; Rowley, E.R.; Priest, H.D.; Michael, T.P.; Lyons, E.; Filichkin, S.A.; et al. The Genome of Black Raspberry (*Rubus occidentalis*). *Plant J.* **2016**, *87*, 535–547. [CrossRef] [PubMed]
28. Wang, L.; Lei, T.; Han, G.; Yue, J.; Zhang, X.; Yang, Q.; Ruan, H.; Gu, C.; Zhang, Q.; Qian, T.; et al. The Chromosome-scale Reference Genome of *Rubus chingii* Hu Provides Insight into the Biosynthetic Pathway of Hydrolyzable Tannins. *Plant J.* **2021**, *107*, 1466–1477. [CrossRef]
29. Wang, Y.; Guan, J.; Zhang, Q. Chromosome-Scale Genome, Together with Transcriptome and Metabolome, Provides Insights into the Evolution and Anthocyanin Biosynthesis of *Rubus rosaefolius* Sm. (Rosaceae). *Hortic. Res.* **2024**, *11*, uhae064. [CrossRef]
30. Price, R.J.; Davik, J.; Fernandéz Fernandéz, F.; Bates, H.J.; Lynn, S.; Nellist, C.F.; Buti, M.; Røen, D.; Šurbanovski, N.; Alsheikh, M.; et al. Chromosome-Scale Genome Sequence Assemblies of the 'Autumn Bliss' and 'Malling Jewel' Cultivars of the Highly Heterozygous Red Raspberry (*Rubus idaeus* L.) Derived from Long-Read Oxford Nanopore Sequence Data. *PLoS ONE* **2023**, *18*, e0285756. [CrossRef]
31. Wu, W.; Wang, L.; Huang, W.; Zhang, X.; Li, Y.; Guo, W. A High-Quality Genome Assembly Reveals Adaptations Underlying Glossy, Wax-Coated Leaves in the Heat-Tolerant Wild Raspberry *Rubus leucanthus*. *DNA Res.* **2024**, *31*, dsae024. [CrossRef]
32. Brůna, T.; Aryal, R.; Dudchenko, O.; Sargent, D.J.; Mead, D.; Buti, M.; Cavallini, A.; Hytönen, T.; Andrés, J.; Pham, M.; et al. A Chromosome-Length Genome Assembly and Annotation of Blackberry (*Rubus argutus*, cv. "Hillquist"). *G3* **2023**, *13*, jkac289. [CrossRef]
33. Shi, Y.; Chen, Z.; Jiang, J.; Wu, W.; Xin, Y.; Zeng, W. Assembly and Comparative Analysis of the Complete Mitogenome of *Rubus chingii* var. *suavissimus*, an Exceptional Berry Plant Possessing Sweet Leaves. *Front. Plant Sci.* **2024**, *15*, 1504687. [CrossRef]
34. Bi, C.; Qu, Y.; Hou, J.; Wu, K.; Ye, N.; Yin, T. Deciphering the Multi-Chromosomal Mitochondrial Genome of *Populus simonii*. *Front. Plant Sci.* **2022**, *13*, 914635. [CrossRef]
35. Wick, R.R.; Schultz, M.B.; Zobel, J.; Holt, K.E. Bandage: Interactive Visualization of de Novo Genome Assemblies. *Bioinformatics* **2015**, *31*, 3350–3352. [CrossRef] [PubMed]
36. Li, H. New Strategies to Improve Minimap2 Alignment Accuracy. *Bioinformatics* **2021**, *37*, 4572–4574. [CrossRef]
37. Li, J.; Ni, Y.; Lu, Q.; Chen, H.; Liu, C. PMGA: A Plant Mitochondrial Genome Annotator. *Plant Commun.* **2025**, *6*, 101191. [CrossRef] [PubMed]
38. Tillich, M.; Lehwark, P.; Pellizzer, T.; Ulbricht-Jones, E.S.; Fischer, A.; Bock, R.; Greiner, S. GeSeq—Versatile and Accurate Annotation of Organelle Genomes. *Nucleic Acids Res.* **2017**, *45*, W6–W11. [CrossRef] [PubMed]
39. Shi, L.; Chen, H.; Jiang, M.; Wang, L.; Wu, X.; Huang, L.; Liu, C. CPGAVAS2, an Integrated Plastome Sequence Annotator and Analyzer. *Nucleic Acids Res.* **2019**, *47*, W65–W73. [CrossRef]
40. Chan, P.P.; Lin, B.Y.; Mak, A.J.; Lowe, T.M. tRNAscan-SE 2.0: Improved Detection and Functional Classification of Transfer RNA Genes. *Nucleic Acids Res.* **2021**, *49*, 9077–9096. [CrossRef]
41. Chen, Y.; Ye, W.; Zhang, Y.; Xu, Y. High Speed BLASTN: An Accelerated MegaBLAST Search Tool. *Nucleic Acids Res.* **2015**, *43*, 7762–7768. [CrossRef]
42. Kearse, M.; Moir, R.; Wilson, A.; Stones-Havas, S.; Cheung, M.; Sturrock, S.; Buxton, S.; Cooper, A.; Markowitz, S.; Duran, C.; et al. Geneious Basic: An Integrated and Extendable Desktop Software Platform for the Organization and Analysis of Sequence Data. *Bioinformatics* **2012**, *28*, 1647–1649. [CrossRef]
43. Zhang, X.; Chen, H.; Ni, Y.; Wu, B.; Li, J.; Burzyński, A.; Liu, C. Plant Mitochondrial Genome Map (PMGmap): A Software Tool for the Comprehensive Visualization of Coding, Noncoding and Genome Features of Plant Mitochondrial Genomes. *Mol. Ecol. Resour.* **2024**, *24*, e13952. [CrossRef] [PubMed]
44. Beier, S.; Thiel, T.; Münch, T.; Scholz, U.; Mascher, M. MISA-Web: A Web Server for Microsatellite Prediction. *Bioinformatics* **2017**, *33*, 2583–2585. [CrossRef]
45. Benson, G. Tandem Repeats Finder: A Program to Analyze DNA Sequences. *Nucleic Acids Res.* **1999**, *27*, 573–580. [CrossRef]
46. Kurtz, S. REPuter: The Manifold Applications of Repeat Analysis on a Genomic Scale. *Nucleic Acids Res.* **2001**, *29*, 4633–4642. [CrossRef]
47. Zhang, D.; Gao, F.; Jakovlić, I.; Zou, H.; Zhang, J.; Li, W.X.; Wang, G.T. PhyloSuite: An Integrated and Scalable Desktop Platform for Streamlined Molecular Sequence Data Management and Evolutionary Phylogenetics Studies. *Mol. Ecol. Resour.* **2020**, *20*, 348–355. [CrossRef] [PubMed]
48. Edera, A.A.; Small, I.; Milone, D.H.; Sanchez-Puerta, M.V. Deepred-Mt: Deep Representation Learning for Predicting C-to-U RNA Editing in Plant Mitochondria. *Comput. Biol. Med.* **2021**, *136*, 104682. [CrossRef]
49. Katoh, K. MAFFT: A Novel Method for Rapid Multiple Sequence Alignment Based on Fast Fourier Transform. *Nucleic Acids Res.* **2002**, *30*, 3059–3066. [CrossRef] [PubMed]

50. Wang, D.; Zhang, Y.; Zhang, Z.; Zhu, J.; Yu, J. KaKs_Calculator 2.0: A Toolkit Incorporating Gamma-Series Methods and Sliding Window Strategies. *Genom. Proteom. Bioinform.* **2010**, *8*, 77–80. [CrossRef]
51. Chen, C.; Wu, Y.; Li, J.; Wang, X.; Zeng, Z.; Xu, J.; Liu, Y.; Feng, J.; Chen, H.; He, Y.; et al. TBtools-II: A "One for All, All for One" Bioinformatics Platform for Biological Big-Data Mining. *Mol. Plant* **2023**, *16*, 1733–1742. [CrossRef]
52. Xie, J.; Chen, Y.; Cai, G.; Cai, R.; Hu, Z.; Wang, H. Tree Visualization By One Table (tvBOT): A Web Application for Visualizing, Modifying and Annotating Phylogenetic Trees. *Nucleic Acids Res.* **2023**, *51*, W587–W592. [CrossRef]
53. Marçais, G.; Delcher, A.L.; Phillippy, A.M.; Coston, R.; Salzberg, S.L.; Zimin, A. MUMmer4: A Fast and Versatile Genome Alignment System. *PLoS Comput. Biol.* **2018**, *14*, e1005944. [CrossRef] [PubMed]
54. He, W.; Yang, J.; Jing, Y.; Xu, L.; Yu, K.; Fang, X. NGenomeSyn: An Easy-to-Use and Flexible Tool for Publication-Ready Visualization of Syntenic Relationships across Multiple Genomes. *Bioinformatics* **2023**, *39*, btad121. [CrossRef]
55. Liu, M.; Fan, R.; Wang, C.; Dai, L.; Chu, S. Complete Analysis and Phylogenetic Analysis of *Polygonatum sibiricum* Mitochondria. *BMC Plant Biol.* **2025**, *25*, 471. [CrossRef] [PubMed]
56. Huang, L.; Wen, J.; Crabbe, M.J.C.; Chen, C.; Ren, Z. Complete Chloroplast Genome Characterization of Three *Plagiomnium* Species and the Phylogeny of Family Mniaceae. *Genetica* **2025**, *153*, 6. [CrossRef]
57. Angiosperm Phylogeny Group; Chase, M.W.; Christenhusz, M.J.; Fay, M.F.; Byng, J.; Judd, W.; Soltis, D.; Mabberley, D.; Sennikov, A.; Soltis, P. An Update of the Angiosperm Phylogeny Group Classification for the Orders and Families of Flowering Plants: APG IV. *Bot. J. Linn. Soc.* **2016**, *181*, 1–20. [CrossRef]
58. Wang, H.; Wu, Z.; Li, T.; Zhao, J. Highly Active Repeat-Mediated Recombination in the Mitogenome of the Aquatic Grass *Hygroryza aristata*. *BMC Plant Biol.* **2024**, *24*, 644. [CrossRef]
59. Wang, Z.; Wang, R.; Sang, Y.; Wang, T.; Su, Y.; Liao, W. Comparative Analysis of Mitochondrial Genomes of Invasive Weed *Mikania micrantha* and Its Indigenous Congener *Mikania cordata*. *Int. J. Biol. Macromol.* **2024**, *281*, 136357. [CrossRef]
60. Xu, C.; Bi, W.; Ma, R.; Li, P.; Liu, F.; Liu, Z. Assembly and Comparative Analysis of the Complete Mitochondrial of *Spodiopogon sagittifolius*, an Endemic and Protective Species from Yunnan, China. *BMC Plant Biol.* **2025**, *25*, 373. [CrossRef]
61. Gao, C.; Wang, S.; Huang, Y.; Deng, Y. Assembly and Comparative Analysis of the Complete Mitochondrial Genome of *Echinacanthus longipes* (Acanthaceae), Endemic to the Sino-Vietnamese Karst Flora. *BMC Genom.* **2025**, *26*, 251. [CrossRef]
62. Mower, J.P. Variation in Protein Gene and Intron Content among Land Plant Mitogenomes. *Mitochondrion* **2020**, *53*, 203–213. [CrossRef]
63. Alverson, A.J.; Rice, D.W.; Dickinson, S.; Barry, K.; Palmer, J.D. Origins and Recombination of the Bacterial-Sized Multichromosomal Mitochondrial Genome of Cucumber. *Plant Cell* **2011**, *23*, 2499–2513. [CrossRef] [PubMed]
64. Skippington, E.; Barkman, T.J.; Rice, D.W.; Palmer, J.D. Miniaturized Mitogenome of the Parasitic Plant *Viscum scurruloideum* Is Extremely Divergent and Dynamic and Has Lost All *Nad* Genes. *Proc. Natl. Acad. Sci. USA* **2015**, *112*, E3515–E3524. [CrossRef] [PubMed]
65. Putintseva, Y.A.; Bondar, E.I.; Simonov, E.P.; Sharov, V.V.; Oreshkova, N.V.; Kuzmin, D.A.; Konstantinov, Y.M.; Shmakov, V.N.; Belkov, V.I.; Sadovsky, M.G.; et al. Siberian larch (*Larix sibirica* Ledeb.) Mitochondrial Genome Assembled Using Both Short and Long Nucleotide Sequence Reads Is Currently the Largest Known Mitogenome. *BMC Genom.* **2020**, *21*, 654. [CrossRef]
66. Han, F.; Bi, C.; Zhao, Y.; Gao, M.; Wang, Y.; Chen, Y. Unraveling the Complex Evolutionary Features of the *Cinnamomum camphora* Mitochondrial Genome. *Plant Cell Rep.* **2024**, *43*, 183. [CrossRef] [PubMed]
67. Dong, S.; Chen, L.; Liu, Y.; Wang, Y.; Zhang, S.; Yang, L.; Lang, X.; Zhang, S. The Draft Mitochondrial Genome of *Magnolia biondii* and Mitochondrial Phylogenomics of Angiosperms. *PLoS ONE* **2020**, *15*, e0231020. [CrossRef]
68. Du, M.-Z.; Liu, S.; Zeng, Z.; Alemayehu, L.A.; Wei, W.; Guo, F.-B. Amino Acid Compositions Contribute to the Proteins' Evolution under the Influence of Their Abundances and Genomic GC Content. *Sci. Rep.* **2018**, *8*, 7382. [CrossRef]
69. Kubo, T.; Newton, K.J. Angiosperm Mitochondrial Genomes and Mutations. *Mitochondrion* **2008**, *8*, 5–14. [CrossRef]
70. Jiang, M.; Ni, Y.; Zhang, J.; Li, J.; Liu, C. Complete Mitochondrial Genome of *Mentha spicata* L. Reveals Multiple Chromosomal Configurations and RNA Editing Events. *Int. J. Biol. Macromol.* **2023**, *251*, 126257. [CrossRef]
71. Fajardo, D.; Schlautman, B.; Steffan, S.; Polashock, J.; Vorsa, N.; Zalapa, J. The American Cranberry Mitochondrial Genome Reveals the Presence of Selenocysteine (tRNA-Sec and SECIS) Insertion Machinery in Land Plants. *Gene* **2014**, *536*, 336–343. [CrossRef]
72. Li, Y.; Hu, X.; Xiao, M.; Huang, J.; Lou, Y.; Hu, F.; Fu, X.; Li, Y.; He, H.; Cheng, J. An Analysis of Codon Utilization Patterns in the Chloroplast Genomes of Three Species of *Coffea*. *BMC Genom. Data* **2023**, *24*, 42. [CrossRef]
73. Liu, D.; Guo, H.; Zhu, J.; Qu, K.; Chen, Y.; Guo, Y.; Ding, P.; Yang, H.; Xu, T.; Jing, Q.; et al. Complex Physical Structure of Complete Mitochondrial Genome of *Quercus acutissima* (Fagaceae): A Significant Energy Plant. *Genes* **2022**, *13*, 1321. [CrossRef] [PubMed]
74. Parvathy, S.T.; Udayasuriyan, V.; Bhadana, V. Codon Usage Bias. *Mol. Biol. Rep.* **2022**, *49*, 539–565. [CrossRef] [PubMed]
75. Bock, H.; Brennicke, A.; Schuster, W. *Rps3* and *Rpl16* Genes Do Not Overlap in Oenothera Mitochondria: GTG as a Potential Translation Initiation Codon in Plant Mitochondria? *Plant Mol. Biol.* **1994**, *24*, 811–818. [CrossRef] [PubMed]

76. Wang, Y.; Cui, G.; He, K.; Xu, K.; Liu, W.; Wang, Y.; Wang, Z.; Liu, S.; Bi, C. Assembly and Comparative Analysis of the Complete Mitochondrial Genome of *Ilex rotunda* Thunb. *Forests* **2024**, *15*, 1117. [CrossRef]
77. Sun, N.; Han, F.; Wang, S.; Shen, F.; Liu, W.; Fan, W.; Bi, C. Comprehensive Analysis of the *Lycopodium japonicum* Mitogenome Reveals Abundant tRNA Genes and Cis-Spliced Introns in Lycopodiaceae Species. *Front. Plant Sci.* **2024**, *15*, 1446015. [CrossRef]
78. Liu, Q.; Wu, Z.; Tian, C.; Yang, Y.; Liu, L.; Feng, Y.; Li, Z. Complete Mitochondrial Genome of the Endangered *Prunus pedunculata* (*Prunoideae*, Rosaceae) in China: Characterization and Phylogenetic Analysis. *Front. Plant Sci.* **2023**, *14*, 1266797. [CrossRef]
79. Zhang, Y.; Guo, X.; Ma, R.; Yu, M.; Xu, J.; Shen, Z. Assembly and Analysis of the Complete Mitochondrial Genome of *Prunus davidiana* (Rosaceae). *Front. Plant Sci.* **2025**, *16*, 1558619. [CrossRef]
80. Sloan, D.B.; Wu, Z.; Sharbrough, J. Correction of Persistent Errors in Arabidopsis Reference Mitochondrial Genomes. *Plant Cell* **2018**, *30*, 525–527. [CrossRef]
81. Bi, C.; Lu, N.; Xu, Y.; He, C.; Lu, Z. Characterization and Analysis of the Mitochondrial Genome of Common Bean (*Phaseolus vulgaris*) by Comparative Genomic Approaches. *Int. J. Mol. Sci.* **2020**, *21*, 3778. [CrossRef]
82. Sun, S.; Xiao, N.; Sha, Z. Complete mitochondrial genomes of four deep-sea echinoids: Conserved mitogenome organization and new insights into the phylogeny and evolution of Echinoidea. *PeerJ.* **2022**, *10*, e13730. [CrossRef]
83. Wang, D.; Liu, F.; Wang, L.; Huang, S.; Yu, J. Nonsynonymous Substitution Rate (Ka) Is a Relatively Consistent Parameter for Defining Fast-Evolving and Slow-Evolving Protein-Coding Genes. *Biol. Direct* **2011**, *6*, 13. [CrossRef]
84. Zhou, P.; Zhang, Q.; Li, F.; Huang, J.; Zhang, M. Assembly and Comparative Analysis of the Complete Mitochondrial Genome of *Ilex metabaptista* (Aquifoliaceae), a Chinese Endemic Species with a Narrow Distribution. *BMC Plant Biol.* **2023**, *23*, 393. [CrossRef]
85. Li, Z.-Z.; Wang, Y.; He, X.-Y.; Li, W.-G. The *Taihangia* Mitogenome Provides New Insights into Its Adaptation and Organelle Genome Evolution in Rosaceae. *Planta* **2025**, *261*, 59. [CrossRef] [PubMed]
86. Ma, Q.; Wang, Y.; Li, S.; Wen, J.; Zhu, L.; Yan, K.; Du, Y.; Ren, J.; Li, S.; Chen, Z.; et al. Assembly and Comparative Analysis of the First Complete Mitochondrial Genome of *Acer truncatum* Bunge: A Woody Oil-Tree Species Producing Nervonic Acid. *BMC Plant Biol.* **2022**, *22*, 29. [CrossRef]
87. Christensen, A.C. Plant Mitochondrial Genome Evolution Can Be Explained by DNA Repair Mechanisms. *Genome Biol. Evol.* **2013**, *5*, 1079–1086. [CrossRef] [PubMed]
88. Yisilam, G.; Liu, Z.; Turdi, R.; Chu, Z.; Luo, W.; Tian, X. Assembly and Comparative Analysis of the Complete Mitochondrial Genome of *Isopyrum anemonoides* (Ranunculaceae). *PLoS ONE* **2023**, *18*, e0286628. [CrossRef]
89. Yong, Y.; Wang, Y.; Wang, D.; Yuan, X.; Zhang, Q. The Organelle Genomes of the Endangered Seagrass *Zostera caespitosa* Reveal Sequence Divergences, Massive Gene Transfer, and Uncommon RNA Editing Types. *Front. Plant Sci.* **2025**, *16*, 1550467. [CrossRef] [PubMed]
90. Wang, X.-C.; Chen, H.; Yang, D.; Liu, C. Diversity of Mitochondrial Plastid DNAs (MTPTs) in Seed Plants. *Mitochondrial DNA Part A* **2018**, *29*, 635–642. [CrossRef]
91. Bi, C.; Paterson, A.H.; Wang, X.; Xu, Y.; Wu, D.; Qu, Y.; Jiang, A.; Ye, Q.; Ye, N. Analysis of the Complete Mitochondrial Genome Sequence of the Diploid Cotton *Gossypium raimondii* by Comparative Genomics Approaches. *BioMed Res. Int.* **2016**, *2016*, 5040598. [CrossRef]
92. Gao, C.; Ren, X.; Mason, A.S.; Liu, H.; Xiao, M.; Li, J.; Fu, D. Horizontal Gene Transfer in Plants. *Funct. Integr. Genom.* **2014**, *14*, 23–29. [CrossRef]
93. Clifton, S.W.; Minx, P.; Fauron, C.M.-R.; Gibson, M.; Allen, J.O.; Sun, H.; Thompson, M.; Barbazuk, W.B.; Kanuganti, S.; Tayloe, C.; et al. Sequence and Comparative Analysis of the Maize NB Mitochondrial Genome. *Plant Physiol.* **2004**, *136*, 3486–3503. [CrossRef] [PubMed]
94. Unseld, M.; Marienfeld, J.R.; Brandt, P.; Brennicke, A. The Mitochondrial Genome of *Arabidopsis thaliana* Contains 57 Genes in 366,924 Nucleotides. *Nat. Genet.* **1997**, *15*, 57–61. [CrossRef] [PubMed]
95. Li, Y.-X.; Li, Z.-H.; Schuiteman, A.; Chase, M.W.; Li, J.-W.; Huang, W.-C.; Hidayat, A.; Wu, S.-S.; Jin, X.-H. Phylogenomics of Orchidaceae Based on Plastid and Mitochondrial Genomes. *Mol. Phylogenetics Evol.* **2019**, *139*, 106540. [CrossRef] [PubMed]
96. Yang, Z.; Ferguson, D.K.; Yang, Y. New Insights into the Plastome Evolution of Lauraceae Using Herbariomics. *BMC Plant Biol.* **2023**, *23*, 387. [CrossRef]
97. Liu, C.; Chen, H.-H.; Tang, L.-Z.; Khine, P.K.; Han, L.-H.; Song, Y.; Tan, Y.-H. Plastid Genome Evolution of a Monophyletic Group in the *Subtribe lauriineae* (*Laureae*, Lauraceae). *Plant Divers.* **2022**, *44*, 377–388. [CrossRef]
98. Dong, W.; Xu, C.; Cheng, T.; Lin, K.; Zhou, S. Sequencing Angiosperm Plastid Genomes Made Easy: A Complete Set of Universal Primers and a Case Study on the Phylogeny of Saxifragales. *Genome Biol. Evol.* **2013**, *5*, 989–997. [CrossRef] [PubMed]
99. Qiu, Y.-L.; Li, L.; Wang, B.; Xue, J.-Y.; Hendry, T.A.; Li, R.-Q.; Brown, J.W.; Liu, Y.; Hudson, G.T.; Chen, Z.-D. Angiosperm Phylogeny Inferred from Sequences of Four Mitochondrial Genes. *J. Syst. Evol.* **2010**, *48*, 391–425. [CrossRef]
100. Pearl, S.A.; Welch, M.E.; McCauley, D.E. Mitochondrial Heteroplasmy and Paternal Leakage in Natural Populations of *Silene vulgaris*, a Gynodioecious Plant. *Mol. Biol. Evol.* **2009**, *26*, 537–545. [CrossRef]

101. Howarth, D.G.; Gardner, D.E.; Morden, C.W. Phylogeny of *Rubus* subgenus *Idaeobatus* (Rosaceae) and Its Implications Toward Colonization of the Hawaiian Islands. *Syst. Bot.* **1997**, *22*, 433–441. [CrossRef]
102. Wang, X.; Wang, J.; Jin, D.; Guo, H.; Lee, T.-H.; Liu, T.; Paterson, A.H. Genome Alignment Spanning Major Poaceae Lineages Reveals Heterogeneous Evolutionary Rates and Alters Inferred Dates for Key Evolutionary Events. *Mol. Plant* **2015**, *8*, 885–898. [CrossRef]

horticulturae

MDPI

Article

Cytochrome P450 CYP76F14 Mediates the Conversion of Its Substrate Linalool in Table Grape Berries

Zhizhong Song [1,2,†], Jinjin Zhang [1,3,†], Matthew Shi [2,4], Dong Li [1,*] and Xiaohua Liu [1,2,*]

[1] College of Horticulture, Ludong University, Yantai 264025, China
[2] Department of Plant Science, University of Cambridge, Cambridge CB2 3EA, UK
[3] Co-Innovation Center for Industry-Academy-Research, Linyi Vocational University of Science and Technology, Linyi 276000, China
[4] Wolfson College, University of Cambridge, Cambridge CB3 9BB, UK
[*] Correspondence: 3128@ldu.edu.cn (D.L.); xl652@cam.ac.uk (X.L.)
[†] These authors contributed equally to this work.

Abstract: Aroma composition serves as a pivotal quality determinant in table grapes (*Vitis vinifera*). While the cytochrome P450 enzyme CYP76F14 is implicated in aroma biosynthesis, its functional role in grape berries remains uncharacterized. A comparative analysis of three aroma-distinct cultivars—Muscat type 'Irsai Oliver', Neutral type 'Yanhong', and Berry-like type 'Venus Seedless'—revealed cultivar-specific linalool accumulation patterns. 'Irsai Oliver' exhibited sustained linalool biosynthesis from the fruit set through to maturity (from Stage 1 to Stage 5), with concentrations peaking at Stage 3 (veraison phase) and remaining elevated until harvest, surpassing the other two cultivars. Transcriptional profiling demonstrated that the *CYP76F14* expression exhibited a similar trend with the accumulation of linalool levels, showing a higher expression in 'Irsai Oliver' across the developmental stages. A structural analysis identified 12 divergent residues in the 'Irsai Oliver' CYP76F14 variant, including E378 and T380 within the conserved substrate recognition site. The site-directed mutagenesis of these residues (CYP76F14-E378G/T380A) reduced the catalytic efficiency by 68–72% compared to the wild-type (in vitro LC-MS/MS assays), confirming their functional significance. This work reveals that cytochrome P450 CYP76F14 mediates the conversion of its substrate linalool in table grape berries, especially of Muscat type grapes, and proposes the CYP76F14 polymorphic variants as molecular markers for aroma-type breeding. The identified catalytic residues (E378/T380) provide targets for enzymatic engineering to modulate the terpenoid profiles in *Vitis* species.

Keywords: table grape; Muscat type; linalool; cytochrome P450; site-directed amino acid mutagenesis

1. Introduction

Aroma serves as a critical sensory indicator of fruit quality in table grapes [1–5]. Cytochrome P450 is a superfamily of enzymes with heme as a secondary group involved in the biosynthesis of metabolites, such as fatty acids, plant sterols, plant hormones, and phenylpropanoids, and plays an important role in plant growth and development [6–8]. Recently, Kunert et al. identified two enzymes belonging to the CYP87A subfamily in the distant plants *Digitalis purpurea* and *Calotropis procera*. They act on cholesterol and the plant sterols (Brassinosteroids and β-sitosterols) to form pregnenolone, which is the first step in the biosynthesis of cardiac sterols [9]. The overexpression of these CYP87A enzyme encoding genes in *Arabidopsis thaliana* resulted in the ectopic accumulation of pregnenolone

https://doi.org/10.3390/horticulturae11060651

in transgenic plants, while silencing CYP87A in the leaves of *Eucommia ulmoides* through RNA interference led to a significant reduction in pregnenolone and cardiac sterols [9].

Unlike most plant monofunctional cytochrome P450 enzymes that catalyze a single substrate, the CYP76 subfamily of enzymes is a class of multifunctional monooxygenases that can catalyze multiple substrates [10]. Among them, the redox partner NADPH cytochrome P450 reductase (CPR) acts as an electron donor, promoting the catalytic activity of CYP76 [7,8,11,12]. In *Arabidopsis*, CYP76C1 can catalyze the conversion of linalool to various terpenoid compounds, such as 8-hydroxylinalool, 8-oxolinalool, and 8-carboxylinalool, and participate in the formation of syringaldehyde and syringanol [7,13]. In wine grapes (*Vitis vinifera*), enzymes from the CYP76F subfamily (VvCYP76F14) undergo a three-step enzymatic process (hydroxylation, dehydrogenation, and carboxylation) to catalyze the substrate linalool to produce 8-carboxylinalool [2,11,14–16], which is a precursor substance for the synthesis of bicyclic monoterpenoid lactones and participates in regulating the formation of wine's aroma.

According to the different types and contents of aroma compounds, table grapes can be classified into three different aroma types, such as Muscat type, Neutral type, and Berry-like type. Among them, linalool and geraniol are the main aroma compounds of Muscat type grapes, and together with other volatile compounds form the characteristic aromatic profile of table grapes [1,2,17]. However, the molecular mechanism behind the aroma formation of Muscat type grapes is still unclear, and the enzymatic activity characteristics of CYP76F14 catalyzing linalool are still unknown. In this study, *CYP76F14* genes were cloned from three different flavor type table grape fruits, and CYP76F14 proteins were expressed and purified in *Escherichia coli* (*E. coli*), and the key amino acid residues that determine CYP76F14 enzymatic activity were analyzed. This study provides a theoretical basis for studying the molecular mechanism of table grape aroma formation.

2. Materials and Methods

2.1. Table Grape Cultivars

Three table grape cultivars of 'Irsai Oliver' (Muscat type), 'Yanhong' (Neutral type), and 'Venus Seedless' (Berry-like type) were provided by the National Grape Germplasm Repository of the National Grape Industry System (Yantai, China). Five-year-old vines were grown under routine field management. Following the methods described by Song et al. [18], berries exhibiting similar maturity stages were harvested at the Stage 1 (cell division phase), Stage 2 (seed hardening phase), Stage 3 (veraison phage), Stage 4 (post-veraison, cell expansion phase), and Stage 5 (mature phase), respectively. Berries with consistent maturity and no damage or disease were randomly harvested and stored in a −80 °C freezer after liquid nitrogen treatment for subsequent analysis. Three biological replicates were performed, each with 20 individual clusters.

2.2. Determination of Linalool Content in Berries

Fresh berries were smashed and subjected to a vacuum freeze-drying treatment. A mixed solvent of ethanol and ethyl acetate with the ration of 6:1 (v/v) was added to the powdered material to accumulate linalool. The content of linalool in the table grape berries was determined using high-performance liquid chromatography combined with high-resolution mass spectrometry (HPLC-HRMS) (Waters, Milford, MA, USA) by Shanghai Bioprofile Technology Co. Ltd. (Shanghai, China), as described by Iln et al. [15] and Song et al. [2,16].

2.3. Isolation and Sequence Analysis of *CYP76F14*

Total RNA was extracted from table grape berries of different aroma types, and the first strand cDNA was synthesized using the PrimeScriptTM RT reagent Kit (TaKaRa, Dalian, China) as a template. The coding region (CDS) sequence of the *CYP76F14* gene in wine grapes was used as a reference sequence [2,16,19]. The *CYP76F14* gene was amplified from three table grape varieties with different aroma types using high fidelity polymerase Prime STARTM HS DNA (TaKaRa, Dalian, China) and sent to Biotechnology (Shanghai) Co., Ltd. (Shanghai, China) for sequencing verification. The amino acid sequence identity of the encoded CYP76F14 proteins was aligned via the help of the ClustalX 2.0.13 software.

2.4. Quantitative Real Time PCR (qRT-PCR)

The specific expression primers for *CYP76F14* were designed using the NCBI/Primer BLAST online server (Forward: 5′-TGTATCCACACCATAT-3′, Reverse: 5′-TCCCAGCTTC-CTCCATCACA-3′). Grape *Ubiquitin* (Forward: 5′-CCTCATCTTCGCTGGCAAAC-3′, Reverse: 5′-GGTGTAGGTCTTCTTCTTGCG-3′) was chosen as an internal reference [2,16,18–20]. The expression levels of *CYP76F14* during the different fruit developmental stages were analyzed by labeling with the fluorescent dye SYBR Green (TaKaRa, Dalian, China) and using the ABI 7500 real-time fluorescence quantitative PCR instrument (ABI, New York, NY, USA). The PCR reaction procedure was as follows: first, pre denature at 95 °C for 30 s. Second, denaturation at 95 °C for 5 s, annealing at 60 °C for 34 s, 40 cycles. Finally, extend at 72 °C for 15 s. The corresponding Ct values were obtained using an ABI 7500 PCR instrument. The relative expression level was calculated using the $2^{-\Delta\Delta Ct}$ method [2,16–21] after normalization to the internal reference *Ubiquitin* from three independent biological repeats, each with four technical replicates.

2.5. Site-Directed Mutagenesis (SM)

To explore the catalytic activities of key amino acid residues in the CYP76F14 proteins with different aroma types, we employed the alanine-scanning method to generate alanine-substituted CYP76F14-SMs, using the QuikChange Lightning Site-Directed Mutagenesis Kit (Agilent Stratagene, New York, NY, USA), as previously described [2,16,19,21]. The *CYP76F14* gene from the Muscat type 'Irsai Oliver' was cloned, and site-directed mutagenesis kits were used to obtain single amino acid mutagenesized mutants of the CYP76F14 protein, referred to as CYP76F14-SMs. The specific mutants obtained were CYP76F14-N46S, CYP76F14-T107R, CYP76F14-N111K, CYP76F14-R175Q, CYP76F14-L222V, CYP76F14-M264I, CYP76F14-S286N, CYP76F14-K325T, CYP76F14-E378G, CYP76F14-T380A, CYP76F14-E383G, and CYP76F14-T386A.

2.6. In Vitro Enzymatic Activity Assay

The crude protein of CYP76F14-MBP or CYP76F14-SM-MBP was purified by the NEBExpress® MBP Fusion and Purification System (New England Biolabs, Hitchin, UK), using the affinity between MBP and amylose resin [2,16,19]. According to the method described by Song et al. [2] and Kunert et al. [9], table grape CPR1 was used as the electron transfer redox partner for the fresh grape CYP76F14 at an appropriate ratio of 2:1 (CYP76F14/CPR1), and linalool was used as the substrate. The purified recombinant enzymes of CYP76F14 or CYP76F14-SMs were subjected to an in vitro enzymatic activity assay. The VvCYP76F14 enzyme assays were performed in a total reaction volume of 5 mL with a 100 mM Na^+/K^+ phosphate buffer (pH 5.0), with varying substrate concentrations, 1 mM NADPH, and adjusted enzyme amounts. The reactions were carried out at 26 °C for 1 h with agitation. The reaction product was collected and analyzed using HPLC-HRMS (Waters, Milford, MA, USA), and boiled protein (nonfunctional) served as a control. The turnover number (K_{cat}) and affinity (K_m) were individually calculated, and the K_{cat}/K_m ratio represents the corresponding enzyme activity [2,16,19]. Each reaction undergoes 6 technical replicates for a total of three biological replicates.

2.7. Statistical Analysis

The statistical graphs were generated using the OriginPro 12.0 software (OriginLab Corporation, Northampton, MA, USA). Significant differences were analyzed using ANOVA Fisher's LSD in the SPSS 13.0 software (SPSS, Chicago, IL, USA). The conditions for the applicability of LSD were checked and the ANOVA showed global significance.

3. Results

3.1. Determination of Linalool Content in Three Different Flavor Type Table Grape Varieties

Three different flavor type table grapes of 'Irsai Oliver' (Muscat type), 'Yanhong' (Neutral type), and 'Venus Seedless' (Berry-like type) were selected in this study. The results of the linalool content determination showed that the linalool content in the Muscat type 'Irsai Oliver' berries was significantly higher than that in the 'Yanhong' and 'Venus Seedless' fruits throughout the entire grape development period, from Stage 1 to Stage 5 (Table 1). In addition, the linalool content in the fruits of all the three different flavor types of grapes continuously increased with the development of the grape fruits, reaching a peak during Stage 3 and continuing until Stage 5 (Table 1).

Table 1. Linalool determination from the different fruit developmental stages of three distinct flavor type varieties.

Cultivar	Flavor Type	Content of Linalool (μg·g^{-1} FW)				
		Stage 1	Stage 2	Stage 3	Stage 4	Stage 5
Irsai Oliver	Muscat	0.46 ± 0.04 a	1.07 ± 0.012 a	1.66 ± 0.14 a	1.68 ± 0.17 a	1.62 ± 0.13 a
Yanhong	Neutral	0.008 ± 0.0009 b	0.013 ± 0.0014 b	0.017 ± 0.0015 b	0.016 ± 0.0014 b	0.016 ± 0.0016 b
Venus Seedless	Berry-like	0.0031 ± 0.00043 c	0.0054 ± 0.00062 c	0.0091 ± 0.0011 c	0.0094 ± 0.0013 c	0.0093 ± 0.0014 c

Note: Stage 1, cell division phase. Stage 2, seed hardening phase. Stage 3, veraison phage. Stage 4, post-veraison (cell expansion phase). Stage 5, mature phase. Data were presented as the means ± SE (n = 3). Letters represent significant differences among the different flavor type cultivars at a significance level of $p \leq 0.01$, as determined using ANOVA followed by Fisher's LSD test.

3.2. Cloning of the CYP76F14 Gene and Analysis of the Encoded Protein Sequence

Using the CDS sequence of the *CYP76F14* gene from wine grapes as a reference [2,16,19], the homologous *CYP76F14* genes were cloned from three flavor type grape varieties. The encoded products all contained the AGTDT (active domain) and FGAGRRICFG (HEME-binding domain) motifs, which are typical conserved domains of the cytochrome P450 family proteins (Figure 1). Amino acid sequence alignment showed that the amino acid sequence identity of the CYP76F14 proteins in the three different flavor type varieties was 99.20%. Compared with the CYP76F14 amino acid sequences of 'Yanhong' and 'Venus Seedless', the CYP76F14 amino acids in the Muscat type 'Irsai Oliver' showed mutations in the N46, T107, N111, R175, L222, M264, S286, K325, E378, T380, E383, and T386 residues (Figure 1).

Figure 1. Amino acid sequence alignment of CYP76F14 proteins from the three distinct flavor type table grape cultivars.

3.3. Relative Expression of CYP76F14 Genes in Three Different Table Grape Varieties

qRT-PCR analysis showed that the expression level of the *CYP76F14* gene in the 'Irsai Oliver' (Muscat type) berries was significantly higher than that in the 'Yanhong' (Neutral type) and 'Venus Seedless' (Berry-like type) berries throughout the entire fruit development period (Figure 2). The expression level of the *CYP76F14* gene in the 'Irsai Oliver' berries continuously increased with the fruit's development, reaching a peak during Stage 3 and continuing until Stage 5 (Figure 2). In particular, the expression trends of the *CYP76F14* genes were closely aligned with the accumulation patterns of the linalool content during different developmental stages in the three table grape varieties, respectively.

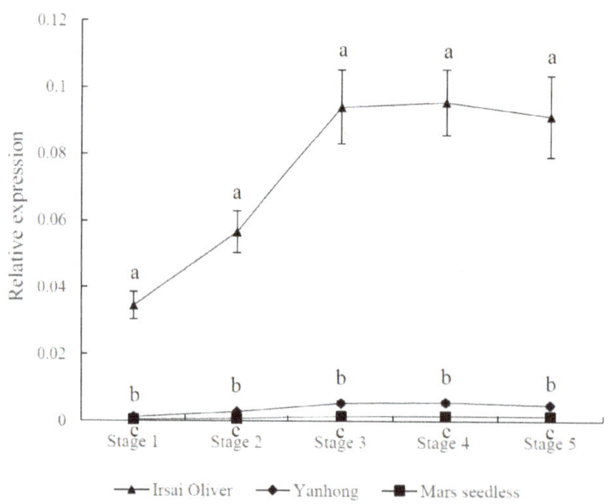

Figure 2. Expression analysis of *CYP76F14* genes during the different fruit developmental stages. Stage 1, cell division phase. Stage 2, seed hardening phase. Stage 3, veraison phage. Stage 4, post-veraison (cell expansion phase). Stage 5, mature phase. Data were presented as the means ± SE (n = 3). Letters represent significant differences among the different flavor type cultivars at a significance level of $p \leq 0.01$, as determined using ANOVA followed by Fisher's LSD test.

3.4. Site-Directed Mutagenesis of CYP76F14 (CYP76F14-SMs) and Heterologous Expression of CYP76F14 Protein in E. coli

The *CYP76F14* gene from the Muscat type 'Irsai Oliver' berries was cloned, and 12 site-directed mutant proteins (CYP76F14-SMs: CYP76F14-N46S, CYP76F14-T107R, CYP76F14-N111K, CYP76F14-R175Q, CYP76F14-L222V, CYP76F14-M264I, CYP76F14-S286N, CYP76F14-K325T, CYP76F14-E378G, CYP76F14-T380A, CYP76F14-E383G, and CYP76F14-T386A) were obtained and sequenced. The recombinant plasmids of pMAL-c6T-*CYP76F14* or pMAL-c6T-*CYP76F14*-SMs were transformed into *E. coli* BL21 strains for expression and purification. The resulting CYP76F14-MBP recombinant proteins were subsequently used for the in vitro enzyme activity assays. The expression and purification of the CYP76F14-MBP recombinant proteins is exhibited in Figure 3.

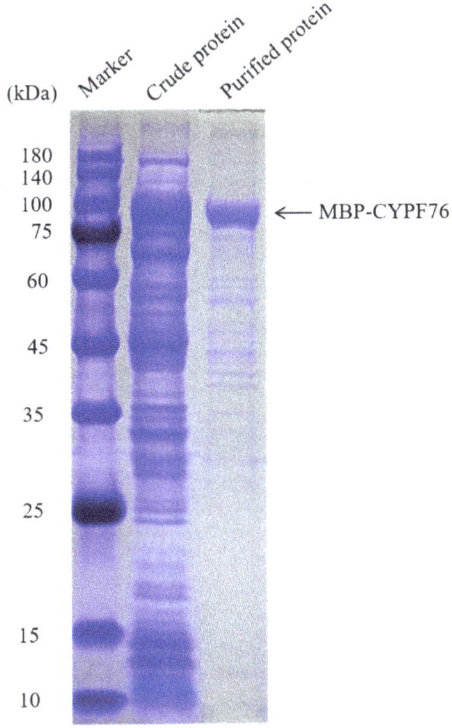

Figure 3. SDS-PAGE analysis of the expression and purification of the CYP76F14 MBP fusion protein. SDS-PAGE analysis of the recombinant CYP76F14 from the Muscat type 'Irsai Oliver' in *E. coli*. The pMAL-c6T-VvCYP76F14s plasmids were expressed in the *E. coli* BL21 strain. The crude protein of MBP-VvCYP76F14 was purified by NEBExpress® MBP Fusion and Purification System. Lane 1 represents the standard protein marker. Lane 2 represents the crude CYP76F14-MBP protein expressed in the *E. coli*. Lane 3 represents purified VvCYP76F14 eluted from the amylose column with maltose.

3.5. In Vitro Enzyme Activity Assay of CYP76F14 and CYP76F14-SMs

According to the method described by Peng et al. [16] and Xia et al. [19], using linalool as the substrate, the purified recombinant enzymes CYP76F14-MBP or CYP76F14-SMs-MBP were subjected to the in vitro enzyme activity assays. The enzymatic kinetic analysis results showed that the enzyme activities of the CYP76F14-E378G and CYP76F14-T380A site-directed mutant proteins were significantly lower than those of the control protein CYP76F14. However, there were no significant differences in the enzyme activities of the other 10 amino acid residue site-directed mutant proteins CYP76F14-SMs (CYP76F14-N46S, CYP76F14-T107R, CYP76F14-N111K, CYP76F14-R175Q, CYP76F14-L222V, CYP76F14-M264I, CYP76F14-S286N, CYP76F14-K325T, CYP76F14-E383G, and CYP76F14-T386A) compared to the control protein (Figure 4 and Table 2).

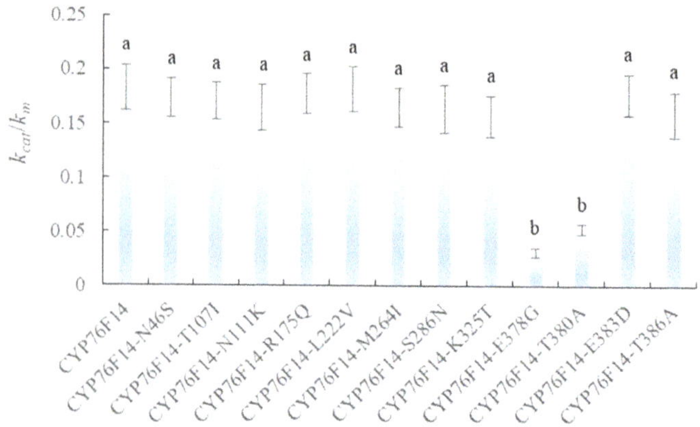

Figure 4. Enzyme kinetics of the table grape CYP76F14 was isolated from the 'Irsai Oliver' berries. CYP76F14 and its site-directed mutant proteins (CYP76F14-SMs) using linalool as the substrate. Data were presented as the means ± SE (n = 3). Letters represent significant differences among CYP76F14 and CYP76F14-SMs at a significance level of $p \leq 0.01$, as determined using ANOVA followed by Fisher's LSD test.

Table 2. In vitro determination of the 'Irsai Oliver' CYP76F14 and CYP76F14-SM mutants.

Protein	k_m (μM)	k_{cat} (S^{-1})	Remaining Amount of Linalool ($mmol \cdot min^{-1} \cdot mg^{-1}$ Protein)
CYP76F14	69.20 ± 5.26 ^{b}	12.93 ± 1.42 ^{a}	8.02 ± 0.89 ^{c}
CYP76F14-N146S	70.52 ± 7.53 ^{b}	12.27 ± 1.33 ^{a}	8.43 ± 0.94 ^{c}
CYP76F14-T107I	76.84 ± 8.27 ^{b}	13.14 ± 1.48 ^{a}	7.79 ± 0.82 ^{c}
CYP76F14-N111K	68.24 ± 6.65 ^{b}	11.26 ± 1.21 ^{a}	7.80 ± 0.91 ^{c}
CYP76F14-R175Q	75.56 ± 8.21 ^{b}	13.45 ± 1.44 ^{a}	8.24 ± 0.78 ^{c}
CYP76F14-L222V	67.25 ± 6.23 ^{b}	12.24 ± 1.51 ^{a}	7.87 ± 0.83 ^{c}
CYP76F14-M264I	71.39 ± 6.96 ^{b}	11.78 ± 1.09 ^{a}	8.11 ± 0.74 ^{c}
CYP76F14-S286N	74.08 ± 7.24 ^{b}	12.15 ± 1.18 ^{a}	8.39 ± 0.90 ^{c}
CYP76F14-K325T	70.57 ± 8.16 ^{b}	11.08 ± 1.02 ^{a}	8.33 ± 0.76 ^{c}
CYP76F14-E378G	269.35 ± 25.18 ^{a}	8.35 ± 0.89 ^{b}	16.67 ± 1.85 ^{b}
CYP76F14-T380A	134.91 ± 14.67 ^{a}	7.15 ± 0.83 ^{b}	23.72 ± 0.94 ^{a}
CYP76F14-E383D	66.95 ± 7.02 ^{b}	11.85 ± 1.16 ^{a}	7.92 ± 0.81 ^{c}
CYP76F14-T386A	83.27 ± 9.12 ^{b}	13.24 ± 1.46 ^{a}	8.11 ± 0.85 ^{c}

Note: Different lowercase letters indicate extremely significant differences ($p < 0.01$) among the different recombinant proteins.

4. Discussion

Aroma is one of the important quality indicators of grape fruits, and the content of linalool varies among the different aroma types of table grapes [1,2,17]. In wine grapes, the enzyme CYP76F14 catalyzes the production of different terpene compounds from linalool, which are precursor substances that determine the aroma of wine [2,15–17]. The physiological function of CYP76F14 in regulating the formation of aromas in table grape fruits is still unclear.

The highly conserved monooxygenase region of the cytochrome P450 features a general activity characteristic motif A(G)G(A)XD(E)T [22,23]. In this study, the typical AGTDT activity domain and HEME binding domain were detected in the monooxygenase region of CYP76F14 proteins from three different aroma types of table grapes, consistent with previous reports in *Arabidopsis* [22] and yeast [24]. Notably, sequence differences in the CYP76F14 protein were found among different aroma types of table grapes, with multiple amino acid mutation sites detected only in the Muscat type 'Irsai Oliver' CYP76F14 protein. Similarly, amino acid sequence differences were observed among CYP76F14 proteins in different wine grape varieties, closely related to the enzyme activity of their in vitro expressions [16,19]. Moreover, similar reports of amino acid residue mutations in cytochrome P450 subfamily proteins exist in potatoes (CYP77A) [25] and *Salvia miltiorrhiza* (CYP76AK) [26].

In recent years, studies have utilized prokaryotic systems to express plant cytochrome P450 proteins and conduct in vitro enzymatic kinetic characteristic analyses. Such studies have been reported in *Arabidopsis* [9,22,23], potatoes [25], *S. miltiorrhiza* [26], and wine grapes [2,16,19]. In this study, site-directed mutagenesis of two amino acid residues (E378G and T380A) of the table grape CYP76F14 protein resulted in a significant decrease in in vitro CYP76F14 enzyme activity, suggesting that these two amino acid residues may be critical for CYP76F14 enzyme activity and may directly participate in the process of CYP76F14 catalyzing the production of other terpene compounds from the substrate linalool, warranting further ontological functional verification. These results again indicate that certain key amino acid residues in plant CYP76F14 proteins directly determine cytochrome P450 enzyme activity [2,15,18], and their specific biological functions and regulatory mechanisms merit further in-depth study.

In summary, this study provides gene resources for revealing the biological function of CYP76F14 in table grapes and lays a theoretical foundation for exploring the use of CYP76F14 as a fingerprint marker in breeding table grape varieties with different aroma types.

5. Conclusions

There is a significant difference in the content of linalool in the fruits of three different flavor-type table grape varieties. The Muscat type 'Irsai Oliver' exhibited sustained linalool accumulation from Stage 1 (cell division phase) to Stage 5 (mature phase), with concentrations peaking at Stage 3 (veraison phage) and remaining elevated until harvest, surpassing other two cultivars of the Neutral type 'Yanhong' and Berry-like aroma type 'Venus Seedless'. Transcriptional profiling demonstrated that the CYP76F14 expression exhibited a similar trend with the linalool levels showing a higher expression in 'Irsai Oliver' across the developmental stages. A structural analysis identified 12 divergent residues in the 'Irsai Oliver' CYP76F14 variant. The site-directed mutagenesis of two residues (CYP76F14-E378G and CYP76F14-T380A) reduced catalytic efficiency by 68–72% compared to the wild-type. E378 and T380 may be two key amino acid residues determining the enzyme activity of 'Irsai Oliver' CYP76F14.

Author Contributions: Conceptualization, Z.S. and X.L.; methodology, Z.S., J.Z. and D.L.; validation, J.Z., D.L. and M.S.; investigation, J.Z., D.L. and M.S.; data curation, D.L. and M.S.; writing—original draft preparation, Z.S.; writing—review and editing, M.S. and X.L.; project administration, X.L.; funding acquisition, Z.S. and D.L. All authors have read and agreed to the published version of the manuscript.

Funding: This work was jointly supported by the following grants: Key R&D Program of Shandong Province-Innovation Capability Enhancement Project for Technology based Small and Medium sized Enterprises (2024TSGC0718), Major Project of Science and Technology of Shandong Province (2022CXGC010605), and the China Scholarship Council Fund (202208370080).

Institutional Review Board Statement: Not applicable. The study utilized publicly available data from the NCBI database.

Informed Consent Statement: Not applicable.

Data Availability Statement: The original contributions presented in the study are included in the article, further inquiries can be directed to the corresponding author.

Acknowledgments: The authors are grateful to Julia M. Davies, Department of Plant Sciences, University of Cambridge for the critical reading and valuable suggestions.

Conflicts of Interest: The authors declare no conflicts of interest.

References

1. Alegre, Y.; Saenz-Navajas, M.P.; Hernandez-Orte, P.; Ferreira, V. Sensory, olfactometric and chemical characterization of the aroma potential of Garnacha and Tempranillo winemaking grapes. *Food Chem.* **2020**, *331*, 127207. [CrossRef] [PubMed]
2. Song, Z.Z.; Tang, M.L.; Xiao, H.L.; Xu, H.H.; Shi, M.; Dark, A.; Xie, Z.Q.; Peng, B. Unraveling the trisubstrate-triproduct reaction mechanisms of wine grape VvCYP76F14 to improve wine bouquet. *Food Chem.* **2025**, *474*, 143077. [CrossRef] [PubMed]
3. Robinson, A.L.; Boss, P.K.; Solomon, P.S.; Trengove, R.D.; Heymann, H.; Ebeler, S.E. Origins of grape and wine aroma. Part 2. Chemical and sensory analysis. *Am. J. Enol. Vitic.* **2014**, *65*, 25–42. [CrossRef]
4. Thomas-Danguin, T.; Ishii-Foret, A.; Atanasova, B.; Etievant, P. Wine bouquet: The perceptual integration of chemical complexity. In *Wine Active Compounds;* Conference Object; Oenoplurimédia: Beaune, France, 2011.
5. Zhai, H.Y.; Li, S.Y.; Zhao, X.; Lan, Y.B.; Zhang, X.K.; Shi, Y.; Duan, C.Q. The compositional characteristics, influencing factors, effects on wine quality and relevant analytical methods of wine polysaccharides: A review. *Food Chem.* **2023**, *403*, 134467. [CrossRef]
6. Noguerol-Pato, R.; Gonzalez-Barreiro, C.; Cancho-Grande, B.; Santiago, J.L.; Martinez, M.C.; Simal-Gandara, J. Aroma potential of Brancellao grapes from different cluster positions. *Food Chem.* **2012**, *132*, 112–124. [CrossRef]
7. Lukic, I.; Lotti, C.; Vrhovsek, U. Evolution of free and bound volatile aroma compounds and phenols during fermentation of Muscat blanc grape juice with and without skins. *Food Chem.* **2017**, *232*, 25–35. [CrossRef] [PubMed]
8. Wang, X.J.; Tao, Y.S.; Wu, Y.; An, R.Y.; Yue, Z.Y. Aroma compounds and characteristics of noble-rot wines of Chardonnay grapes artificially botrytized in the vineyard. *Food Chem.* **2017**, *226*, 41–50. [CrossRef]
9. Kunert, M.; Langley, C.; Lucier, R.; Ploss, K.; López, C.Z.R.; Guerrero, A.D.S.; Rothe, E.; O'Connor, S.E. Promiscuous CYP87A enzyme activity initiates cardenolide biosynthesis in plants. *Nat. Plants* **2023**, *9*, 1607–1617. [CrossRef]
10. Parker, M.; Capone, D.L.; Francis, I.L.; Herderich, M.J. Aroma Precursors in Grapes and Wine: Flavor Release during Wine Production and Consumption. *J. Agric. Food Chem.* **2017**, *66*, 2281–2286. [CrossRef]
11. Picard, M.; Tempere, S.; de Revel, G.; Marchand, S. A sensory study of the ageing bouquet of red Bordeaux wines: A three-step approach for exploring a complex olfactory concept. *Food Qual. Prefer.* **2015**, *42*, 110–122. [CrossRef]
12. Ghaste, M.; Narduzzi, L.; Carlin, S.; Vrhovsek, U.; Shulaev, V.; Mattivi, F. Chemical composition of volatile aroma metabolites and their glycosylated precursors that can uniquely differentiate individual grape cultivars. *Food Chem.* **2015**, *188*, 309–319. [CrossRef]
13. Lin, J.; Massonnet, M.; Cantu, D. The genetic basis of grape and wine aroma. *Hortic. Res.* **2019**, *6*, 81. [CrossRef] [PubMed]
14. Giaccio, J.; Capone, D.L.; Håkansson, A.E.; Smyth, H.E.; Elsey, G.M.; Sefton, M.A.; Taylor, D.K. The formation of wine lactone from grape-derived secondary metabolites. *J. Agric. Food Chem.* **2011**, *59*, 660–664. [CrossRef] [PubMed]
15. Ilc, T.; Halter, D.; Miesch, L.; Lauvoisard, F.; Kriegshauser, L.; Ilg, A.; Baltenweck, R.; Hugueney, P.; Werck-Reichhart, D.; Duchêne, E.; et al. A grapevine cytochrome P450 generates the precursor of wine lactone, a key odorant in wine. *N. Phytol.* **2017**, *213*, 264–274. [CrossRef]
16. Peng, B.; Ran, J.G.; Li, Y.Y.; Tang, M.L.; Xiao, H.L.; Shi, S.P.; Ning, Y.Z.; Dark, A.; Guan, X.Q.; Song, Z.Z. Site-directed mutagenesis of VvCYP76F14 (cytochrome P450) unveils its potential for selection in wine grape varieties linked to the development of wine bouquet. *J. Agric. Food Chem.* **2024**, *72*, 3683–3694. [CrossRef] [PubMed]
17. Yang, C.X.; Wang, Y.J.; Liang, Z.C.; Fan, P.G.; Wu, B.H.; Yang, L.; Wang, Y.N.; Li, S.H. Volatiles of grape berries evaluated at the germplasm level by headspace-SPME with GC–MS. *Food Chem.* **2009**, *114*, 1106–1114. [CrossRef]
18. Liu, W.; Xiao, H.; Shi, M.; Tang, M.; Song, Z. D299T mutation in CYP76F14 led to a decrease in wine bouquet precursor production in wine grape. *Genes* **2024**, *15*, 1478. [CrossRef]

19. Xia, G.; Shi, M.; Xu, W.; Dark, A.; Song, Z. Cytochrome P450 VvCYP76F14 dominates the production of wine bouquet precursors in wine grapes. *Front. Plant Sci.* **2024**, *15*, 1450251. [CrossRef]
20. Xie, Z.; Peng, B.; Shi, M.; Yang, G.; Song, Z. Table grape ferritin1 is implicated in iron accumulation, iron homeostasis, and plant tolerance to iron toxicity and H_2O_2 induced oxidativestress. *Horticulturae* **2025**, *11*, 146. [CrossRef]
21. Song, Z.Z.; Peng, B.; Gu, Z.X.; Tang, M.L.; Li, B.; Liang, M.X.; Wang, L.M.; Guo, X.T.; Wang, J.P.; Sha, Y.F.; et al. Site-directed mutagenesis identified the key active site residues of alcohol acyltransferase PpAAT1 responsible for aroma biosynthesis in peach fruits. *Hortic. Res.* **2021**, *8*, 32. [CrossRef]
22. Hofer, R.; Boachon, B.; Renault, H.; Gavira, C.; Miesch, L.; Iglesias, J.; Ginglinger, J.F.; Allouche, L.; Miesch, M.; Grec, S.; et al. Dual function of the cytochrome P450 CYP76 family from *Arabidopsis thaliana* in the metabolism of monoterpenols and phenylurea herbicides. *Plant Physiol.* **2014**, *166*, 1149–1161. [CrossRef] [PubMed]
23. Renault, H.; Bassard, J.E.; Hamberger, B.; Werck-Reichhart, D. Cytochrome P450-mediated metabolic engineering: Current progress and future challenges. *Curr. Opin. Plant Biol.* **2014**, *19*, 27–34. [CrossRef] [PubMed]
24. Bathe, U.; Frolov, A.; Porzel, A.; Tissier, A. CYP76 oxidation network of abietane diterpenes in lamiaceae reconstituted in yeast. *J. Agric. Food Chem.* **2019**, *67*, 3437–13450. [CrossRef] [PubMed]
25. Grausem, B.; Widemann, E.; Verdier, G.; Nosbusch, D.; Aubert, Y.; Beisson, F.; Schreiber, L.; Franke, R.; Pinot, F. CYP77A19 and CYP77A20 characterized from Solanum tuberosum oxidize fatty acids in vitro and partially restore the wild phenotype in an Arabidopsis thaliana cutin mutant. *Plant Cell Environ.* **2014**, *37*, 2102–2115. [CrossRef]
26. Li, B.; Li, J.; Chai, Y.; Huang, Y.; Li, L.; Wang, D.; Wang, Z. Targeted mutagenesis of CYP76AK2 and CYP76AK3 in *Salvia miltiorrhiza* reveals their roles in tanshinones biosynthetic pathway. *Int. J. Biol. Macromol.* **2021**, *189*, 455–463. [CrossRef]

horticulturae

MDPI

Article

PavSPL Expression Dynamics in Fruits and Seeds and in Relation to Endocarp Lignification Status During the Transition from Development to Ripening in Sweet Cherry

Matías Zavala [1], Marcela Menares [1], Orlando Acevedo [2], Mirna Melo [1], Carlos Nuñez [3], Camila Arancibia [1], Romina Pedreschi [1,4], José Manuel Donoso [5], Lee A. Meisel [2], Jonathan E. Maldonado [3] and Nathalie Kuhn [1,*]

1 Facultad de Ciencias Agronómicas y de los Alimentos, Escuela de Agronomía, Pontificia Universidad Católica de Valparaíso, Quillota 2340025, Chile
2 Instituto de Nutrición y Tecnología de los Alimentos, Universidad de Chile, Santiago 820808, Chile
3 Facultad de Química y Biología, Departamento de Biología, Universidad de Santiago de Chile, Santiago 9170022, Chile
4 Millennium Institute Center for Genome Regulation (CRG), Santiago 8331150, Chile
5 Centro Regional INIA Rayentué, Instituto de Investigaciones Agropecuarias, Rengo 2940000, Chile
* Correspondence: nathalie.kuhn@pucv.cl

Abstract: The transition to ripening in non-climacteric species is governed by several signals, including hormones that enhance or counteract the abscisic acid (ABA)-promoting effect. The SQUAMOSA Promoter-binding protein-Like (SPL) transcription factors are involved in ripening through the modulation of anthocyanin biosynthesis. In sweet cherry fruits, several miR156-targeted *PavSPLs* are expressed before and during ripening. Recently, some *PavSPLs* were found in the transition from development to ripening in cultivars contrasting in maturity time. Additionally, several forms of miR156 were expressed in sweet cherry seeds of an early-season cultivar. In this work, we addressed the relevance of endocarp lignification and *PavSPLs* expression for the transition to ripening. First, we characterized early- and late-season sweet cherry cultivars, 'Celeste' and 'Regina', focusing on fruit and seed development, endocarp lignification, and *PavSPL* expression profile. Fruit growth dynamics revealed an earlier onset of color development and lignification in 'Celeste', while 'Regina' exhibited a prolonged lag phase and delayed embryo development. Transcript profiling at the light green stage showed a higher expression of *PavSPL* genes in fruits and identified cultivar-specific expressions, especially between 'Regina' and 'Celeste' seeds. Co-expression networks linked *PavSPLs* to genes involved in lignin and anthocyanin biosynthesis. We focused on *PavSPL2* and *PavSPL9*, which were targeted by mtr-miR156a and gma-miR156f. Both *PavSPLs* and miRNAs were expressed in fruits and seeds at the yellow stage, an advanced point in the transition to ripening in sweet cherry. Exogenous application of auxin-related compounds in the mid-season cultivar 'Lapins' modulated endocarp lignification and pigmentation. Notably, *p*-IBA treatment, which enzymatically targets the lignin pathway, transiently increased anthocyanin accumulation and reduced lignin deposition, effects that correlated with the downregulation of *PavSPL* gene expression. These findings highlight the interplay between lignification, color evolution, and pigment biosynthesis during the transition from development to ripening in sweet cherry fruits, and suggest a role for *PavSPL* genes in this transition.

Keywords: anthocyanins; endocarp; gene expression; lignin; maturity; miRNA

1. Introduction

Sweet cherry (*Prunus avium* L.) is a profitable crop whose production worldwide increased from 1.9 to 2.96 million metric tons between 2000 and 2023 [1]. This species is classified as a non-climacteric fruit, as it does not present significant ethylene production or a respiration burst at the onset of ripening [2]. Since the sweet cherry genome has been sequenced and annotated, this species is a good model for unveiling this type of ripening [3]. Sweet cherry depends on abscisic acid (ABA) for the triggering of ripening [4–7], which is similar to other non-climacteric species like strawberry and grapevine [8,9]. Other hormones modulate ripening in non-climacteric species. For instance, auxin exerts a negative ripening control in grapevine, whose content is higher in fruits with a high seed content [10]. However, Clayton-Cuch et al. reported that auxin can promote fruit ripening in sweet cherry [11]. This is supported by the finding that indole-3-acetic acid (IAA) treatment increased ABA concentration in sweet cherry fruits [7]. Hence, auxin in conjunction with ABA can positively modulate the ripening process in this species. On the other hand, cytokinins (CKs) increase fruit size and delay the fruit coloring process [12], and some CK pathway genes are regulated at the epigenomic level by ABA during ripening in sweet cherry [13]. Finally, gibberellin (GA) is a negative ripening regulator in sweet cherry, since the application of gibberellic acid (GA_3) delayed ripening [14–16], suggesting a role in the control of maturity time. It is worth noting that GA_3 treatment is a common agronomical practice, usually applied during the light green or straw yellow stages. This suggests that this moment is key for determining ripening initiation. In addition, cultivars with differences in maturity time have different lengths for these stages, as well as differences in hormone content and secondary metabolic profile [17]. Therefore, the evidence suggests that some signals occurring before the pink-stage-related ABA increase could modulate ripening initiation.

The growth of sweet cherry fruit is characterized by a double sigmoid curve, including three key phases. During stage I, intense cell division takes place; then a growth arrest occurs at stage II (also known as lag phase), where embryo development and pit hardening due to endocarp lignification occurs [18]; finally, during stage III a second exponential growth takes place [19]. At stage III, fruit ripening initiates, including the coloring process associated with changes in the expression of ABA and anthocyanin biosynthesis genes in the fruit skin [20]. However, other processes such as de-greening due to chlorophyll degradation occur some weeks earlier [21]. The lignification–ripening connection comes from the idea that a fully lignified endocarp indicates that the fruit is ripe and ready to disperse [22]. Thus, different signals are expected to coordinate the endocarp and mesocarp processes. In this regard, plant hormones control lignification, as shown in model organisms. For example, in *Arabidopsis*, auxin has been described as regulating lignin biosynthesis [23]. Stone cell development in pear (*Pyrus pyrifolia*) fruits is also affected by auxin [24].

Early-, mid-, and late-season sweet cherry cultivars differ in stage II duration, where lignification occurs [15,17,25]. As stage II is shorter in early-season cultivars, the embryo cannot reach its full size, since its development coincides with fruit ripening and senescence [25]. Thus, variations in the harvest time between cultivars could be associated with differences in the degree of embryo development. This is relevant considering that seeds have at least a one order of magnitude higher hormone content than the mesocarp, including auxin and GA_3 [26]; hence, differences in seed development between cultivars may involve differences in hormone production, including auxin and GA, which in turn are key for controlling ripening initiation, as mentioned above. On the other hand, the endocarp lignification occurring at stage II could be a critical period, since profuse remodeling at the transcriptomic level occurs in fruits and seeds pre- versus post-lignification, including genes encoding transcription factors with hormone-specific responsiveness, as shown in *Prunus persica* [27].

Transcription factors are key in the regulation of plant processes. The SQUAMOSA Promoter-binding protein-Like (*SPL*) genes are part of a family of transcription factors acting as direct upstream regulators [28]. *SPLs* are regulated at the post-transcriptional level by miR156s [29], which repress their expression, resulting in the regulation of processes such as floral induction, shoot branching, and pigment biosynthesis [30–32]. *SPL* transcripts are part of a regulatory module, *SPL*/miR156, where their expression in a given process/tissue is controlled by specific miRNA156. Regulatory small RNA miR156 is part of a very conserved miRNA family, consisting of 20 nucleotides [33]. Overexpressing miR156 resulted in a prolonged juvenile stage and delayed flowering [34,35], showing a key role in regulating plant developmental transitions.

In fruit trees, SPLs are key in phase transitions such as the vegetative-to-adult reproductive phase, as Song et al. reported in pear (*Pyrus* spp.) [36]. There are 16 *PavSPL* encoding sequences in sweet cherry [29], with 12 of them predicted as targets of miR156 [37]. These *PavSPLs* are involved in flowering time and abiotic stress response. In the case of miR156 sequences, they were present during bud dormancy in sweet cherry [38].

The role of the *SPL*/miR156 module in the transition from development to ripening has recently gained attention. Functional analysis in *Pyrus pyrifolia* showed that this module is involved in anthocyanin accumulation during light-induced red peel coloration [39]. In that work, the abundance of several miR156s increased upon light-induced anthocyanin pathway activation, whereas *PpySPL* expression decreased, suggesting that SPLs are negative regulators of the coloring process and miRNA156 are positive regulators. This in line with findings showing that the expression of miR156 increases during fruit ripening in *Solanum lycopersicum* [40]. In blueberry (*Vaccinium corymbosum*), lines overexpressing the gene VcMIR156a increased anthocyanin accumulation, supporting their pro-ripening role, whereas *VcSPL12* positively regulated chlorophyll accumulation, since it physically binds to the promoter of the chlorophyll biosynthetic gene *VcDVR* [41].

PavSPL genes and several miR156 sequences are present at color initiation in sweet cherry [42]. Their expression decreases in fruits during ripening, with differences starting from the de-greening to yellow stages [43], suggesting a role in the transition from development to ripening. *PavSPL* expression is affected by hormonal treatments. For instance, ABA decreased the transcript abundance of several *PavSPLs*, whereas GA—a ripening-antagonizing hormone—increased *PavSPL* gene expression [43], supporting the idea that they are negative ripening regulators. In a recent report, *PavSPL* genes were found to be expressed in fruits and seeds at the light green stage [44], where the seeds of the early-season cultivar 'Celeste' were highly abundant in several miR156 variants, consistent with the pro-ripening role of these miRNAs. Maldonado et al. showed several fruit–seed associations, suggesting that seeds may modulate the transition from development to ripening in sweet cherry [44].

As several *PavSPLs* have been found to have high expression levels in fruits, whereas miR156s have high abundance in seeds [44], here we hypothesize that the sweet cherry *PSPL*/miR156 module could have a role in inter-tissue post-transcriptional regulation. Long-distance movement of miRNAs in plant tissues has previously been discussed [45]. We argue that this could be addressed, at least from an exploratory perspective, by disturbing the lignification of the endocarp, which is located between the seed and fruit mesocarp tissues. This is supported by the fact that endocarp lignification is related to the ripening process, as mentioned above. Therefore, the aim of this work was to characterize the expression profile of sweet cherry miR156s and *PavSPLs* in seeds and fruits, and then to analyze the expression profile of selected *PavSPL* genes in response to treatments affecting lignification during the transition from development to ripening in sweet cherry.

To address this, we first analyzed the gene expression of sweet cherry *PavSPLs* in the fruits and seeds of early- and late-season cultivars. Then, we explored the effect of lignin modulators on fruit endocarp lignification and on ripening markers (chlorophyll, carotenoid, and anthocyanin contents) over two seasons in field conditions. The modulators included auxin-related compounds and a compound targeting the activity of a key lignification enzyme. Finally, the effect of these modulators on *PavSPL* gene expression was evaluated.

To our knowledge, the *SPL* response to lignin modulators at the level of gene expression has not been assessed in stone fruits, nor the physiological effect of disturbing lignification, specifically in relation to ripening markers in non-climacteric species at the transition from development to ripening.

2. Methods

2.1. Plant Material

2.1.1. Plant Material for Early- and Late-Season Cultivar Characterization

Sweet cherry (*Prunus avium* L.) fruits selected from trees growing in the Experimental Station INIA Los Tilos, located in Buin, Región Metropolitana, Chile (33°42′ S, 70°42′ W) were used during the 2022–2023 season. Two cultivars contrasting in maturity time were selected, the late-season 'Regina', and the early-season 'Celeste' (denomination Sumpaca®, 13S.24.28; licensed to McGrath Nurseries Ltd. in Hamilton, New Zealand). The ten-year-old trees were grafted on 'Cap-6P' rootstock. Plants were arranged in a 4 × 2 frame, with regular phytosanitary and fertirrigation management, and with hydrogen cyanamide treatment to break dormancy. Gibberellic acid was not applied in the orchard during the 2022–2023 season.

2.1.2. Plant Material for Mid-Season Cultivar Characterization

Sweet cherry fruits from trees established in 2006 and 2011 in the Experimental Station La Palma, Valparaíso, Chile (32°52′ S, 71°12′ W) were used in the 2023–2024 and 2024–2025 seasons, respectively. 'Lapins', a mid-season cultivar grafted on 'Colt' rootstock was selected. The plant material in Sections 2.1.1 and 2.1.2 corresponds to registered varieties and it was commercially obtained. The trees were in a 4 × 2.5 array, with similar productive management as mentioned in Section 2.1.1, except that the trees were covered with plastic bags when gibberellic acid (GA_3) was applied, to avoid interactions with the treatments we applied to the plant material.

2.2. Experimental Design

2.2.1. Experimental Design for Early- and Late-Season Cultivar Characterization in the 2022–2023 Season

Three plants were randomly selected from 'Celeste' and 'Regina' cultivars, which were part of the *arboretum* of INIA Los Tilos. We considered each tree as a biological replicate; thus, three replicates (n = 3) were utilized. Sampling for qPCR analysis was at the straw yellow (SY) stage, which was 40 and 41 days after full bloom (DAFB) in 'Celeste' and 'Regina', respectively, with 0 DAFB when 50% flowers were open. For the qPCR analyses, eight representative fruits without apparent defects were sampled from each tree, separated into fruit (mesocarp) and seed tissues, and immediately frozen in liquid nitrogen until processing. Once in the laboratory, fruits and seeds were ground separately using liquid nitrogen. The endocarp tissue was carefully separated with tweezers and discarded. Ground samples were stored at −80 °C until the RNA extractions. In this work, 'fruits' corresponds to mesocarp-enriched tissues, whereas 'seeds' includes embryo and its surrounding tissues (endosperm and seed coat). Both, 'seed' and 'fruit' samples, were used for qPCR analysis. The fruit equatorial diameter of the same trees as used for sampling was measured during the season using 20 fruits per tree; lignification was evaluated at 21, 28, 35, and 42 DAFB in eight fruits per tree; and seed and embryo lengths were evaluated at 21, 28, 35, 42, and 49 DAFB by sampling three fruits per tree. The ripening parameter soluble solids content (SSC) was evaluated at the red (R) stage in five fruits per tree at 59 and 60 DAFB in 'Celeste' and 'Regina', respectively.

2.2.2. Experimental Design for the Treatments with Lignification Modulators in the 2023–2024 Season in the Mid-Season Cultivar 'Lapins'

Four treatments with lignification modulators were performed in a randomized complete block design. For this, four trees of the 'Lapins' cultivar were randomly selected in a row, with the four blocks located in different rows, where the tree was the experimental unit. As four trees were selected (one of each row), four replicates per treatment (n = 4) were utilized. The treatments were NPA (naphtylphtalamic acid; Sigma-Aldrich, St. Louis, MO, USA), IAA (indol-3-acetic acid; Sigma-Aldrich, St. Louis, MO, USA), and *p*-IBA (*p*-iodobenzoic acid, Thermo Fisher Scientific, Waltham, MA, USA). All the treatments were at 40 μM and the control treatment corresponded to untreated plants. The treatments were applied at 25 DAFB, (T0), corresponding to 20% light green (LG)/80% green (G). Eight representative fruits without visible defects were collected from each replicate and frozen in liquid nitrogen. Grinding at the laboratory was carried out with a constant supply of liquid nitrogen, where every sample consisted of entire fruits (including endocarps and seeds). The samples for pigment content analyses correspond to 28 (T3), 46 (T21), 56 (T31), and 79 (T54) DAFB, with T3 corresponding to 50% LG/50% G; T21 to 90% LG/10% SY; T31 to 90% SY/10% pink (P); and T54 to 70% R/30% P. For qPCR analyses, only samples of the control and *p*-IBA treatments at T21 were selected. The samples were kept at −80 °C until they were used for qPCR and pigment content analyses. Sampling for the effect of the treatments on endocarp lignification was at 29 DAFB (T4). Eight fruits were sampled and fixed in FAA solution (10/50/5, formaldehyde (37%)/absolute ethanol/anhydrous acetic acid). The equatorial diameter of the fruit was recorded with 15 fruits per tree during the entire season. Ripening parameters, firmness, and SSC were evaluated at 71 DAFB in ten fruits per tree.

2.2.3. Experimental Design for the Treatments with Lignification Modulators in the 2024–2025 Season in the Mid-Season Cultivar 'Lapins'

Five treatments with lignification modulators were performed in a randomized complete block design with three blocks. For this, five trees of the 'Lapins' cultivar were randomly selected per row, where each tree was the experimental unit. As three trees were selected (one for each row), three replicates per treatment (n = 3) were utilized. The treatments were *p*-IBA (Sigma-Aldrich, St. Louis, MO, USA) at 40 μM and 60 μM, and NAA (1-naphthaleneacetic acid; Sigma-Aldrich, St. Louis, MO, USA) at 40 μM and 100 μM. The water solutions were 1% Tween 20 (Thermo Fisher Scientific, Waltham, MA, USA). The control treatment corresponded to untreated plants. The treatments were applied at 25 DAFB (T0, where the tree phenology was 100% LG) through spraying until run-off. Eight representative fruits without visual defects were sampled for pigment analysis and immediately frozen in liquid nitrogen, another eight representative and healthy fruits were sampled and fixed in FAA solution (10/50/5, formaldehyde (37%)/absolute ethanol/anhydrous acetic acid). For endocarp lignification assessment and pigment analysis, the sample dates were 25 DAFB (before the treatments), 35 DAFB (T10, 90% LG/10% SY), and 50 DAFB

(T25, 70% SY/30% P). At 68 DAFB, the fruit equatorial diameter of five fruits per tree was recorded.

2.3. Growth and Ripening Parameters Assessment

Non-destructive equatorial diameter measurement was performed in fruits of 'Celeste', 'Regina', and 'Lapins' with a caliper, using millimeters (mm) as the unit. SSC and firmness were recorded using a Pocket Brix-Acidity Meter (PAL-BX | ACID3, ATAGO Co., Ltd., Tokyo, Japan) and a durometer device (Durofel T.R. Turoni, Forlì, Italy), respectively. The durometer was placed on two opposite cheeks, according to San Martino et al. [46]. From these two quantifications, the average was calculated. Seed and embryo length were measured by longitudinally cutting the fruits and using mm as units. Then, the Q index was calculated by dividing the length of embryo by the length of seed [47]. Finally, to determine the color distribution of the fruits at harvest in the 2023–2024 season, a color chart (CTIFL, Paris, France) was used, where one is light red, two is red, three is red, and four is light mahogany.

2.4. RNA-Seq Expression Levels of *PavSPL* Genes and Gene Co-Expression Network Analysis

RNA-seq data from sweet cherry cultivars 'Regina' and 'Celeste' were used to determine *PavSPL* expression levels in fruits and seeds [44]. Samples were taken at the LG stage (34 DAFB) from the same trees as those indicated in Section 2.1.1. For the sequence processing, mapping, and RNAseq procedure visit the mentioned publication. For heatmap and clustering visualization of *PavSPL* gene expression, Morpheus software of the Broad Institute (https://software.broadinstitute.org/morpheus (accessed on 5 March 2025)) was utilized. Samples and genes were clustered by the 'One minus Pearson correlation' and average linkage method. Using the gene ontology categories of *Prunus avium* predicted genes ('Tieton' cv.; [48]), subgroups of genes related to flavonoid, anthocyanin, and lignin metabolism were selected. Then, their gene expression was extracted from [44] datasets (fruits and seeds of 'Celeste' and 'Regina'). The gene expression of the selected gene datasets, as well as *PavSPL* gene expression, were used as input to the GENIE3 algorithm v.1.31.0 [49] using the R environment v.4.4.2 [50], for the inference of putative gene co-expression/interactions. The GENIE3 tool was run with standard parameters using 1000 decision trees, Random Forest as the tree-based method, and seeds of 123 for reproducibility. The output scores were used to create subnetworks based on the top 10% scores of *PavSPL*–gene pairs. Cytoscape v.3.10.21 was utilized to visualize the co-expression network maps [51].

2.5. RNA Extraction

For relative expression analyses using qPCR, total RNA was isolated from 0.5 g of mesocarp and seed tissue of 'Celeste' and 'Regina', using a Zymo extraction kit (Zymo Research, Irvine, CA, USA). In the case of 'Lapins' samples RNA was isolated from 0.5 g of whole fruit tissue using the CTAB method [52]. The integrity was evaluated by denaturant electrophoresis with MOPS buffer, using an agarose gel (1.5%) in order to detect two ribosomal bands (28S and 14S). The genomic DNA traces were eliminated using TURBOTM DNase (Thermo Fisher Scientific, Waltham, MA, USA) according to the manufacturer's instructions. For all extractions, the purity values of the ratios A260/A230 and A260/A280 were close to 2.0.

2.6. cDNA Synthesis

For cDNA synthesis, reverse transcription was performed with 0.8 µg of RNA using a BIO-RAD iScript cDNA Synthesis kit (Bio-Rad Laboratories, Inc., Irvine, CA, USA), according to the manufacturer's instructions. For microRNA (miRNA), reverse transcription was performed with 1 μg of RNA using a Mir-X miRNA First-Strand Synthesis Kit (Takara Bio, Kusatsu, Japan).

2.7. Selection of Candidate Genes for qPCR Expression Analyses

To select candidate genes, small RNA-seq and mRNA-seq datasets of fruits and seeds of 'Celeste' and 'Regina' were utilized [44]. From small RNA-seq data, [44] obtained a list of 25 miRNAs functionally annotated using miRbase [53]. Then, these 25 miRNAs were used to search in the online tool TarDB (http://www.biosequencing.cn/TarDB/ (accessed on 25 June 2024)). This database delivered the ID of 17 CDSs. Using a BLASTp tool on NCBI (https://blast.ncbi.nlm.nih.gov/Blast.cgi, (accessed on 25 June 2024)), 15 sweet cherry ortholog genes were obtained. The *Prunus avium* cv. 'Tieton' Genome v2.0 assembly and annotation was used to identify the putative function of these orthologs [48]. We selected three sweet cherry *SQUAMOSA Promoter-binding protein-Like* (*PavSPL*) genes from the list of orthologs delivered by TarDB search [44], FUN_003506-T1 (*PavSPL16*), FUN_003521-T1 (*PavSPL2*), and FUN_037655-T1 (*PavSPL9*). These genes are included in the list of sweet cherry *SPLs* (Table S4). The three genes were orthologs (lowest E value) of the output sequences retrieved by TarDB (lowest *p* value) using as query miRNAs sequences zma-miR156j, mtr-miR156a, and gma-miR156f, respectively (here designated as miR156j, miR156a, and miR156f). Then, we selected only the *PavSPLs* with one allele in the cultivar 'Tieton' Genome v2.0 assembly; thus, *PavSPL16* was excluded from the qPCR analyses. The psRNATarget tool (https://www.zhaolab.org/psRNATarget/ (accessed on 25 June 2024)) was used to predict the miR156-targeted *PavSPL* genes [54], showing that *PavSPL2* and *PavSPL9* were predicted targets of miR156a and miR156f, whereas miR156j was not targeting these genes; thus, miR156j was excluded from the expression analyses.

2.8. Phylogenetic Tree and Motif Discovery

BLASTx on NCBI was performed with *PavSPL2* and *PavSPL9* CDSs against *Prunus* spp. The best six hits sorted by E value and annotated as SPL were used for alignment. The nucleotide sequences were aligned with the online software ClustalW v.2.0.12, included on the website https://www.phylogeny.fr/ (accessed on 15 January 2025). We added the best hit against the *Vitis vinifera* genome in this alignment. MEGA software v.12 was utilized to construct the phylogenetic tree [55]. A neighbor-joining tree was generated using the bootstrap method for testing phylogeny, with 1000 replications, of which 977 were valid and 23 failed. For motif discovery, amino acid sequences were used as input on the online tool Motif Search (https://www.genome.jp/tools-bin/search_motif_lib (accessed on 15 January 2025)), selecting the Pfam database.

2.9. Primer Design for Gene Expression Analyses

To quantify mRNA abundance, specific qPCR primers were designed using the NCBI primer BLAST tool (https://www.ncbi.nlm.nih.gov/tools/primer-blast/ (accessed on 25 July 2024)), as shown in Table S1. The site of primer alignment is shown in Figure S1. For miRNA abundance, primers were designed from the miRbase sequence database (https://www.mirbase.org/ (accessed on 25 July 2024)) [53], using the entire sequence of the mature miRNA (Table S2). Sweet cherry *Actin V1* (XM_021976055.1) gene [7] was selected for the normalization of gene expression (Table S3), following the recommendations for the use of reference genes in *Prunus* spp. species [56]. Primers were evaluated using the online OligoAnalyzer tool from Integrated DNA Technologies (https://www.idtdna.com (accessed on 25 July 2024)). Every primer was tested for specificity using a dissociation curve analysis. Primer efficiency was determined using the LinRegPCR software v11.0 [57]. Primer efficiency was considered for relative transcript abundance calculations, as indicated by Pfaffl [58].

2.10. Quantification of Relative Gene Expression by qPCR

The qPCR analyses were performed on a QIAGEN Rotor-Gene Q system (QIAGEN, Hilden, Germany), using the conditions recommended in "Minimum information established for qRT-PCR experiments" (MIQE) [59] and "Golden Rules of Quantitative PCR" [60]. Takyon SYBR Green qPCR Master Mix (Eurogentec, Seraing, Belgium) was used, according to the manufacturer's indications. For miRNA, TB Green Advantage® qPCR Premix (Takara, Japan) was used, according to the manufacturer's instructions. For the mRNA and miRNA abundance reactions, 10 and 20 ng of cDNA were used, respectively. Three technical replicates were performed for each biological replicate.

2.11. Endocarp Lignification Assessment

Samples fixed in FAA solution and stored at 4 °C were washed three times with phosphate-buffered saline (PBS) solution (pH 7.0). Then, the fruits were longitudinally cut in half and immersed in a solution with one volume of concentrated HCl (37%) and two volumes of 3% phloroglucinol with absolute ethanol (phloroglucinol-HCl or Wiesner staining). After 90 s, the samples were washed with 70% ethanol [61]. For visualization, a 7×–45× stereo Binocular Microscope (AmScope, Irvine, CA, USA) was utilized.

2.12. Pigment Concentration Estimation

For pigment extraction, 30 mg of ground frozen tissue was mixed with 1.0 mL of 95% ethanol. The samples were incubated on an orbital shaker for 1 h, followed by overnight incubation at 4 °C. Subsequently, the samples were centrifuged at 16,000× *g* for 10 min to recover the supernatant. Chlorophyll (a and b) and carotenoid concentrations were quantified using a microplate spectrophotometer (Thermo Fisher Scientific, USA) by measuring the absorbance at 470, 648, 664, and 750 nm wavelengths. Total chlorophyll and carotenoid contents were calculated according to Lichtenthaler [62] and expressed as µg/g dry weight. Anthocyanin content was determined using the pH differential method described by Giusti and Wrolstad [63]. This method involved absorbance measurements of samples diluted in buffer solution at pH 1.0 and 4.5. Absorbance was measured at 520, 700, 900, and 975 nm. The content of anthocyanins was calculated following the equation described by Dyankova and Doneva [64], and contents were expressed as µg/g dry weight of cyanidin 3-*O*-glucoside.

2.13. Statistical Analysis

Data were subjected to a Shapiro–Wilk test to check the normality. The homogeneity of variances was evaluated with Levene's test. For relative expression comparisons, Student's *t*-test was used, with a significance level set at $p < 0.05$. SSC data comparisons between 'Celeste' and 'Regina' were analyzed using Student's *t*-test ($p < 0.05$). One-way ANOVA followed by Tukey's post hoc analysis ($p < 0.05$) was used to compare treatments performed on the 'Lapins' cultivar in both seasons. When indicated, Tukey's post hoc test was used with a significance level of $p < 0.05$. INFOSTAT software, 2020 version obtained from https://www.infostat.com.ar/ (accessed on 25 July 2024), was used for statistical analyses (InfoStat, Córdoba, Argentina) and GraphPad Prism (GraphPad Software, Boston, MA, USA) software v.8.0.2 was used for visualizing the results.

3. Results

The early- and late-season cultivars 'Celeste' and 'Regina' were characterized during the 2022–2023 season. The fruit growth curves from 21 to 63 DAFB revealed that 'Regina' had a more pronounced lag phase, especially from 34 to 42 DAFB, while the pink coloration started earlier in the early-season cultivar 'Celeste' (Figure 1A). To determine differences between both cultivars regarding stone formation, lignin staining of the endocarp—the fruit tissue between mesocarp and seed—was performed. Differences were observed at 21 DAFB, with the lignification process of 'Celeste' starting earlier; additionally, the lignin staining was more marked in the early-season cultivar (Figure 1B). The ripening marker SSC was measured at the R stage, where 'Regina' had significantly less Brix° than 'Celeste' (Figure 1C).

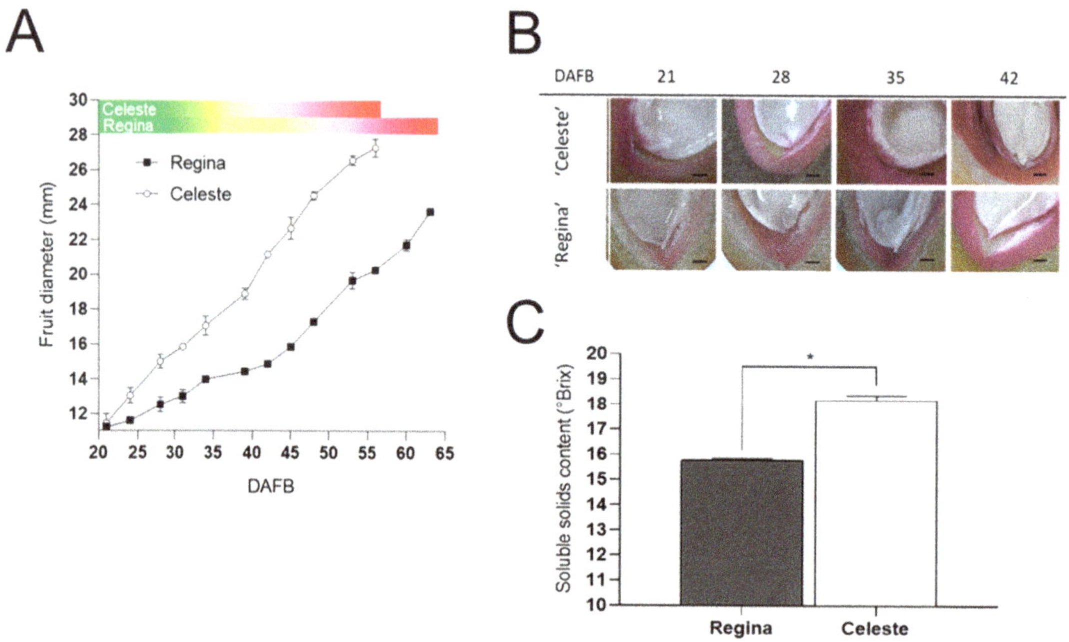

Figure 1. Fruit growth, endocarp lignification, and soluble solids content (SSC) in the early-season cv. 'Celeste' and the late-season cv. 'Regina'. (**A**) Equatorial diameter on different days after full bloom (DAFB). Phenology is presented as colored bars. (**B**) Endocarp lignin deposition as revealed by phloroglucinol-HCl staining. The black bars represent 1 mm. (**C**) SSC at the red (R) stage in 'Celeste' and 'Regina'. In A and C, data are presented as the mean ± standard deviation (SD). In (**C**), the asterisk indicates statistical differences, as revealed by Student's *T*-test with $p < 0.05$.

Regarding the seeds, at 28 DAFB, the embryo was observed, but only in the late-season cultivar 'Regina' (Figure 2A). At 42 DAFB, the embryo of 'Regina' was larger; finally, at 49 DAFB, 'Regina' had full embryo development, as the space inside the seed was completely occupied. The seed length was measured between 21 and 49 DAFB, with 'Regina' having bigger seeds throughout the season (Figure 2B). Finally, the Q index was assessed (Figure 2C), evaluating the relation between embryo length and seed length, where 'Regina' had higher values at the beginning of the season (Figure 2A).

To determine differences between the cultivars at the molecular level, we used the expression levels of *PavSPLs* in the sweet cherry fruits and seeds at the transition from development to ripening of the early cultivar 'Celeste' and the late cultivar 'Regina', obtained in a previous transcriptome analysis performed at the LG stage [44]. First, we identified the *PavSPLs* already characterized in earlier works (Table S4) and searched for their expression level in the datasets. We observed clear differences between seeds and fruits, with *PavSPLs* being in general more abundant in fruits (Figure 3). Nevertheless, some *PavSPLs* showed differences between cultivars, including both *PavSPL16* (FUN_003506-T1 and T2), *PavSPL17* (FUN_021909-T1), and *PavSPL9* (FUN_021909-T2; an allelic variant of *PavSPL17* in the cv. 'Tieton' Genome v2.0 annotation), especially in seed tissues (CS *versus* RS).

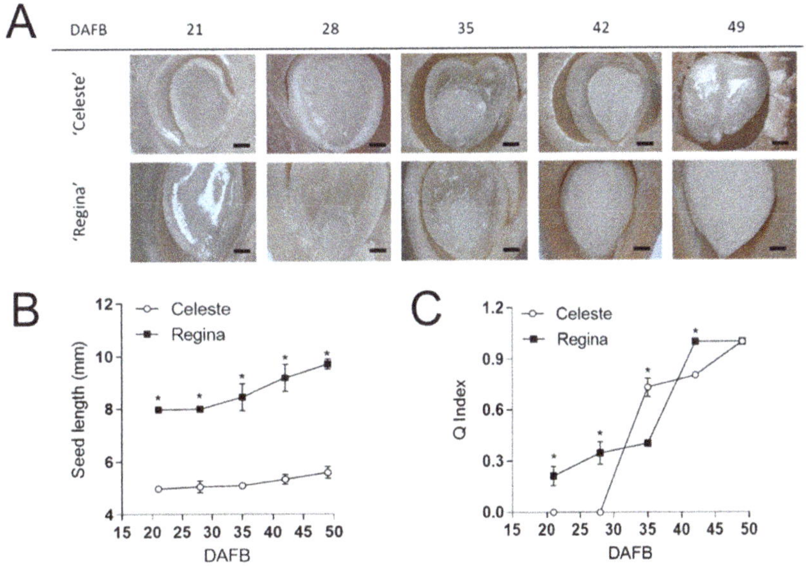

Figure 2. Seed structure evolution in the early-season cv. 'Celeste' and the late-season cv. 'Regina'. (**A**) Pictures showing seeds at different DAFB. The black bars represent 1 mm. (**B**) Quantification of seed length during development. (**C**) Q index estimation at different DAFB, corresponding to the ratio between embryo and seed lengths. The results in (**B**,**C**) are presented as mean ± SD. Zero values represent non-visible embryo; the asterisk indicates statistical differences at each date, as revealed by Student's *T*-test with $p < 0.05$.

The co-expression network shows connections with several transcripts found in the seed and fruit datasets of both cultivars. The transcripts included flavonoid, anthocyanin, and lignin metabolism genes that connected with *PavSPLs* (Figure 3B). *PavSPL8* and *PavSPL2* were co-expressed with several lignin related genes, whereas *PavSPL9*, *PavSPL17* (and its related allelic variant), *PavSPL12*, *PavSPL3*, and *PavSPL1* were connected with anthocyanin genes.

Based on the TarDB search using sequences of miR156 identified in fruit and seed samples [44], we obtained two candidates for expression analysis, *PavSPL2* (FUN_003521-T1) and *PavSPL9* (FUN_037655-T1). We analyzed in silico the predicted site for miR156 binding, and found that both *PavSPLs* were targeted for cleavage by miR156a and miR156f (Table 1). *PavSPL2* and *PavSPL9* were used for expression analyses. The sequence FUN_021909-T2 (also annotated as *PavSPL9*; Table S4) was not analyzed, as it corresponded to *PavSPL17* allelic variant.

Table 1. *PavSPL* sequences complementary to miR156a and miR156f.

miRNA	Target	miRNA Aligned Fragment	Alignment	Target-Aligned Fragment	Expectation	Target Start	Target End	Inhibition
miR156f	*PavSPL9*	UUGACAGAAGAGAGAGAGCACA	:::::::::::::::::::::	UGUGCUCUCUCUCUUCUGUCAA	0.0	784	805	Cleavage
miR156f	*PavSPL2*	UUGACAGAAGAGAGAGAGCACA	:::::::::::::::::::::	CGUGCUCUCUCUCUUCUGUCAA	0.0	1150	1171	Cleavage
miR156a	*PavSPL9*	UGACAGAAGAGAGAGAGCACA	::::::::::::::::::::	UGUGCUCUCUCUCUUCUGUCA	0.0	784	804	Cleavage
miR156a	*PavSPL2*	UGACAGAAGAGAGAGAGCACA	::::::::::::::::::::	CGUGCUCUCUCUCUUCUGUCA	0.0	1150	1170	Cleavage

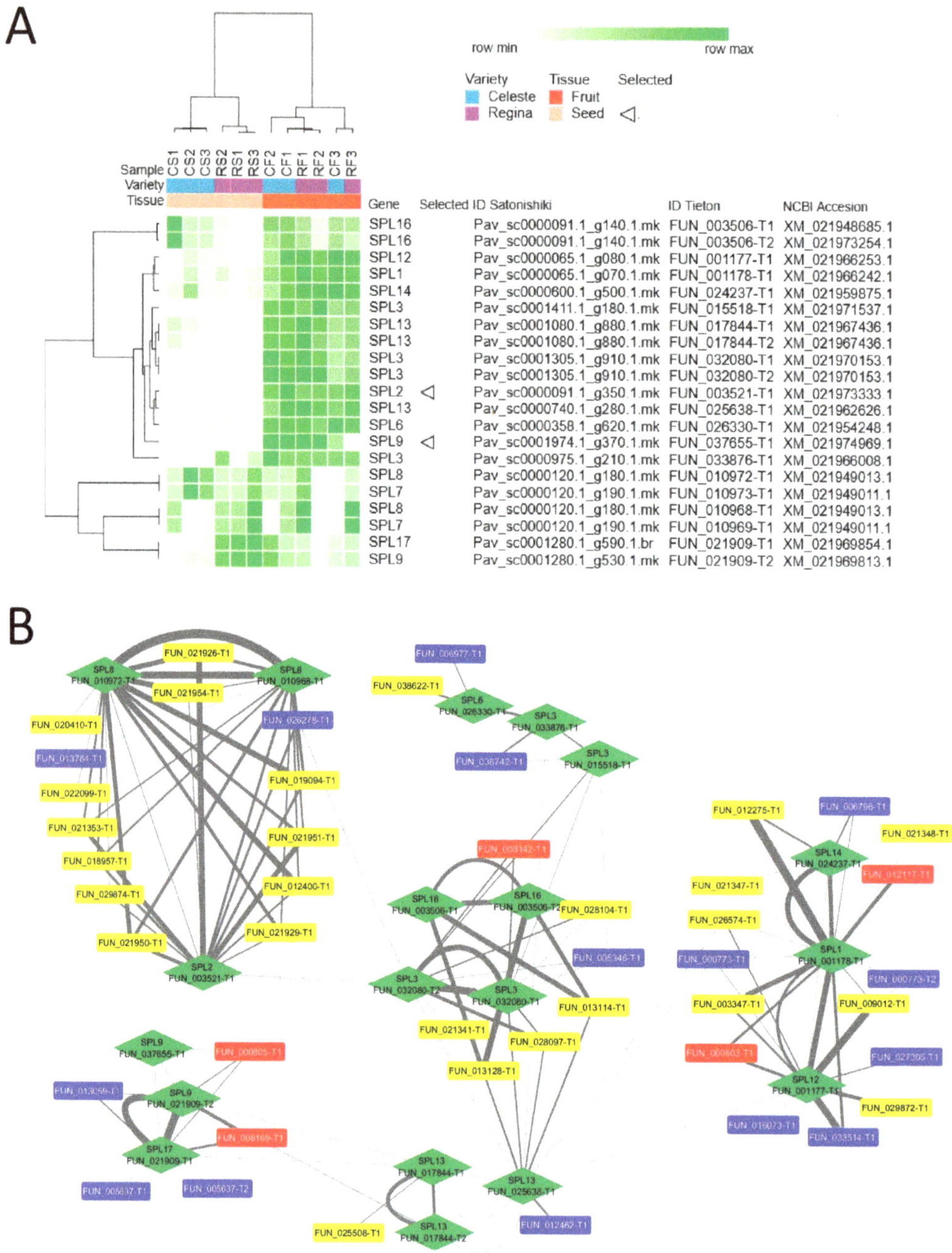

Figure 3. Expression levels of *PavSPL* genes in seeds and fruits at the light green (LG) stage and co-expression of *PavSPLs* with secondary metabolism genes. (**A**) Heatmap for *PavSPL* expression in seeds and fruits of the cvs. 'Celeste' and 'Regina'. For normalization, the trimmed mean of M-values were used. Hierarchical clustering was performed using one minus Pearson correlation with average, and clustering genes and samples. Names of the genes are related to NCBI accessions (Table S4). (**B**) Inferred co-expression/interaction networks of certain *PavSPL* genes (green) with flavonoid (blue), anthocyanin (red), and lignin (yellow) metabolism genes. Triangles indicate genes utilized in subsequent analysis.

To characterize *PavSPL9* and *PavSPL2*, we assessed the presence of the SBP motif, which was found in both genes (Figure 4A). Then, we performed a phylogeny analysis and obtained that *PavSPL9* and *PavSPL2* grouped in different clades, having phylogenetic proximity with other *Prunus* spp. and *Vitis vinifera SPL* genes (Figure 4B).

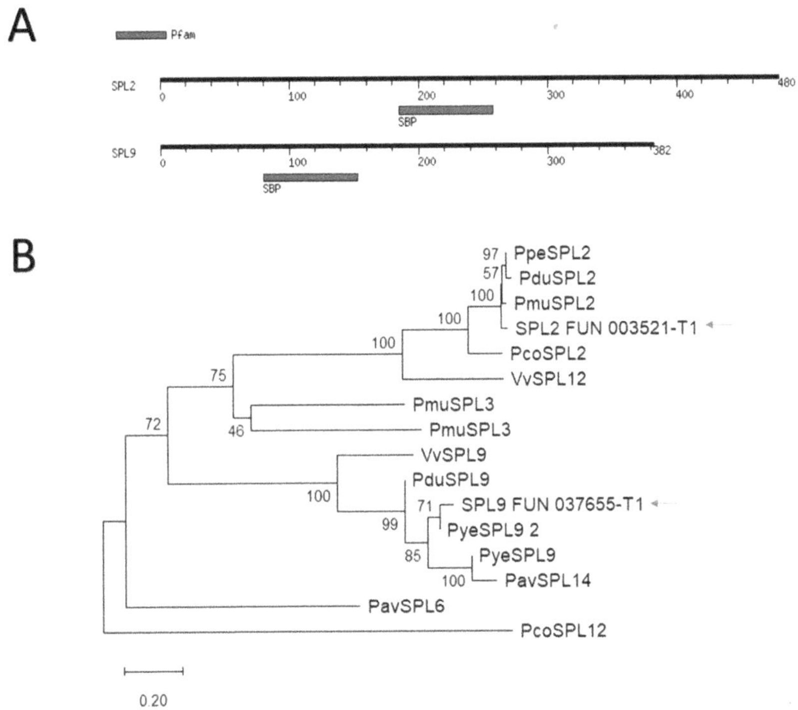

Figure 4. *PavSPL2* and *PavSPL9* phylogeny and identification of the SBP motif. (**A**) SBP (for SQUAMOSA-PROMOTER BINDING PROTEIN) domain identified in the motif search performed using amino acidic sequences (blue colored boxes). (**B**) A bootstrap consensus neighbor-joining tree with 977 bootstrap replicates was generated. Blue arrows indicate genes utilized in subsequent gene expression analysis. Abbreviations of species are according to the Standard Nomenclature for Gene Designation in Rosaceae species [65]; Vv for *Vitis vinifera*.

We analyzed the *PavSPL2* and *PavSPL9* gene expression and miR156a and miR156f abundance in fruits and seeds at the SY stage of each cultivar. SY was chosen since we aimed to determine if they were expressed at a more advanced developmental point in the transition from development to ripening. Both *PavSPLs* were detected in all fruit samples (Figure 5A,B). In contrast, *PavSPL2* was not detected in the seed tissues of the late-season cultivar 'Regina'. Then, we analyzed the profile of miR156a and miR156f in the same samples (Figure 5C,D) and found that miR156a and miR156f were expressed in all the fruit and seed samples.

We evaluated the effect of lignification modulators in a mid-season cultivar during the 2023–2024 season. The IAA, NPA, and *p*-IBA treatments were performed at the initiation of the LG stage (25 DAFB). We measured color, pigments, and lignin staining (Figure 6).

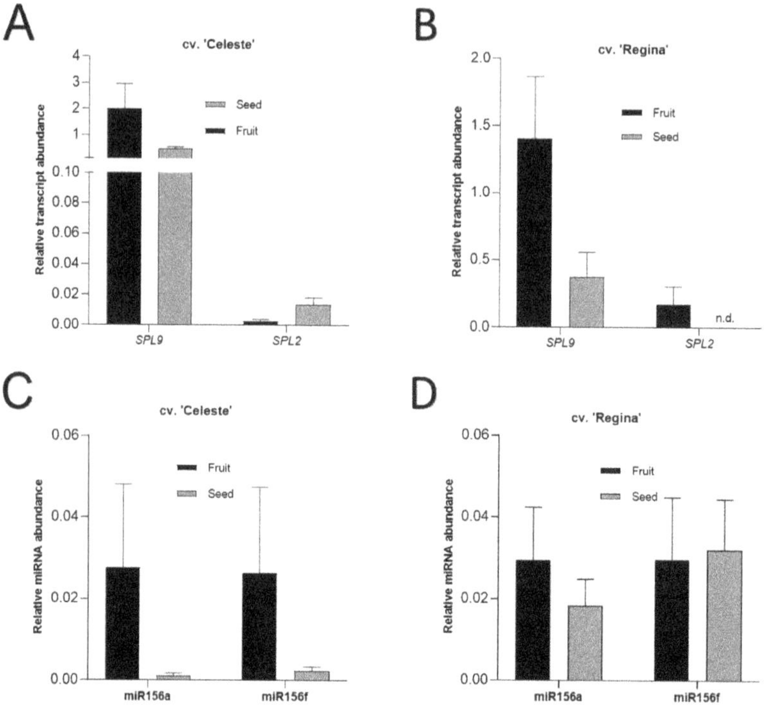

Figure 5. Expression profile of sweet cherry *SQUAMOSA Promoter-binding protein-Like* (*PavSPL*) genes and miR156a and miR156f in seeds and fruits at the straw yellow (SY) stage. *PavSPL9* and *PavSPL2* transcript abundance of cvs. 'Celeste' and 'Regina' are presented in (**A**,**B**), respectively. The miRNA abundances of miR156a and miR156f are presented in (**C**,**D**), respectively. The *Actin V1* (XM_021976055.1) gene was used as normalize for relative expression. *PavSPL* relative expression values were amplified by 10^3. The axis of 'Celeste' *PavSPL* was split in order to visualize the *PavSPL2* expression in fruits. Data are presented as mean ± standard error of the mean (SEM). Non-detected genes are represented as n.d.

We found that at 46 DAFB, three weeks after the treatments, an increase in the fruit coloration occurred in the IAA and *p*-IBA treatments (Figure 6A). Regarding lignin deposition, the *p*-IBA and NPA treatments presented less staining in the endocarp (Figure 6B). Chlorophyll and carotenoids reduced steadily during the season, whereas anthocyanins abruptly increased from 56 to 79 DAFB (Figure 6C). At 46 DAFB, there were significant differences when the pigments of all treatments were compared (Figure 6C). When we further analyzed these differences at this time point, we found that chlorophyll significantly increased in the *p*-IBA and IAA treatments (Figure 6D). Anthocyanins showed a trend of higher content in the *p*-IBA treatment compared to the control, though this was not significant. Fruit size significantly increased in the NPA treatment (Figure S2), whereas firmness and SSC did not change (Figure S3). The effect of the treatments at 46 DAFB was transient, as there were no differences between treatments at 56 DAFB (Figure 6C).

The effect of *p*-IBA and auxin treatments on lignification was assessed in the following season (2024–2025) in the same cultivar (Figure 7). The NAA analog was used instead of IAA, and both compounds, *p*-IBA and NAA, were tested at two doses. Thus, the five treatments further explored the effect of treatments that produced significant differences in the 2023–2024 season. Differences between treatments were observed at 48 DAFB (T23), where *p*-IBA increased red coloration in the transition to P stage (Figure 7A).

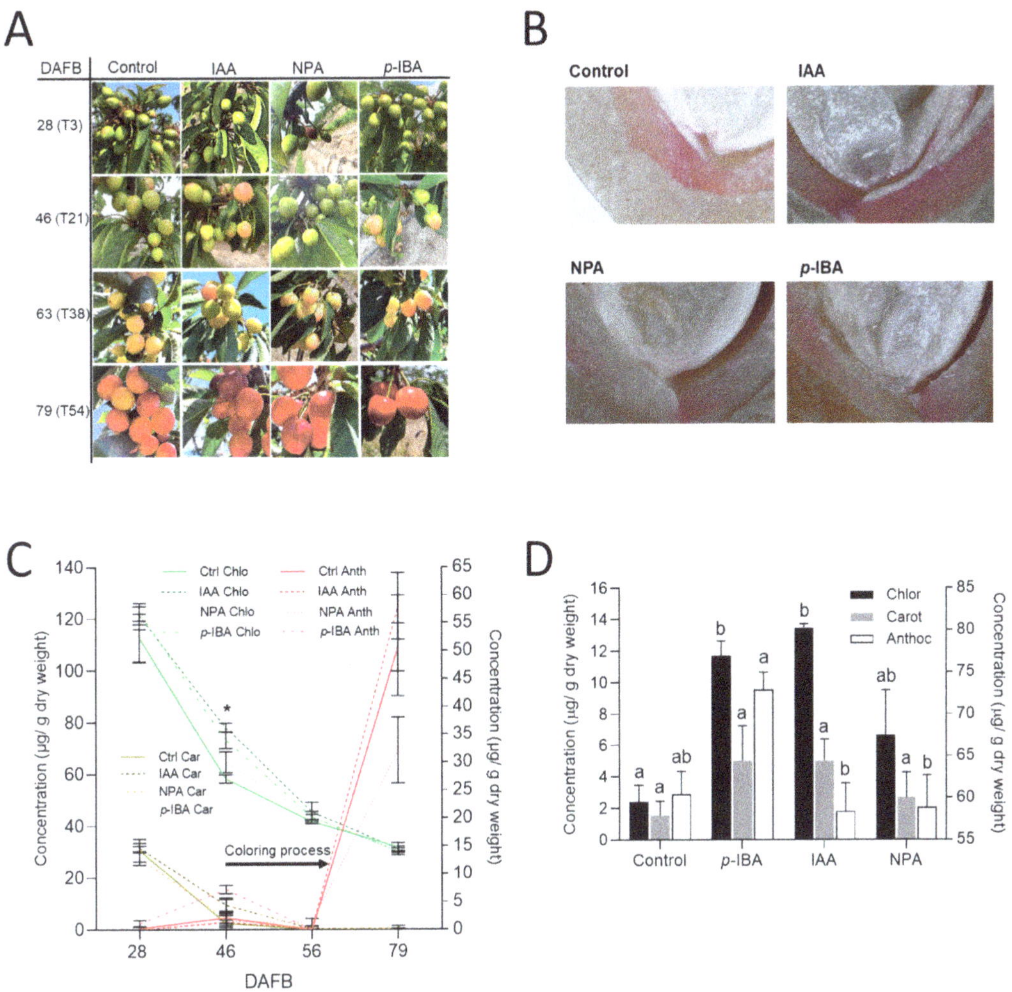

Figure 6. Effect of treatments that modulated lignification on lignin deposition and coloring in the mid-season cv. 'Lapins' during the 2023–2024 season. Control corresponded to untreated trees, and 40 µM NPA, *p*-IBA and IAA treatments were applied at 25 DAFB (T0). (**A**) Sweet cherry fruit color at 46 DAFB (T21), corresponding to light green (LG) stage, i.e., 90% LG/10% straw yellow (SY). (**B**) Endocarp lignin deposition at 29 DAFB (T4), as revealed by phloroglucinol-HCl staining. (**C**) Quantification of total chlorophylls, carotenoids, and anthocyanins at different DAFB; * indicates the date with significant differences according to one-way ANOVA (comparisons at each date separately). (**D**) Quantification of total chlorophylls, carotenoids and total anthocyanins at 46 DAFB (T21). Data are presented as the mean ± SD, and different letters indicate statistical differences, as revealed by ANOVA Tukey's post hoc test with $p < 0.05$, where each pigment was calculated independently.

These differences were not maintained over time, as the other treatments started to develop red color; thus, all the treatments were similar at 50 DAFB (T25). Regarding pigment content, significant differences were observed in anthocyanins versus control at 35 DAFB (T10) in the *p*-IBA treatment, whereas the chlorophyll and carotenoid content was not affected by any treatment (Figure 7B). Endocarp lignification analyses showed that at 25 DAFB (T0) there was no lignin deposition, whereas at 35 DAFB it was possible to observe lignification (Figure 7C). At this time point, the treatments presented differences, where both *p*-IBA treatments, and the NAA 100 μM treatment, had less lignin deposition (Figure 7C). However, these differences were transient, since at 50 DAFB all treatments had similar lignin staining. No differences were found in fruit diameter at 68 DAFB between the treatments (Figure S4).

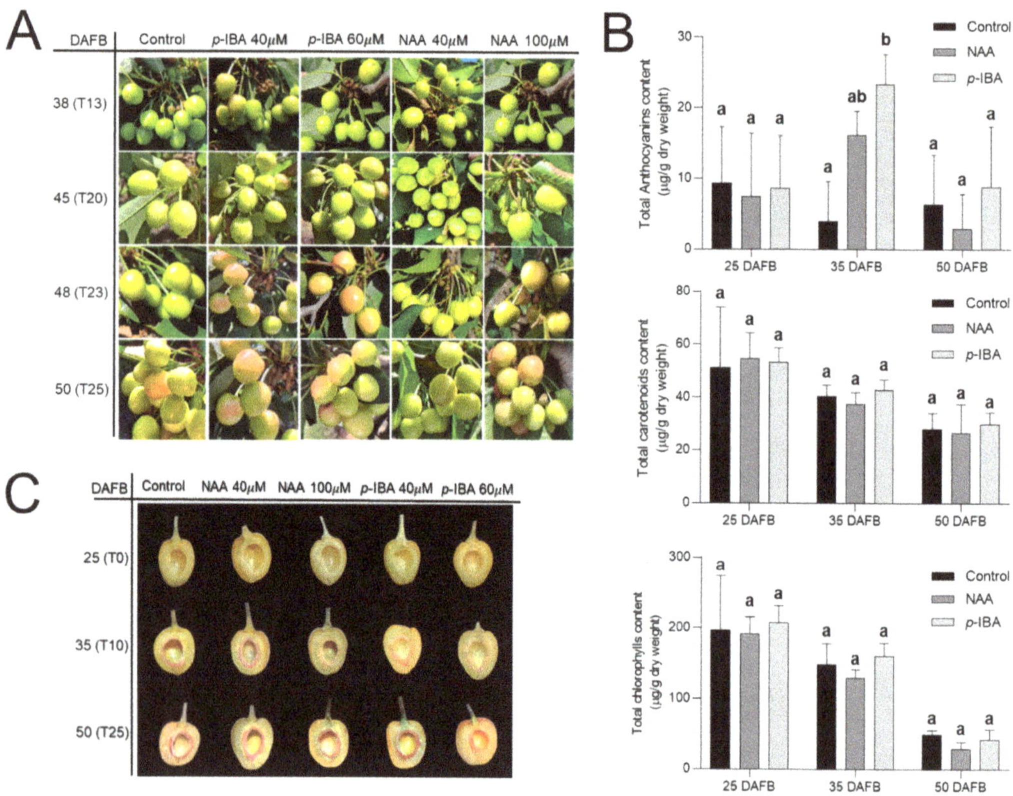

Figure 7. Effect of treatments that modulate lignification on lignin deposition and coloring in the mid-season cv. 'Lapins' during the 2024–2025 season. Control corresponded to untreated trees, and 40 μM *p*-IBA, 60 μM *p*-IBA, 40 μM NAA, and 100 μM NAA were applied at 25 DAFB (T0). (**A**) Sweet cherry fruit color at 38, 45, 48, and 50 DAFB. (**B**) Total anthocyanins, carotenoids, and chlorophylls in the fruits at 25 DAFB (T0, 100% LG), 35 DAFB (T10, 90% LG/10% SY), and 50 DAFB (T25, 70% SY/30% P); *p*-IBA and NAA μM 40 were included. (**C**) Endocarp lignin deposition revealed by phloroglucinol-HCl staining at 25, 35, and 50 DAFB. In B, data are presented as the mean ± SD, and different letters indicate statistical differences, as revealed by one-way ANOVA and Tukey's post-hoc test with $p < 0.05$.

Overall, lignification disturbance led to differences in pigment accumulation and color. In particular, anthocyanin content increased with *p*-IBA, which was significant in the 2024–2025 season. Thus, we aimed to associate these physiological effects with *PavSPL* expression, since they were co-expressed with lignin and anthocyanin genes (Figure 3B). We found that *p*-IBA treatment reduced the transcript abundance of both *PavSPL* genes, and this was significant in the case of *SPL9* (Figure 8).

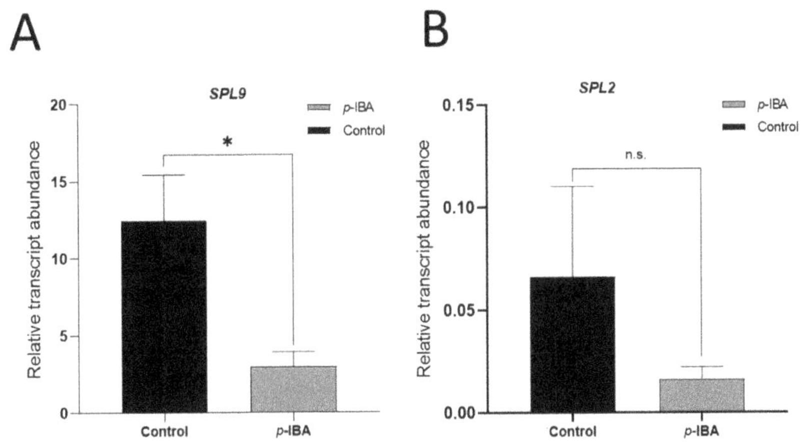

Figure 8. Effect of *p*-IBA on *PavSPL* gene expression in the mid-season cv. 'Lapins' during the 2023–2024 season. Control corresponds to untreated trees, and *p*-IBA 40µM treatment was applied at 25 DAFB (T0). The expression was evaluated by qPCR at 46 DAFB (T21), corresponding to the light green stage (LG). For quantification of *PavSPL9* (**A**) and *PavSPL2* (**B**) relative transcript abundance, the *Actin V1* (XM_021976055.1) gene was used as a normalizer. Data are presented as means ± SEM. Asterisk indicates statistical differences, as revealed by the Student's *T*-test with $p < 0.05$. n.s., not significant.

4. Discussion

We characterized cultivars contrasting in harvest time, regarding fruit and seed development, and endocarp lignification (Figure 1). Differences were found in the stage II progression (Figure 1A). This is consistent with previous reports [15,17], where stage II had different lengths depending on the cultivar. Stage II can affect the timing of the following stage, by delaying the initiation of stage III, where SSC and anthocyanin accumulation occur. The shorter stage II in 'Celeste' in fact resulted in earlier sugar accumulation, as this cultivar presented more SSC (Figure 1C).

The endocarp lignification process also differed between cultivars (Figure 1B). According to previous reports, lignification may be a turning point in the onset of ripening at the molecular level, as profuse remodeling occurs at the transcriptomic level in fruits and seeds of *Prunus persica* before and after lignification [27]. The modulation of lignification is not so well understood, but it is possible that fruit or seed signals influence lignin deposition. Kondo et al. reported that the seed had higher levels of GA_3 and auxin compared to the mesocarp [26]. On the other hand, the IAA flux is basipetal [66], suggesting that the IAA produced in the seeds moves to the fruit and pedicels. Auxin modulates lignification in *Arabidopsis thaliana* [23] and pear [24]. Thus, cultivars with differences in hormone content in fruits or seeds, especially auxin, might present differences in lignification. In this regard, Ponce et al. showed that fruits of an early-season cultivar had a higher mesocarp IAA content than a mid-season cultivar [17]. Whether this IAA comes from the fruit should be explored in future works. On the other hand, the increased lignin in the endocarp observed in the early-season phenotype could hinder the movement of GAs, which delay ripening [15].

Braak reported that the embryo of early-season cultivars does not reach full development since the ripening processes are advanced [25]. In this regard, we observed differences between 'Celeste' and 'Regina' in embryo and seed length (Figure 2). 'Regina' embryo was larger and reached full development. Possibly there are differences in hormone content associated with this differential degree of seed development between cultivars, though this must be addressed in future works. This could also influence lignification processes.

Then, we evaluated whether contrasting cultivars differed at the molecular level by focusing on *PavSPL* genes. *SPL*/miR156 module has been shown to be relevant for ripening initiation in other fruit species. In *Pyrus* spp., this module was implicated in the anthocyanin synthesis in response to light during fruit development [39]. In line with this, the authors showed that *PpySPL10* and *PpySPL13* proteins interacted with *PpyMYB10*, key in the regulation of anthocyanin structural genes. In blueberry, several *SPLs* expressed during ripening, and overexpression of miR156a led to more anthocyanin accumulation [41]. Regarding non-climacteric species, in fruits of litchi (*Litchi chinensis* Sonn.), *SPLs* co-express with anthocyanin genes [67].

We found that several *PavSPLs* were expressed at the transition from development to ripening, specifically at the LG (Figure 3), which was consistent with Wang et al. [42], though in this work an advanced stage was analyzed, corresponding to the initial R stage. However, Sun et al. showed a decrease in the expression of *PavSPLs* from LG onwards [43]. The decrease could be associated with hormonal regulation, since GA and ABA increased and decreased *PavSPL* expression, respectively [43]. GA is abundant at LG in sweet cherry [17], whereas ABA increases at the beginning of the P stage [5]. Thus, the reduction in *PavSPL* expression during ripening could be controlled by ABA, which in turn is key for anthocyanin accumulation [21]. Regarding GA, Liu et al. found that *PavSPL14* protein (here *PavSPL17* according to NCBI nomenclature; Table S4) interacted with *PavDWARF8*, a DELLA protein, part of the GA pathway [29].

PavSPLs are expressed in fruits of sweet cherry [29,43,44]; however, their expression in other tissues may also be important for fruit processes including ripening. Maldonado et al. showed that some *PavSPLs* also were expressed in the developing seed at the LG stage, consistent with our findings (Figure 3), though we profiled the complete gene family [44]. This allowed us to determine that most *PavSPLs* had higher expression in fruits than in seeds. This is consistent with findings in litchi fruits, where *PavSPLs* were more abundant in the pericarp than in the seed [67].

Interestingly, some *PavSPLs* showed differences between both cultivars, especially in seeds (Figure 3). For instance, *PavSPL8* (FUN_010968-T1) and *PavSPL17* (FUN_021909-T1) had more expression in 'Regina' seeds, whereas both alleles of *PavSPL16* (FUN_003506-T1 and T2) were more abundant in 'Celeste' seeds. Whether this is related to differences in the hormone content of seeds should be explored in future works, but an influence of hormones is expected, according to the effect of ABA, GA, and methyl jasmonate (MeJA) treatments reported by Sun et al. [43]. In line with this, Maldonado et al. found differences in the content of several hormones (IAA, GAs, CKs, and JA) between 'Celeste' and 'Regina' seeds [44].

PavSPLs are targeted by several *miR156* forms in sweet cherry [29,42]. Here, we identified two *PavSPL* genes, possibly targeted by *miR156*, using different in silico approaches (TarDB search using *miR156 sequences*; cleavage site predictions). We obtained complete complementarity between *PavSPLs* and *miR156a* and *miR156f* (Table 1). It is worth noting that *PavSPL9* (FUN_037655-T1) corresponded to *PavSPL16,* while *PavSPL2* (FUN_003521-T1) corresponded to *PavSPL4,* according to Liu et al. [29], but here we followed NCBI nomenclature, as in Wang et al. [42], as shown in Table S4. In the case of *PavSPL9,* we had additional indications that this gene was targeted by *miR156,* since it presented negative correlations (opposing expression patterns across the fruit and seed samples) with the hairpin *bdi-MIR156a* [44].

In the phylogenetic tree (Figure 4), *PavSPL9* grouped with *PavSPL14* (here *PavSPL17*). It is worth noting that, in NCBI, there is another *PavSPL* annotated as *PavSPL9*

(FUN_021909-T2); but this sequence corresponds to *PavSPL15*, which arose from *PavSPL17* in a gene duplication event [29]; this explains the two allelic variants in the cv. 'Tieton' Genome v2.0 annotation.

PavSPL2 and *PavSPL9* were also present at the SY stage (Figure 5). *PavSPL2* was not detected in seed tissues of 'Regina', which coincided with a higher abundance of both miR156 variants in the seeds of this cultivar compared to 'Celeste' seeds, though these differences were not significant. Maldonado et al. [44] found that miR156 forms were highly abundant in seeds of the early cultivar 'Celeste', which could be related to the earlier stage analyzed (LG).

In line with the idea that lignified endocarp could act as a barrier for signal communication between the seed and the mesocarp, or it being a site of signal production itself, we aimed to disturb the lignification process and determine if *PavSPL* gene expression was altered, as well as the ripening process. In line with this, *PavSPL2* and *PavSPL9* were co-expressed with lignin and anthocyanin genes, respectively (Figure 3B). For this, we applied modulators of lignification in the 2023–2024 season (Figure 6). We used an enzymatic inhibitor of the CINNAMATE-4-HYDROXILASE [68], *p*-IBA, at 40 μM. We used this concentration as it affected lignification without producing a stress response [68]. We also utilized IAA, since it modulated ripening in sweet cherry [11]; in addition, Qu et al. [23] reported that IAA regulates lignin biosynthesis. We used a low dose (40 μM) in relation to a previous report in sweet cherry that utilized 1 mM [42], with fruit endogenous levels in the order of 2–4 μg per g of dry tissue [44] and 0.2 μg per g of fresh tissue [7]. Finally, we tested NPA, an inhibitor of polar auxin transport, in order to indirectly alter the IAA levels in the fruits [69]. NPA was also reported to modulate ripening in grapevine at a 40 μM dose [70].

As shown in Figure 6, the phloroglucinol-HCl staining was reduced in NPA and *p*-IBA treatments; on the other hand, it was more intense in the IAA treatment. In pear, the low dose (200 μM IAA) reduced stone cell content, while higher concentrations (>500 μM IAA) increased the stone cell content [24]. As such, our low dose treatment resembled the high dose effect found in pear, though it is worth noting that the lignification processes are likely different between both species and cellular types, especially considering that pear is climacteric. NPA caused the opposite effect to IAA in terms of lignification, and this is possibly related to a disturbance in auxin homeostasis, as polar auxin transport should be affected at this dose [66,70]. At 46 DAFB, IAA and *p*-IBA had a transient positive effect on the fruit coloration (Figure 6). It seems that this was due to different mechanisms, since *p*-IBA targets enzymatic activity, whereas IAA regulates several pathways involved in ripening, including ABA. Hence, though lignification increased with IAA, more color might be attributed to IAA regulation of ABA levels, as reported by Wang et al. [7]. On the other hand, the *p*-IBA-related color increase could be associated with lower lignin production, which opens the possibility of inter-tissue communication, under the idea that lignified endocarp represents a barrier between seeds and fruits, though this should be explored in future works. It is worth noting that *p*-IBA 40 μM was not associated with stress, since toxic responses usually occur within hours or a few days, and here we did not observe color differences at T3. Therefore, we propose that the effect on color of *p*-IBA could be related to lignification disturbance. Regarding pigments, the increase in anthocyanin content followed the chlorophyll decrease, as previously reported [21]. IAA did not produce the significant anthocyanin increase that was expected for more colored fruits. Regarding *p*-IBA, a trend of higher anthocyanin content compared to the control was observed, but it was not significant. Possibly this was due to sampling, since high heterogenicity was observed between fruits. Additionally, the low anthocyanin concentration at the P stage compared to R stage may have led to higher deviations between replicates. Finally, field and not controlled conditions could have influenced the treatment effect. Therefore, we repeated the assay the following season, 2024–2025, by focusing on treatments that improved color and with two concentrations.

We found a transient color increase with both concentrations of *p*-IBA versus control at 48 DAFB, accompanied with a significantly higher concentration of anthocyanins in the fruits in the 40 µM *p*-IBA treatment at 35 DAFB (Figure 7). NAA also improved color but only at 100 µM. Hence, similarly to the 2023–2024 season, less lignification was accompanied by more color in *p*-IBA, whereas more lignin deposition was related to color increase in NAA, suggesting different mechanisms of modulating color development. Interestingly, 40 µM NAA reduced lignin staining, which reinforced that the effect of auxin on the lignification process is dose-dependent, as shown in pear stone cells [24]. Similarly, 40 µM NAA reduced fruit color at 50 DAFB. Thus, only the 100 µM NAA dose resembled the 40 µM IAA treatment of the 2023–2024 season, regarding the positive effect on color and lignin deposition. The effect of IAA and *p*-IBA on the chlorophyll content, as observed in the 2023–2024 season, was not replicated; possibly, this could have been influenced by other environmental conditions. As in 2023–2024, the effect on lignification, color, and anthocyanins was transient for all the treatments, including *p*-IBA. Regarding NAA, this was different from the treatments using 100 µM that produced significantly increased anthocyanin content in sweet cherry [11]. This was possibly related to the application time, as the spraying was at the SY stage in Clayton-Cuch et al. [11], whereas we performed the treatments at LG in both seasons, where the ripening-related genes, including ABA signaling genes, are less expressed compared to SY [4]. Additionally, GA pathways exert a negative regulation on the ABA pathway when applied at LG [15,16]. Future works should also explore the effect of *p*-IBA at the SY.

It is worth noting that we detected an effect of *p*-IBA and IAA on the intensity of the phloroglucinol staining at T4 in the 2023–2024 season, and at T10 in the *p*-IBA and NAA treatments of the 2024–2025 season; thereafter, the color effect occurred around two weeks later. This is consistent with the work of Clayton-Cuch et al. [11], showing an effect of NAA on anthocyanin content three weeks after the treatment. Under the hypothesis that both processes—lignification and color development—were connected, a temporality should exist, where a treatment disturbs the lignification process and this, in turn, affects other processes that may lead to changes in the evolution of coloring.

We found that the treatments did not modulate ripening parameters differently than the color and pigment content, at least under our experimental conditions, since firmness and SSC were not affected, nor fruit diameter (Figures S2–S4). This suggests that color based on anthocyanins is partially independent of other ripening parameters. In line with this, desynchronization of color and firmness has been shown in other fruit crops. In 'Hass' avocado (*Persea americana*), when skin color develops coupled with firmness, certain markers are present, including ABA [50,71]. It would be interesting to apply these treatments to cultivars contrasting in maturity time, as they have a ten times difference in ABA fruit levels [17], in order to detect effects other than color.

The *PavSPL* gene expression was evaluated in response to *p*-IBA treatment (Figure 8). The downregulation of both genes—significant in the case of *PavSPL9*—was expected, as the *p*-IBA treated fruits presented more color and *PavSPLs* have been proposed to antagonize ripening, whereas miR156 is pro-ripening [39,41]. Several *PavSPLs* decrease their expression during sweet cherry fruit ripening and in response to ABA [43]; hence, it seems that the expression of these transcription factors has to be downregulated to allow the transition to ripening. In litchi and sweet cherry, SPLs are upregulated by GA applied to fruits [43,67]. Therefore, one possibility is that less lignin in the endocarp in the *p*-IBA treatment makes the endocarp more permeable to the flux of signals from the seed, including hormones such as IAA and GA, which modulate color development [7,16]. IAA from the seed could promote coloring through the ABA pathway, and ABA, in turn, could downregulate fruit *PavSPLs*; miR156 from the seed might contribute to *PavSPL* downregulation through post-transcriptional regulation (Figure 9). Thereafter, GA would increase the *PavSPL* levels, explaining the transient effect of *p*-IBA on color. This hypothesis is tempting but requires functional demonstration, as well as tracing of seed signals, including miRNAs and hormones.

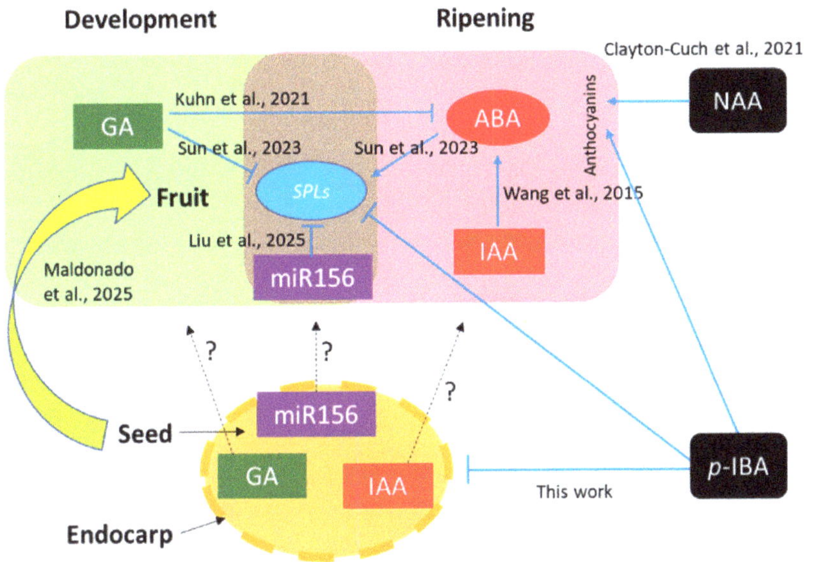

Figure 9. Working model on the regulation of the sweet cherry transition to ripening, integrating hormone signals, the *SPL*/miR156 module, and the role of seed and endocarp. Works cited were performed in sweet cherry, and positive or negative effects are indicated in blue. Black dotted arrows refer to inter-tissue movement and interrogation signs denote that this should be demonstrated in future works. Large yellow arrow represents fruit–seed associations through multiomics analyses. Finally, the effect of exogenous treatments is indicated, where NAA and *p*-IBA promoted fruit color formation [7,11,16,29,43,44].

5. Conclusions

In this work, we contributed to the molecular understanding of the non-climacteric transition to ripening. Here, we characterized the expression of several sweet cherry SPL family members in the transition from development to ripening. This work also showed that *PavSPLs*, abundant in fruits, but also present in seeds, are co-expressed with anthocyanin and lignin genes. They were also found to be targeted by miR156 expressed in seeds and fruits, and downregulated in response to a treatment that impaired lignification and triggered color formation, though transiently. These treatments revealed that the endocarp lignification process is relevant to the transition to ripening. Future studies should further explore the effects of compounds that disrupt lignification, with the aim of developing growth regulators useful for orchard management. Finally, the *SPL*/miR156 module is a promising candidate for identifying genetic factors that could be used in marker-assisted breeding programs.

Supplementary Materials: The following supporting information can be downloaded at https://www.mdpi.com/article/10.3390/horticulturae11060706/s1, Figure S1: Position of the PavSPL Forward (Fw) and Reverse (Rv) primers on the sweet cherry cultivar 'Tieton' Genome v. 2.0; Figure S2:

Effect of treatments that modulate lignification on fruit growth in the mid-season cultivar 'Lapins' at different DAFB. Figure S3: Effect on SSC and firmness of the lignification modulators treatments on the mid-season cultivar 'Lapins' at 71 DAFB (2023-2024 season); Figure 4: Effect on fruit size of the lignin modulators treatments on the mid-season cultivar 'Lapins' at 68 DAFB (2024-2025 season); Table S1: qPCR primer sequences with melting temperature (TM); Table S2: qPCR miRNA primer sequences with melting temperature (TM); Table S3: qPCR normalizer gene primer sequences with melting temperature (TM); Table S4. Accessions, IDs and *PavSPL* names.

Author Contributions: Conceptualization, N.K., J.E.M. and M.Z.; methodology, C.A., M.Z., O.A., M.M. (Mirna Melo) and M.M. (Marcela Menares); software, J.E.M., C.N.; validation, M.Z., C.N. and J.E.M.; formal analysis, R.P., M.Z. and M.M. (Marcela Menares); investigation, N.K., J.E.M., M.Z.; resources, L.A.M., N.K., J.E.M., R.P. and J.M.D.; data curation, J.E.M.; writing—original draft preparation, N.K., M.M. (Marcela Menares) and M.Z.; writing—review and editing, L.A.M., N.K., J.E.M., R.P. and J.M.D.; visualization, N.K., M.Z., J.E.M., C.N.; supervision, N.K., J.E.M.; project administration, N.K.; funding acquisition, L.A.M., N.K., J.E.M., R.P. All authors have read and agreed to the published version of the manuscript.

Funding: This research was funded by Fondecyt Iniciación ANID 11221186, Fondecyt Regular ANID 1220223; Fondecyt Regular ANID 1231491; Fondecyt Regular ANID 1220097; SIA sa77210103. In addition, this research was partially supported by the supercomputing infrastructure of the NLHPC (ECM-02) (Powered@NLHPC). Romina Pedreschi was supported by ANID Millennium Science Initiative Program ICN2021_044. The APC was funded by Departamento de Investigación, Pontificia Universidad Católica de Valparaíso, and Proyecto InES de Género INGE220007, Pontificia Universidad Católica de Valparaíso.

Data Availability Statement: The original contributions presented in the study are included in the article/Supplementary Material, further inquiries can be directed to the corresponding author.

Conflicts of Interest: The authors declare no conflict of interest.

Abbreviations

Abscisic acid (ABA); indole-3-acetic acid (IAA); gibberellin (GA); cytokinin (CK); gibberellic acid (GA_3); SQUAMOSA Promoter-binding protein-Like (SPL); microRNA (miRNA); straw yellow (SY); days after full bloom (DAFB); soluble solids content (SSC); *p*-IBA (*p*-iodobenzoic acid); green (G); light green (LG); pink (P); red (R); NAA (1-naphthaleneacetic acid); NPA (naphtylphtalamic acid); standard deviation (SD); standard error of the mean (SEM).

References

1. Statista. Available online: https://www.statista.com/statistics/577489/world-cherry-production/ (accessed on 19 March 2025).
2. Vignati, E.; Lipska, M.; Dunwell, J.M.; Caccamo, M.; Simkin, A.J. Fruit development in sweet cherry. *Plants* **2022**, *11*, 1531. [CrossRef] [PubMed]
3. Dirlewanger, E.; Quero-Garcia, J.; Le Dantec, L.; Lambert, P.; Ruiz, D.; Dondini, L.; Arús, P. Comparison of the genetic determinism of two key phenological traits, flowering and maturity dates, in three *Prunus* species: Peach, apricot and sweet cherry. *Heredity* **2012**, *109*, 280–292. [CrossRef]
4. Kuhn, N.; Ponce, C.; Arellano, M.; Time, A.; Multari, S.; Martens, S.; Meisel, L.A. ABA influences color initiation timing in *P. avium* L. fruits by sequentially modulating the transcript levels of ABA and anthocyanin-related genes. *Tree Genet. Genomes* **2021**, *17*, 20. [CrossRef]
5. Luo, H.; Dai, S.; Ren, J.; Zhang, C.; Ding, Y.; Li, Z.; Leng, P. The role of ABA in the maturation and postharvest life of a nonclimacteric sweet cherry fruit. *J. Plant Growth Regul.* **2014**, *33*, 373–383. [CrossRef]
6. Ren, J.; Chen, P.; Dai, S.J.; Li, P.; Li, Q.; Ji, K.; Leng, P. Role of abscisic acid and ethylene in sweet cherry fruit maturation: Molecular aspects. *N. Z. J. Crop Hortic. Sci.* **2011**, *39*, 161–174. [CrossRef]
7. Wang, Y.; Chen, P.; Sun, L.; Li, Q.; Dai, S.; Sun, Y.; Ji, K.; Li, Q.; Leng, P. Transcriptional regulation of *PaPYLs*, *PaPP2Cs* and *PaSnRK2s* during sweet cherry fruit development and in response to abscisic acid and auxin at onset of fruit ripening. *Plant Growth Regul.* **2015**, *75*, 455–464. [CrossRef]

8. Medina-Puche, L.; Blanco-Portales, R.; Molina-Hidalgo, F.J.; Cumplido-Laso, G.; García-Caparrós, N.; Moyano-Cañete, E.; Rodríguez-Franco, A. Extensive transcriptomic studies on the roles played by abscisic acid and auxins in the development and ripening of strawberry fruits. *Funct. Integr. Genom.* **2016**, *16*, 671–692. [CrossRef] [PubMed]
9. Wheeler, S.; Loveys, B.; Ford, C.; Davies, C. The relationship between the expression of abscisic acid biosynthesis genes, accumulation of abscisic acid and the promotion of *Vitis vinifera* L. berry ripening by abscisic acid. *Aust. J. Grape Wine Res.* **2009**, *15*, 195–204. [CrossRef]
10. Gouthu, S.; Deluc, L.G. Timing of ripening initiation in grape berries and its relationship to seed content and pericarp auxin levels. *BMC Plant Biol.* **2015**, *15*, 46. [CrossRef]
11. Clayton-Cuch, D.; Yu, L.; Shirley, N.; Bradley, D.; Bulone, V.; Böttcher, C. Auxin treatment enhances anthocyanin production in the non-climacteric sweet cherry (*Prunus avium* L.). *Int. J. Mol. Sci.* **2021**, *22*, 10760. [CrossRef]
12. Zhang, C.; Whiting, M.D. Improving 'Bing' sweet cherry fruit quality with plant growth regulators. *Sci. Hortic.* **2011**, *127*, 341–346. [CrossRef]
13. Kuhn, N.; Arellano, M.; Ponce, C.; Hodar, C.; Correa, F.; Multari, S.; Meisel, L.A. RNA-seq and WGBS analyses during fruit ripening and in response to ABA in sweet cherry (*Prunus avium*) reveal genetic and epigenetic modulation of auxin and cytokinin genes. *J. Plant Growth Regul.* **2025**, *44*, 1165–1187. [CrossRef]
14. Kappel, F.; MacDonald, R.A. Gibberellic acid increases fruit firmness, fruit size, and delays maturity of 'Sweetheart' sweet cherry. *J. Am. Pomol. Soc.* **2002**, *56*, 219–222.
15. Kuhn, N.; Ponce, C.; Arellano, M.; Time, A.; Sagredo, B.; Donoso, J.M.; Meisel, L.A. Gibberellic acid modifies the transcript abundance of ABA pathway orthologs and modulates sweet cherry (*Prunus avium*) fruit ripening in early- and mid-season varieties. *Plants* **2020**, *9*, 1796. [CrossRef] [PubMed]
16. Kuhn, N.; Maldonado, J.; Ponce, C.; Arellano, M.; Time, A.; Multari, S.; Meisel, L.A. RNAseq reveals different transcriptomic responses to GA_3 in early and midseason varieties before ripening initiation in sweet cherry fruits. *Sci. Rep.* **2021**, *11*, 13075. [CrossRef]
17. Ponce, C.; Kuhn, N.; Arellano, M.; Time, A.; Multari, S.; Martens, S.; Meisel, L.A. Differential phenolic compounds and hormone accumulation patterns between early-and mid-maturing sweet cherry (*Prunus avium* L.) cultivars during fruit development and ripening. *J. Agric. Food Chem.* **2021**, *69*, 8850–8860. [CrossRef] [PubMed]
18. Gibeaut, D.M.; Whiting, M.D.; Einhorn, T. Time indices of multiphasic development in genotypes of sweet cherry are similar from dormancy to cessation of pit growth. *Ann. Bot.* **2017**, *119*, 465–475. [CrossRef]
19. Zhang, C.; Whiting, M.D. Plant growth regulators improve sweet cherry fruit quality without reducing endocarp growth. *Sci. Hortic.* **2013**, *150*, 73–79. [CrossRef]
20. Alkio, M.; Jonas, U.; Declercq, M.; Van Nocker, S.; Knoche, M. Transcriptional dynamics of the developing sweet cherry (*Prunus avium* L.) fruit: Sequencing, annotation and expression profiling of exocarp-associated genes. *Hortic. Res.* **2014**, *1*, 11. [CrossRef]
21. Shen, X.; Zhao, K.; Liu, L.; Zhang, K.; Yuan, H.; Liao, X.; Li, T. A role for PacMYBA in ABA-regulated anthocyanin biosynthesis in red-colored sweet cherry cv. Hong Deng (*Prunus avium* L.). *Plant Cell Physiol.* **2014**, *55*, 862–880. [CrossRef]
22. Dardick, C.; Callahan, A.M. Evolution of the fruit endocarp: Molecular mechanisms underlying adaptations in seed protection and dispersal strategies. *Front. Plant Sci.* **2014**, *5*, 284. [CrossRef]
23. Qu, G.; Peng, D.; Yu, Z.; Chen, X.; Cheng, X.; Yang, Y.; Zhou, B. Advances in the role of auxin for transcriptional regulation of lignin biosynthesis. *Funct. Plant Biol.* **2021**, *48*, 743–754. [PubMed]
24. Wang, Y.; Wang, Q.; Zhang, F.; Han, C.; Li, W.; Ren, M.; Hou, C.; Tao, S. *PbARF19*-mediated auxin signaling regulates lignification in pear fruit stone cells. *Plant Sci.* **2024**, *344*, 112103. [CrossRef] [PubMed]
25. Braak, J.P. The effect of flowering date and temperature on embryo development in sweet cherry (*Prunus avium* L.). *Neth. J. Agric. Sci.* **1978**, *26*, 13–30.
26. Kondo, S.; Hayata, Y.; Iwasaki, N. Effects of indole-3-acetic acid and gibberellins on fruit development and maturation of sweet cherry. *Acta Hortic.* **2000**, *514*, 75–82. [CrossRef]
27. Bonghi, C.; Trainotti, L.; Botton, A.; Tadiello, A.; Rasori, A.; Ziliotto, F.; Ramina, A. A microarray approach to identify genes involved in seed-pericarp cross-talk and development in peach. *BMC Plant Biol.* **2011**, *11*, 107. [CrossRef]
28. Yamaguchi, A.; Wu, M.F.; Yang, L.; Wu, G.; Poethig, R.S.; Wagner, D. The microRNA-regulated SBP-box transcription factor SPL3 is a direct upstream activator of *LEAFY*, *FRUITFULL*, and *APETALA1*. *Dev. Cell* **2009**, *17*, 268–278. [CrossRef]
29. Liu, X.; Sun, W.; Liu, H.; Wang, L.; Manzoor, M.A.; Wang, J.; Zhang, C. PavSPLs are key regulators of growth, development, and stress response in sweet cherry. *Plant Sci.* **2025**, *350*, 112279.
30. Gou, J.Y.; Felippes, F.F.; Liu, C.J.; Weigel, D.; Wang, J.W. Negative regulation of anthocyanin biosynthesis in *Arabidopsis* by a miR156-targeted SPL transcription factor. *Plant Cell* **2011**, *23*, 1512–1522. [CrossRef]
31. Wang, J.W.; Czech, B.; Weigel, D. miR156-regulated SPL transcription factors define an endogenous flowering pathway in *Arabidopsis thaliana*. *Cell* **2009**, *138*, 738–749. [CrossRef]

32. Wu, G.; Poethig, R.S. Temporal regulation of shoot development in *Arabidopsis thaliana* by miR156 and its target SPL3. *Development* **2006**, *133*, 3539–3547. [CrossRef] [PubMed]
33. Wang, Y.; Luo, Z.; Zhao, X.; Cao, H.; Wang, L.; Liu, S.; Liu, Z. Superstar microRNA, miR156, involved in plant biological processes and stress response: A review. *Sci. Hortic.* **2023**, *316*, 112010. [CrossRef]
34. Duan, Z.; Qin, Y.; Xia, X.; Yin, W. Overexpression of *Populus euphratica peu-MIR156j* gene enhancing salt tolerance in *Arabidopsis thaliana*. *J. Beijing For. Univ.* **2011**, *33*, 1–7.
35. Sun, C.; Zhao, Q.; Liu, D.D.; You, C.X.; Hao, Y.J. Ectopic expression of the apple Md-miRNA156h gene regulates flower and fruit development in *Arabidopsis*. *Plant Cell Tissue Organ Cult.* **2013**, *112*, 343–351. [CrossRef]
36. Song, M.; Wang, R.; Zhou, F.; Wang, R.; Zhang, S.; Li, D.; Yang, Y. SPLs-mediated flowering regulation and hormone biosynthesis and signaling accompany juvenile-adult phase transition in *Pyrus*. *Sci. Hortic.* **2020**, *272*, 109584. [CrossRef]
37. Liu, H.; Yu, H.; Tang, G.; Huang, T. Small but powerful: Function of microRNAs in plant development. *Plant Cell Rep.* **2018**, *37*, 515–528. [CrossRef]
38. Soto, E.; Sanchez, E.; Nuñez, C.; Montes, C.; Rothkegel, K.; Andrade, P.; Almeida, A.M. Small RNA differential expression analysis reveals miRNAs involved in dormancy progression in sweet cherry floral buds. *Plants* **2022**, *11*, 2396. [CrossRef]
39. Qian, M.; Ni, J.; Niu, Q.; Bai, S.; Bao, L.; Li, J.; Teng, Y. Response of miR156-SPL module during the red peel coloration of bagging-treated Chinese sand pear (*Pyrus pyrifolia* Nakai). *Front. Physiol.* **2017**, *8*, 550. [CrossRef]
40. Mohorianu, I.; Schwach, F.; Jing, R.; Lopez-Gomollon, S.; Moxon, S.; Szittya, G.; Dalmay, T. Profiling of short RNAs during fleshy fruit development reveals stage-specific sRNAome expression patterns. *Plant J.* **2011**, *67*, 232–246. [CrossRef]
41. Li, H.; Wang, S.; Zhai, L.; Cui, Y.; Tang, G.; Huo, J.; Bian, S. The miR156/SPL12 module orchestrates fruit colour change through directly regulating ethylene production pathway in blueberry. *Plant Biotechnol. J.* **2024**, *22*, 386–400. [CrossRef]
42. Wang, Y.; Li, W.; Chang, H.; Zhou, J.; Luo, Y.; Zhang, K.; Wang, B. Sweet cherry fruit miRNAs and effect of high CO_2 on the profile associated with ripening. *Planta* **2019**, *249*, 1799–1810. [CrossRef] [PubMed]
43. Sun, Y.; Wang, Y.; Xiao, Y.; Zhang, X.; Du, B.; Turupu, M.; Li, T. Genome-wide identification of the SQUAMOSA promoter-binding protein-like (SPL) transcription factor family in sweet cherry fruit. *Int. J. Mol. Sci.* **2023**, *24*, 2880. [CrossRef]
44. Maldonado, J.E.; Acevedo, O.; Melo, M.; Núñez, C.; Zavala, M.; Menares, M.; Kuhn, N. Seed-fruit multiomics integration of sweet cherry cultivars with different maturity time shows alternative molecular landscapes at the transition from development to ripening, unveiling a role of small RNAs, SPLs, lignin and inositol pathways. *Sci. Hortic.* **2025**, *343*, 114099.
45. Marín-González, E.; Suárez-López, P. "And yet it moves": Cell-to-cell and long-distance signaling by plant microRNAs. *Plant Sci.* **2012**, *196*, 18–30.
46. San Martino, L.; Manavella, F.A.; García, D.A.; Salato, G. Phenology and fruit quality of nine sweet cherry cultivars in South Patagonia. In *V International Cherry Symposium*; ISHS: Leuven, Belgium, 2008; Volume 795, pp. 841–848.
47. Hesse, C.O.; Kester, D.E. Germination of embryos of *Prunus* related to degree of embryo development and method of handling. *Proc. Am. Soc. Hortic. Sci.* **1955**, *65*, 251–264.
48. Wang, J.; Liu, W.; Zhu, D.; Hong, P.; Zhang, S.; Xiao, S.; Wang, L.; Liu, Q. Chromosome-scale genome assembly of sweet cherry (*Prunus avium* L.) cv. Tieton obtained using long-read and Hi-C sequencing. *Hortic. Res.* **2020**, *7*, 36. [CrossRef]
49. Huynh-Thu, V.A.; Irrthum, A.; Wehenkel, L.; Geurts, P. Inferring regulatory networks from expression data using tree-based methods. *PLoS ONE* **2010**, *5*, e12776. [CrossRef]
50. Núñez-Lillo, G.; Ponce, E.; Arancibia-Guerra, C.; Carpentier, S.; Carrasco-Pancorbo, A.; Olmo-García, L.; Pedreschi, R. A multiomics integrative analysis of color de-synchronization with softening of 'Hass' avocado fruit: A first insight into a complex physiological disorder. *Food Chem.* **2023**, *408*, 135215. [CrossRef]
51. Shannon, P.; Markiel, A.; Ozier, O.; Baliga, N.S.; Wang, J.T.; Ramage, D.; Ideker, T. Cytoscape: A software environment for integrated models of biomolecular interaction networks. *Genome Res.* **2003**, *13*, 2498–2504. [CrossRef]
52. Meisel, L.; Fonseca, B.; González, S.; Baeza-Yates, R.; Cambiazo, V.; Campos, R.; Silva, H. A rapid and efficient method for purifying high quality total RNA from peaches (*Prunus persica*) for functional genomics analyses. *Biol. Res.* **2005**, *38*, 83–88.
53. Griffiths-Jones, S. miRBase: The microRNA sequence database. In *MicroRNA Protocols*; Humana Press: Totowa, NJ, USA, 2006; pp. 129–138.
54. Dai, X.; Zhuang, Z.; Zhao, P.X. psRNATarget: A plant small RNA target analysis server. *Nucleic Acids Res.* **2018**, *46*, W49–W54. [CrossRef] [PubMed]
55. Kumar, S.; Stecher, G.; Suleski, M.; Sanderford, M.; Sharma, S.; Tamura, K. MEGA12: Molecular Evolutionary Genetics Analysis Version 12 for adaptive and green computing. *Mol. Biol. Evol.* **2024**, *41*, msae263. [CrossRef] [PubMed]
56. Tong, Z.; Gao, Z.; Wang, F.; Zhou, J.; Zhang, Z. Selection of reliable reference genes for gene expression studies in peach using real-time PCR. *BMC Mol. Biol.* **2009**, *10*, 71. [CrossRef]
57. Ruijter, J.M.; Ramakers, C.; Hoogaars, W.M.H.; Karlen, Y.; Bakker, O.; Van den Hoff, M.J.B.; Moorman, A. Amplification efficiency: Linking baseline and bias in the analysis of quantitative PCR data. *Nucleic Acids Res.* **2009**, *37*, e45. [CrossRef]

58. Pfaffl, M.W. A new mathematical model for relative quantification in real-time RT–PCR. *Nucleic Acids Res.* **2001**, *29*, e45. [CrossRef] [PubMed]
59. Bustin, S.A.; Benes, V.; Garson, J.A.; Hellemans, J.; Huggett, J.; Kubista, M.; Mueller, R.; Nolan, T.; Pfaffl, M.W.; Shipley, G.L.; et al. The MIQE guidelines: Minimum information for publication of quantitative real-time PCR experiments. *Clin. Chem.* **2009**, *55*, 611–622. [CrossRef]
60. Udvardi, M.K.; Czechowski, T.; Scheible, W.R. Eleven golden rules of quantitative RT-PCR. *Plant Cell* **2008**, *20*, 1736–1737. [CrossRef]
61. Mitra, P.P.; Loqué, D. Histochemical staining of *Arabidopsis thaliana* secondary cell wall elements. *J. Vis. Exp.* **2014**, *87*, 51381. [CrossRef]
62. Lichtenthaler, H.K. Chlorophylls and carotenoids: Pigments of photosynthetic biomembranes. In *Methods Enzymology*; Academic Press: Cambridge, MA, USA, 1987; Volume 148, pp. 350–382.
63. Giusti, M.M.; Wrolstad, R.E. Characterization and measurement of anthocyanins by UV-visible spectroscopy. *Curr. Protoc. Food Anal. Chem.* **2001**, *1*, F1.2.1–F1.2.13. [CrossRef]
64. Dyankova, S.; Doneva, M. Extraction and characterization of anthocyanin colorants from plant sources. *Agric. Sci. Technol.* **2016**, *8*, 85–89. [CrossRef]
65. Jung, S.; Bassett, C.; Bielenberg, D.G.; Cheng, C.H.; Dardick, C.; Main, D.; Schaffer, R.J. A standard nomenclature for gene designation in the Rosaceae. *Tree Genet. Genomes* **2015**, *11*, 108. [CrossRef]
66. Else, M.A.; Stankiewicz-Davies, A.P.; Crisp, C.M.; Atkinson, C.J. The role of polar auxin transport through pedicels of *Prunus avium* L. in relation to fruit development and retention. *J. Exp. Bot.* **2004**, *55*, 2099–2109. [CrossRef] [PubMed]
67. Xu, Z.; Wu, J.; Jing, X.; Khan, F.S.; Chen, Y.; Chen, Z.; Zhao, X.; Wei, Y. Genome-wide identification of litchi SPL gene family and expression analysis in pericarp anthocyanin biosynthesis. *Horticulturae* **2024**, *10*, 762. [CrossRef]
68. Van de Wouwer, D.; Vanholme, R.; Decou, R.; Goeminne, G.; Audenaert, D.; Nguyen, L.; Morsa, S.; Lammertyn, F.; Van Doorsselaere, J.; Boerjan, W. Chemical genetics uncovers novel inhibitors of lignification, including p-iodobenzoic acid targeting CINNAMATE-4-HYDROXYLASE. *Plant Physiol.* **2016**, *172*, 198–220. [CrossRef]
69. Serrani, J.C.; Carrera, E.; Ruiz-Rivero, O.; Gallego-Giraldo, L.; Peres, L.E.P.; García-Martínez, J.L. Inhibition of auxin transport from the ovary or from the apical shoot induces parthenocarpic fruit-set in tomato mediated by gibberellins. *Plant Physiol.* **2010**, *153*, 851–862. [CrossRef]
70. Serrano, A.; Kuhn, N.; Restovic, F.; Meyer-Regueiro, C.; Madariaga, M.; Arce-Johnson, P. The glucose-related decrease in polar auxin transport during ripening and its possible role in grapevine berry coloring. *J. Plant Growth Regul.* **2023**, *42*, 365–375. [CrossRef]
71. Arancibia-Guerra, C.; Núñez-Lillo, G.; Cáceres-Mella, A.; Carrera, E.; Meneses, C.; Kuhn, N.; Pedreschi, R. Color desynchronization with softening of 'Hass' avocado: Targeted pigment, hormone and gene expression analysis. *Postharvest Biol. Technol.* **2022**, *194*, 112067. [CrossRef]

horticulturae

MDPI

Article

MdCDPK24 Encoding Calcium-Dependent Protein Kinase Enhances Apple Resistance to *Colletotrichum gloeosporioides*

Jiajun Shi [1,†], Yuxin Ma [1,†], Dajiang Wang [2,*] and Feng Wang [3,*]

1 College of Horticulture, Shenyang Agricultural University, Shenyang 110866, China
2 Research Institute of Pomology, Chinese Academy of Agricultural Sciences, Key Laboratory of Horticulture Crops Germplasm Resources Utilization, Ministry of Agriculture, Xingcheng 125100, China
3 College of Plant Protection, Shenyang Agricultural University, Shenyang 110866, China
* Correspondence: wangdajiang@caas.cn (D.W.); wangfeng@syau.edu.cn (F.W.)
† These authors contributed equally to this work.

Abstract

Calcium-dependent protein kinases (CDPKs) are unique serine/threonine kinases that play significant roles in response to environmental stresses in plants. In this study, we comprehensively characterized the *CDPK* gene family in the apple cultivar 'Hanfu' at the genome-wide level, and 38 *MdCDPKs* were identified. They were unevenly distributed across 14 chromosomes. Based on phylogenetic analysis, the MdCDPKs were classified into four subfamilies. Conserved domain analysis indicated that MdCDPKs contain the catalytic kinase domain and the Ca^{2+} binding domain. During *Colletotrichum gloeosporioides* infection, the expression level of *MdCDPK24* was significantly upregulated. Subsequently, *MdCDPK24* was fused to *GFP* to generate the *MdCDPK24-GFP* construct, and confocal microscopy imaging confirmed its cytoplasmic localization in *Nicotiana benthamiana* leaves. Using agrobacterium-mediated transformation, we generated the overexpression of *MdCDPK24* transgenic calli. *MdCDPK24*-overexpressing calli demonstrated significantly reduced disease severity against *C. gloeosporioides* infection, indicating its positive role in apple bitter rot resistance. The analysis of the *CDPK* gene family in the apple cultivar 'Hanfu' provides a new insight into the identification of *CDPK* genes involved in biotic stress. *MdCDPK24* represents a promising candidate for genetic manipulation to enhance apple bitter rot resistance.

Keywords: apple; *CDPK*; *Colletotrichum gloeosporioides*; apple bitter rot

1. Introduction

Apple (*Malus domestica*) is a globally significant economic crop, valued for its favorable organoleptic properties and nutritional composition [1]. China maintains the world's largest apple cultivation area and production output. Fungal pathogen infections during cultivation have substantially reduced per-unit-area yield and compromised fruit quality, emerging as primary constraints to sustainable apple production in China [2]. Consequently, effective disease management remains a critical agricultural priority. Apple bitter rot (caused by *Colletotrichum gloeosporioides*) is among the most devastating fungal diseases due to its rapid spread. Conventional chemical controls have limited efficacy, while prolonged pesticide application raises environmental contamination concerns [3]. Therefore, elucidating the molecular mechanisms of apple bitter rot resistance and identifying associated disease resistance genes presents significant theoretical and practical value for sustainable agriculture.

Environmental stresses are perceived by diverse sensors and transduced through distinct signaling networks, inducing downstream defense responses such as stress-related protein and metabolite synthesis [4–6]. Calcium ions (Ca^{2+}), as a universal second messenger, play a critical role in plant signal transduction pathways [7]. In plants, various stresses trigger a rapid transient increase in intercellular Ca^{2+} concentration [8,9]. Multiple Ca^{2+} sensors detect and interpret Ca^{2+} signatures, subsequently transducing them into diverse downstream effects [10]. Calcium-dependent protein kinases (CDPKs/CPKs) represent a major class of Ca^{2+}-sensitive serine/threonine (Ser/Thr) kinases, rapidly detect transient intracellular Ca^{2+} signals, recognize specific substrates, phosphorylate them, and thereby propagate and amplify the signals through downstream signaling cascades [11]. The *CDPK* gene family has been identified in many plants. *Arabidopsis thaliana* contains 34 *CDPK* genes by analysis of the genome sequence. *CDPK* genes were also identified in *Oryza sativa* (29 rice *CDPKs*), *Zea mays* (40 maize *CDPKs*), *Solanum tuberosum* (25 potato *CDPKs*), *Pyrus bretschneideri* (30 pear *CDPKs*), and *Populus trichocarpa* (30 polar *CDPKs*) [12–16]. CDPK proteins have four conserved domains: an N-terminal variable region, a Ser/Thr kinase catalytic domain, an autoinhibitory domain, and a calmodulin-like domain containing EF-hands for Ca^{2+} binding [17]. The N-terminal variable domain exhibits intra-species length and sequence heterogeneity while undergoing constitutive and stimulus-induced in vivo phosphorylation mediated by autophosphorylation or upstream kinases [18]. The catalytic domain contains a conserved Ser/Thr kinase domain with high sequence homology. Mutations within this domain abolish catalytic activity, impairing the phosphorylation of substrate proteins [19]. Crucially, CDPK activation by Ca^{2+} binding is a prerequisite for kinase functionality [17]. At the basal Ca^{2+} level, the binding between the autoinhibitory domain and the catalytic domain suppresses protein kinase activity. Under Ca^{2+} influx conditions, Ca^{2+} binds to the EF-hand domain, inducing conformational changes that relieve autoinhibition and initiate autophosphorylation, thereby activating CDPKs [20,21].

Studies across multiple plant species demonstrate that *CDPKs* mediate plant responses to diverse stimuli, including drought, salinity, and pathogen stress. Transgenic lines overexpressing *PtrCDPK10* increase ascorbate peroxidase activity and decrease ROS levels, elevating trifoliate orange drought stress tolerance [22]. In *Phyllostachys edulis*, *PheCPK1* functions as a negative regulator of drought stress responses by repressing stress-responsive gene expression and impairing ROS scavenging capability [23]. Under salt stress, the *AtCPK12*-RNAi mutant accelerates salt-induced damage, including Na^{+} accumulation and ROS production in roots [24]. *ZmCPK11* enhances maize salt tolerance by maintaining foliar Na^{+}/K^{+} homeostasis, which mitigates salt-induced chlorophyll degradation and photosystem II impairment [25]. Under abiotic stress, *CDPKs* enhance plant stress tolerance by modulating ROS levels and ion homeostasis. Accumulating evidence shows that *CDPK* plays an essential role in plant–pathogen interactions. *AcCDPK1* and *AcCDPK5* from onion function as positive regulators of *Phytophthora nicotianae* resistance [26]. OsCPK17 stabilizes the receptor-like cytoplasmic kinase OsRLCK176, thereby directly enhancing rice immunity signaling [27]. During pepper's response to *Ralstonia solanacearum* infection, CaWRKY27b is phosphorylated by CaCDPK29 and acts as a transcriptional activator of CaWRKY40 to promote plant immune responses [28]. During biotic stress, plants strengthen disease resistance via stabilizing disease-related proteins and upregulating resistance-gene expression. However, the role of apple *CDPK* genes in *C. gloeosporioides* resistance remains largely unclear.

In this study, we identified *CDPK* genes in 'Hanfu' apple and performed bioinformatics analysis of the *MdCDPK* family. Furthermore, we analyzed the expression levels of *MdCDPKs* during *C. gloeosporioides* infection using RT-qPCR to identify *C. gloeosporioides*-responsive *MdCDPK* genes. The selected *MdCDPK* gene induced by apple bitter rot was overexpressed in apple to assess disease resistance. Our findings provide a valuable *CDPK* candidate for the targeted enhancement of apple bitter rot resilience.

2. Materials and Methods

2.1. Plant Materials and Microbial Strains

Apple cultivar 'Orin' calli were cultured in Murashige and Skoog medium with 0.4 mg/L 6-Benzylaminopurine and 1.5 mg/L 2,4-Dichlorophenoxyacetic acid at 24 °C. After two weeks in the dark, calli were subcultured onto fresh medium. *N. benthamiana* was grown in a controlled-environment chamber under a 16 h light/8 h dark photoperiod at 26 °C. *C. gloeosporioides* was grown in potato dextrose agar (PDA) medium in the dark at 28 °C. After a one-week incubation, fungal mycelium was subcultured onto newly prepared PDA plates.

2.2. Identification of CDPK Genes in Apple

Protein sequences of AtCDPK were obtained from the TAIR database (https://www.arabidopsis.org/, accessed on 10 August 2025). All protein sequences in 'Hanfu' apple were downloaded from GDR (https://www.rosaceae.org/, accessed on 10 August 2025) [29]. Local blast was performed using all AtCDPK proteins as queries for the identification of CDPK proteins from apple. All putative candidates were manually verified with the Conserved Domain Database (CDD, https://www.ncbi.nlm.nih.gov/cdd/, accessed on 10 August 2025) to confirm the presence of the protein kinase domain and the calmodulin-like domain [30]. The amino acid number, isoelectric point, and molecular weight of apple CDPK proteins were analyzed using the ExPASy tool (https://web.expasy.org/protparam/, accessed on 10 August 2025) [31].

2.3. Chromosomal Locations of CDPKs in Apple

Chromosome location information of *MdCDPK* genes was obtained from GDR. Then, the mapping of *MdCDPK* genes to the chromosomes was performed using the Gene Location Visualize tool with TBtools II software v2.326 [32]. Input parameters included chromosome ID, chromosome length, gene ID, gene start position, and gene end position. Processing these inputs generated a visualization of *MdCDPK* gene locations on the chromosome.

2.4. Phylogenetic Analysis

Protein sequences of AtCDPK and MdCDPK were used for phylogenetic tree analysis. A phylogenetic tree was constructed using MEGA X software utilizing the Neighbor-Joining method and the Poisson model [33]. All protein sequences were aligned using MUSCLE in MEGA X software with UPGMA clustering and default gap penalties. Following alignment, divergent sequences were trimmed from the termini. The curated dataset was subjected to phylogenetic reconstruction in MEGA X via the Neighbor-Joining method under the following parameters: bootstrap analysis (1000 replicates), uniform substitution rates, and pairwise deletion for gap/missing data treatment.

2.5. Conserved Motif and Domain Analysis

Multiple conserved motifs of MdCDPK proteins were identified by the MEME Suite online tool [34]. MdCDPK protein sequences were submitted through the online portal, followed by selection of the 'Classic Motif' discovery model, while retaining all other parameters at default settings. MdCDPK motif analysis results are available for download upon completion through the web interface. The conserved domain of MdCDPK proteins was analyzed by the CDD from the NCBI databases. Consistent with the motif identification protocol, MdCPK protein sequences were submitted to the batch CD-search web platform using default parameters. Conserved domain analysis results were subsequently retrieved from the specified output directory.

2.6. Gene Expression Analysis

Total RNA was isolated from 'Hanfu' leaves or apple calli using a CTAB protocol [35]. Gene expression with RT-qPCR was performed as previously described with minor modifications [36]. RT-qPCR reactions were conducted with UltraSYBR Mixture (CWBIO) on an ABI QuantStudio™ 6 Flex system. Each 10 μL reaction contained 10 μL UltraSYBR mix, 1 μL gene-specific primers (10 μM), 0.5 μL cDNA template, and 3.5 μL nuclease-free water. The thermal cycling conditions were as follows: initial denaturation, 95 °C for 10 min; 40 cycles, 95 °C for 15 s, 60 °C for 1 min; melt curve analysis, 60–95 °C. The results were calculated by normalization to *MdEF-1α* (GenBank: DQ341381 in the NCBI). Specific primers are listed in Table S2. Each experiment was performed with 3 biological replicates.

2.7. Promoter Cis-Acting Element Analysis

The promoter region (2000 bp) of the *MdCDPK24* gene was extracted from the 'Hanfu' apple genome. The *MdCDPK24* promoter sequence was analyzed for *cis*-acting elements using the PlantCARE website (https://bioinformatics.psb.ugent.be/webtools/plantcare/html/, accessed on 10 August 2025). The *MdCDPK24* promoter sequence was inserted into the designated input field of the bioinformatics tool PlantCARE. The analysis request was submitted after providing a valid email address for automated result delivery. *Cis*-acting elements in the *MdCDPK24* promoter were integrated from PlantCARE online tool analysis. The results are listed in Table S3.

2.8. Subcellular Localization Analysis

The CDS sequence of *MdCDPK24* without the stop codon was cloned into the pRI101-GFP. Then, the pRI101-*MdCDPK24-GFP* construct was introduced into *Agrobacterium tumefaciens* EHA105. MdCDPK24-GFP and NF-YA4-mCherry (a nuclear marker) were expressed in 4-week-old *N. benthamiana* using a previously described protocol [37]. Briefly, EHA105 was cultured overnight at 28 °C in Luria–Bertani medium supplemented with 50 μg/mL kanamycin and 25 μg/mL rifampicin. Then, 1 mL of the culture was transferred to 50 mL of fresh Luria–Bertani medium and incubated with shaking until an optical density was reached (OD_{600} = 0.5). The culture was harvested by centrifugation at 5000 rpm for 5 min and resuspended in infiltration buffer to an OD_{600} of 0.5. Suspensions were infiltrated into the abaxial surface of *N. benthamiana* leaves using a blunt syringe. The fluorescence signal was obtained using a confocal microscope (Leica DMi8, Wetzlar, Germany).

2.9. Generation of Transgenic Apple Calli and Fungal Treatment

For overexpressing *MdCDPK24* in 'Orin' calli, the full-length coding region of *MdCDPK24* was amplified from 'Hanfu' leaves. Then, *MdCDPK24* was cloned into the pRI101-AN to generate pRI101-*MdCDPK24*. The construct was transferred into *A. tumefaciens* EHA105 and then transformed into calli using the previously described method [37]. For *C. gloeosporioides* inoculation treatment, mycelia were inoculated onto wild-type (WT, 'Orin') and *MdCDPK24*-overexpressing transgenic calli. Following inoculation, calli were maintained in darkness at 25 °C. Disease symptoms were photographed at 10 days.

2.10. Transient Overexpression of Genes in Fruit

'Hanfu' fruits were brought from the market. The transiently transformed apple fruit was performed using a previously established method [37]. EHA105 culture refers to subcellular localization analysis. The culture was resuspended in infiltration buffer to an

OD_{600} of 0.3. Suspensions were infiltrated into apple fruit using a blunt syringe. Then, *C. gloeosporioides* mycelia were inoculated on the surface of apple fruits. Disease symptoms were photographed at 5 d.

2.11. Statistical Analysis

Statistical analysis in this study was conducted with SPSS software 26. Data are shown as means ± standard deviations. All datasets underwent normality testing and satisfied the criteria for normal distribution. Data appropriate for one-way ANOVA were evaluated using homogeneity tests. Significance of differences was determined using one-way ANOVA and Duncan's test (RT-qPCR analysis of *MdCDPKs* during *C. gloeosporioides* infection) or Student's *t*-test (RT-qPCR analysis of transgenic calli and analysis of apple bitter rot incidence following inoculation in apple). Different lower-case letters or asterisks indicate statistically significant differences ($p < 0.05$).

3. Results

3.1. Identification and Chromosomal Distribution of CDPK Gene Family in Apple

To identify *CDPK* genes in apple, an apple genome database (HFTH1 Whole Genome v1.0) was searched using 34 AtCDPK proteins as a query sequence. A total of 38 *MdCDPK* genes were systematically designated as *MdCDPK1* to *MdCDPK38* based on their chromosomal positions (Table S1). These genes exhibited open reading frame lengths ranging from 1488 to 5109 bp, corresponding to proteins of 495–1702 amino acids. The encoded proteins showed molecular masses of 55.63–190.26 kDa and predicted isoelectric points of 5.19–9.31. Chromosomal mapping revealed uneven distribution of all 38 *MdCDPKs* across 14 chromosomes (Figure 1). Chromosome 5 harbored the highest gene number (five), followed by chromosomes 10/12/14 (four each), chromosomes 2/3/7/15 (three each), chromosomes 6/9/11 (two each), and chromosomes 1/4/17 (one each).

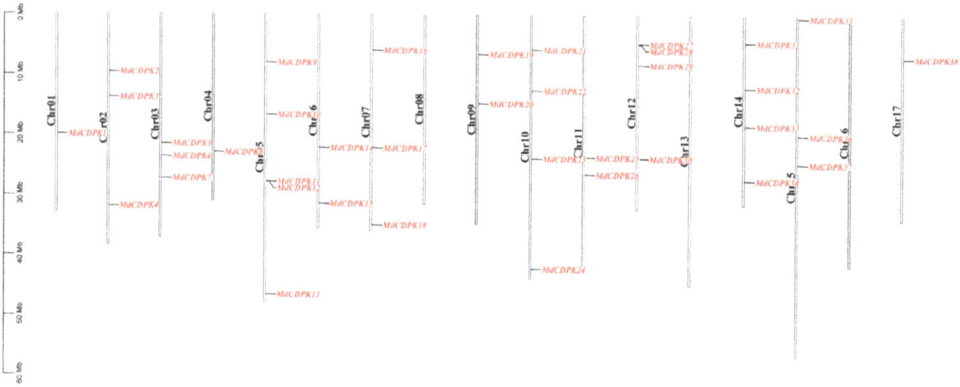

Figure 1. Chromosomal distribution of apple *CDPK* genes.

3.2. Phylogenetic Analysis of MdCDPKs

To analyze the evolutionary relationship, we constructed a phylogenetic tree using 72 CDPK sequences from apple and *A. thaliana* (Figure 2). Based on branch topology alignment and conserved domains of AtCDPK proteins, the 38 MdCDPKs were phylogenetically classified into four subfamilies (I-IV). Subfamily I contained 11 CDPKs from apple (MdCDPK4, MdCDPK9, MdCDPK11, MdCDPK12, MdCDPK16, MdCDPK20, MdCDPK21, MdCDPK23, MdCDPK27, MdCDPK28, and MdCDPK31). Subfamily IV consisted of 10 apple CDPKs (MdCDPK1, MdCDPK2, MdCDPK3, MdCDPK10, MdCDPK17, Md-

CDPK18, MdCDPK22, MdCDPK32, MdCDPK36, and MdCDPK37). Subfamily III contained nine CDPKs from apple (MdCDPK8, MdCDPK6, MdCDPK15, MdCDPK19, MdCDPK26, MdCDPK29, MdCDPK30, MdCDPK34, and MdCDPK38). Subfamily II was the smallest group and comprised eight apple CDPKs (MdCDPK5, MdCDPK7, MdCDPK13, MdCDPK14, MdCDPK24, MdCDPK25, MdCDPK33, and MdCDPK35).

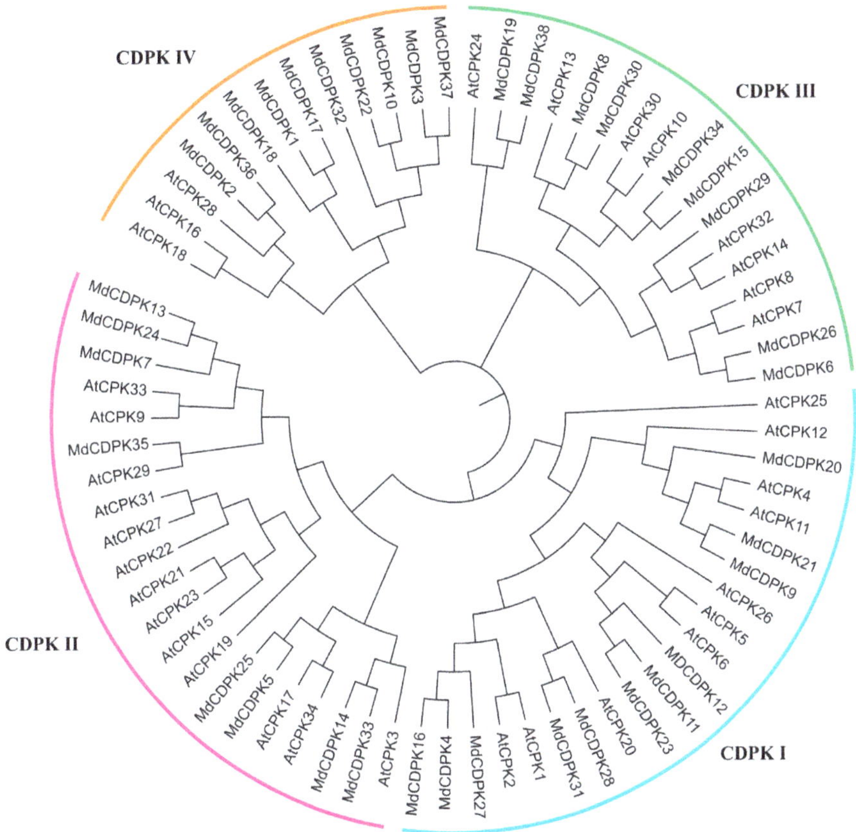

Figure 2. Phylogenetic relationships of CDPK proteins from apple and *Arabidopsis thaliana*. The phylogenetic tree was constructed using MEGA X software by the Neighbor-Joining method with 1000 bootstrap replicates. The subfamily of CDPK proteins is indicated by the distinct color lines.

3.3. Conserved Motif and Domain Analysis of MdCDPKs

The conserved motif of 38 full-length MdCDPKs was analyzed using the MEME program (Figure 3a). All MdCDPK proteins shared conserved motifs 1 to 3. The conserved domain of 38 MdCDPKs was analyzed based on the Conserved Domain Database using the National Center for Biotechnology Information (NCBI) online website (Figure 3a). All MdCDPK proteins contained two conserved domains, including a STKc_CAMK domain (calcium- and calmodulin-stimulated Ser/Thr kinases domain). Most MdCDPKs possessed the PTZ00184 superfamily domain. MdCDPK2, MdCDPK31, and MdCDPK36 contained the FRQ1 domain. The PTZ00183 superfamily domain only existed in the MdCDPK12 protein.

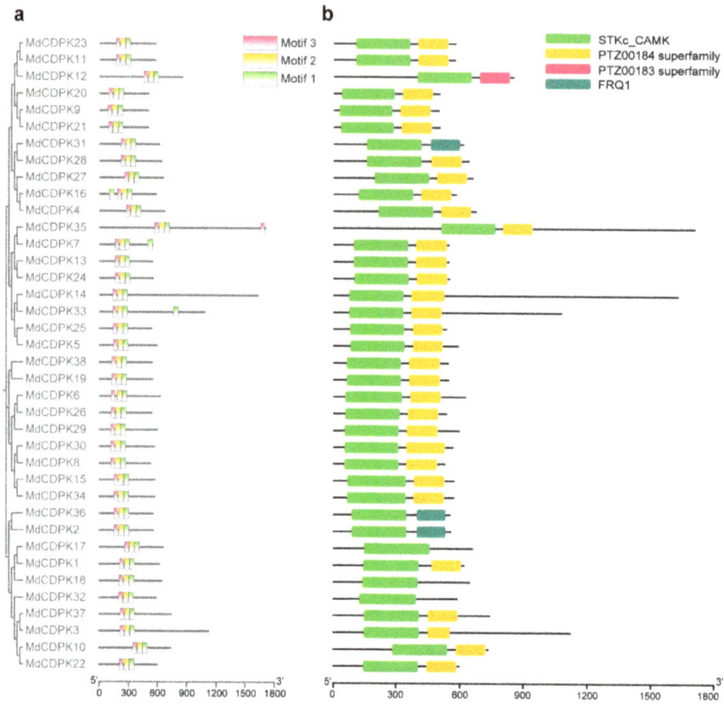

Figure 3. Conserved motif and domain analysis of CDPK members in apple. (**a**) Conserved motif distribution of MdCDPKs. Three motifs are shown by different colored boxes on the right. (**b**) Conserved domain distribution of MdCDPK proteins. Different colored boxes indicate different domains on the right.

3.4. Expression Analysis of MdCDPKs During C. gloeosporioides Infection

To elucidate the transcriptional dynamics of *MdCDPK*s following *C. gloeosporioides* infection, the expression patterns of *MdCDPK* genes in 'Hanfu' leaves treated with *C. gloeosporioides* were examined by RT-qPCR. Following RT-qPCR analysis of 38 *MdCDPK* genes, we selected 9 genes exhibiting altered relative expression during 0–48 h (Figure 4). The relative expression levels of *MdCDPK3*, *MdCDPK8*, *MdCDPK25*, and *MdCDPK31* showed a slight downward trend within 48 h post-inoculation with *C. gloeosporioides*. Only *MdCDPK24* exhibited significantly increased expression, progressively rising from 0 to 48 h. These results suggest that *MdCDPK24* is closely associated with the *C. gloeosporioides* infection process, warranting further investigation into its potential role in pathogen response. To understand the potential functions of MdCDPK24, subcellular localization of MdCDPK24-GFP was investigated in *Nicotiana benthamiana* leaves. In *N. benthamiana* cells transfected with pRI101-GFP, GFP fluorescence was uniformly distributed throughout the cells. *N. benthamiana* leaves expressing MdCDPK24-GFP showed cytoplasm-specific fluorescence, demonstrating cytoplasmic localization of the MdCDPK24-GFP fusion protein (Figure 5).

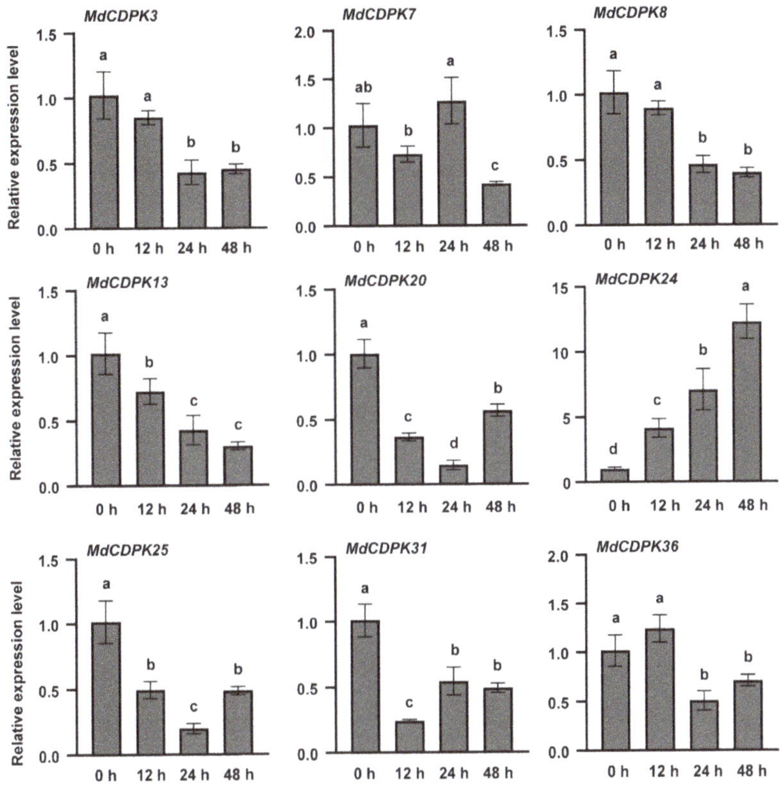

Figure 4. Expression patterns of nine *MdCDPK* genes under *C. gloeosporioides* treatment by RT-qPCR. Data are shown as means. Vertical bars represent standard deviations. Different letters indicate significant differences ($p < 0.05$) using one-way ANOVA and Duncan's test.

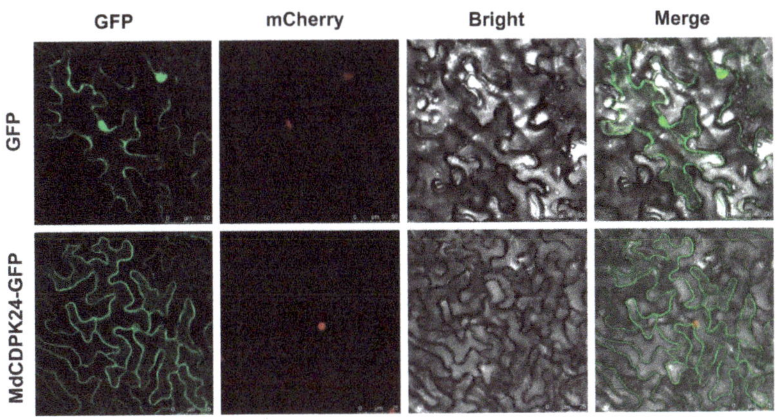

Figure 5. Subcellular localization of MdCDPK24-GFP. NF-YA4-mCherry is a nuclear marker. The scale bars of the control group (GFP) are 50 μm. The scale bars of the experimental group (MdCDPK24-GFP) are 75 μm.

3.5. *MdCDPK24* Positively Regulates Apple Bitter Rot Resistance

To further investigate whether *MdCDPK24* is involved in resistance to *C. gloeosporioides*, we conducted stable overexpression of *MdCDPK24* (*MdCDPK24*-OE) in the 'Orin' calli. The relative expression of *MdCDPK24* significantly increased in *MdCDPK24*-OE calli compared with the control (Figure 6a). We inoculated *MdCDPK24*-OE calli with *C. gloeosporioides* for 10 days and found that the overexpression of *MdCDPK24* significantly reduced the mycelium growth (Figure 6b). We also performed transient *MdCDPK24* overexpression tests in the fruits of 'Hanfu' apple. Following the transient overexpression of *MdCDPK24*, apple fruit was inoculated with *C. gloeosporioides* (Figure 6c). We found that the lesion diameters on *MdCDPK24*-overexpressing (pRI101-*MdCDPK24*) fruits were significantly smaller than those on empty vector (pRI101-AN) control fruits at 5 d (Figure 6d). Longitudinal section analysis revealed significantly shallower lesion depths in fruits expressing pRI101-*MdCDPK24* compared to the control group at 5 d (Figure 6e,f). These results demonstrate that *MdCDPK24* overexpression reduces *C. gloeosporioides* lesion expansion and suppresses pathogen development.

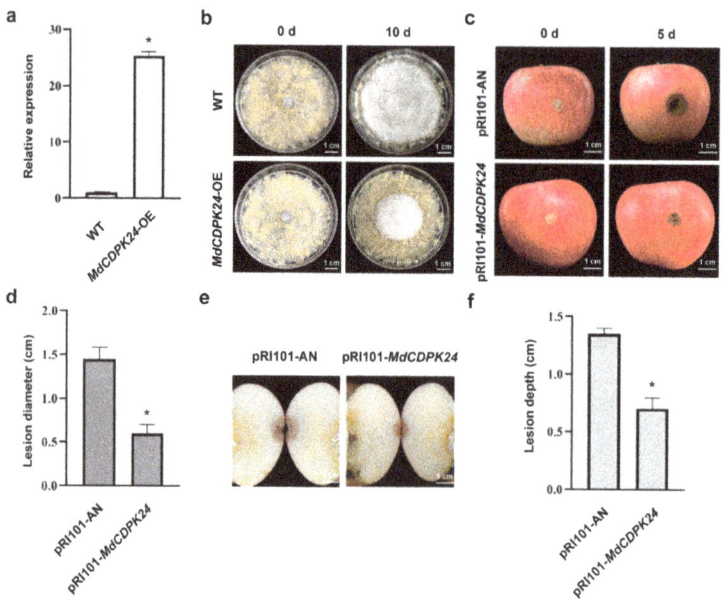

Figure 6. *MdCDPK24* overexpression increases resistance to *C. gloeosporioides* infection. (**a**) Expression levels of *MdCDPK24* in wild-type (WT) and *MdCDPK24* overexpression calli. Data are shown as means. Vertical bars represent standard deviations. An asterisk indicates significant differences (* $p < 0.05$) using Student's *t*-test. (**b**) Phenotype of wild-type (WT) and *MdCDPK24* overexpression calli after inoculation with *C. gloeosporioides* for 10 days. Scale bars = 1 cm. (**c**) Phenotype of fruit transiently overexpressing *MdCDPK24* (pRI101-*MdCDPK24*) or empty vector (pRI101-AN) after inoculation with *C. gloeosporioides* for 5 d. Scale bars = 1 cm. (**d**) Lesion diameter of fruit transiently overexpressing *MdCDPK24* (pRI101-*MdCDPK24*) or empty vector (pRI101-AN) after inoculation with *C. gloeosporioides* for 5 d. Data are shown as means. Vertical bars represent standard deviations. Asterisk indicates significant differences (* $p < 0.05$) using Student's *t*-test. (**e**) Longitudinal section of fruit transiently overexpressing *MdCDPK24* (pRI101-*MdCDPK24*) or empty vector (pRI101-AN) after inoculation with *C. gloeosporioides* at 5 d. Scale bars = 1 cm. (**f**) Lesion depth of fruit transiently overexpressing *MdCDPK24* (pRI101-*MdCDPK24*) or empty vector (pRI101-AN) after inoculation with *C. gloeosporioides* at 5 d. Data are shown as means. Vertical bars represent standard deviations. An asterisk indicates significant differences (* $p < 0.05$) using Student's *t*-test.

4. Discussion

CDPKs serve as primary calcium receptors and central regulators of plant development and stress responses. Utilizing genomic data from 'Hanfu' apple, we identified the *CDPK* gene family members. Phylogenetic analysis and conserved domain characterization revealed evolutionary relationships within this family. Pathogen inoculation with *C. gloeosporioides* induced *MdCDPK24* expression. Subsequently, subcellular localization of MdCDPK24 was performed. We characterized the role of *MdCDPK24* in conferring resistance to apple bitter rot caused by *C. gloeosporioides*.

Since the whole genome sequencing of multiple plant species has been completed, the *CDPK* gene family has been systematically identified, cloned, and analyzed in several plants. Genome-wide analyses have identified 34, 29, and 30 *CDPKs* in *A. thaliana* [12], rice [13], and pear [14]. In our study, we found 38 *CDPKs* in 'Hanfu' apple (Table S1). The apple, pear, rice, and *A. thaliana* genomes harbor comparable numbers of *CDPK* genes. Plants may require approximately 30 CDPK genes to transduce Ca^{2+} signaling cascades. The conserved domain is closely related to gene function. The conserved CDPK consists of three domains: the variable N-terminal domain, the central kinase catalytic domain, and the C-terminal CDPK activation domain [38]. In apple, CDPK proteins exhibit highly conserved amino acid sequences, particularly in their catalytic domain (Figure 3b). When CDPK is activated by Ca^{2+}, the catalytic domain phosphorylates multiple substrates, propagating calcium signals downstream to initiate environmental adaptation responses. AtCPK12 phosphorylates two abscisic acid-responsive transcription factors (ABF1 and ABF4) to regulate seed germination and post-germination growth [39]. OsCPK17 regulates rice cold stress tolerance through the phosphorylation of sucrose–phosphate synthase OsSPS4 and aquaporins OsPIP2;1/OsPIP2;6, thereby modulating sugar metabolism and membrane channel activity [40]. *StCDPK7* is transcriptionally activated during the *Phytophthora infestans* infection of potato and phosphorylates phenylalanine ammonia lyase proteins [41]. These studies suggest that CDPKs in apple potentially regulate both developmental processes and stress responses through substrate phosphorylation mediated by their conserved kinase domains.

CDPK genes are ubiquitously distributed in diverse plant species and play critical roles in mediating abiotic stress responses. ABA signaling orchestrates S-type anion channel SLAH3 activation via spatial regulation within plasma membrane nanodomains, where CPK21 phosphorylates SLAH3 but is antagonized by ABI1 phosphatase. The regulatory components of ABA receptor 1/pyrabactin resistance-like protein 9 (RCAR1/PYL9) modulate this CPK21-SLAH3-ABI1 complex assembly, demonstrating that nanodomain partitioning controls phosphorylation-dependent anion channel gating during drought stress responses [42]. *CPK4* and *CPK11* act as positive regulators in calcium-mediated abscisic acid (ABA) signaling, where ABA-induced kinase activity phosphorylates transcription factors ABF1/ABF4 to modulate ABA-responsive gene expression. Genetic evidence demonstrates that *cpk4*/*cpk11* mutants exhibit ABA- and salt-insensitive phenotypes during germination and seedling development, establishing their essential role in whole-plant salt stress adaptation [43]. These findings establish that CDPKs are central regulators of drought and salt stress responses. Notably, ABA orchestrates key CDPK-mediated signaling cascades during plant stress adaptation. Analysis of the *MdCDPK24* promoter identified multiple ABRE elements, indicating that this kinase may mediate ABA-responsive signaling during abiotic stress responses (Table S3). Moreover, we identified multiple methyl jasmonate (MeJA) response elements, critical for mediating plant responses to both biotic and abiotic stresses. Exogenous ABA and MeJA application may transcriptionally induce *MdCDPK24* expression to enhance multi-stress resistance in apple. Optimal dosage for field-scale deployment requires systematic optimization.

A phylogenetic tree enables the inference of gene evolutionary histories and the prediction of gene functions [44]. Consequently, phylogenetic analysis constitutes a fundamental component of gene family research. A phylogenetic tree analysis of 38 MdCDPKs and 34 AtCDPKs showed that the MdCDPK proteins were classified into four subfamilies (Figure 2). Proteins grouped within the same subfamily typically share conserved motifs or domains, which indicates close evolutionary relationships. Therefore, MdCDPKs and AtCDPK homologs within the same phylogenetic clade likely exhibit conserved functions in environmental stress responses. MdCDPK24 clusters with AtCDPK3 in the same clade position. AtCPK3 phosphorylates actin-depolymerization factor 4 (ADF4) to remodel the actin cytoskeleton, thereby enabling stomatal immunity and resistance to pathogenic bacteria [45]. Phylogenetic tree analysis implicated MdCDPK24 in biotic stress responses, though functional validation remains necessary. Determining subcellular localization is fundamental to characterizing gene function [46]. In *A. thaliana*, CDPK isoforms display distinct subcellular localization, including cytoplasm, nucleus, cell membrane, and peroxisome membrane compartments [47]. In tomato, CDPKs are predominantly localized to the cytoplasm based on predictive analyses, with a minority exhibiting plasma membrane association [48]. Subcellular localization results showed that the protein of MdCDPK24 was a cytoplasmic localization protein (Figure 5). Multiple enzymes regulating plant growth and stress responses—including superoxide dismutase [49], catalase [50], POD, and CAD [51]—localize in the cytoplasm, thus creating spatial proximity conducive to functional interactions with CDPK.

CDPKs also mediate biotic stress signaling pathways in plants. CPK5 acts as a master regulator of plant immunity through phosphorylation of the NADPH oxidase and respiratory burst oxidase homolog D (RBOHD) upon pathogen-associated molecular pattern (PAMP) perception, initiating a self-propagating ROS-CPK5 signaling circuit. This mutual activation cascade enables rapid systemic defense, potentiating salicylic acid-dependent resistance against *Pseudomonas syringae* pv. *tomato* DC3000 and coordinating defense-related transcriptional reprogramming in distal tissues [52]. The E3 ubiquitin ligases ARABIDOPSIS TÓXICOS EN LEVADUR (ATL31 and ATL6) enhance plant immunity by targeting CPK28 for proteasomal degradation upon flg22 elicitation, thereby stabilizing the BOTRYTIS-INDUCED KINASE1 (BIK1). This ubiquitin-proteasome pathway attenuates CPK28-mediated BIK1 turnover, revealing a dual-layer regulatory mechanism for immune signal amplification [53]. Mutational analyses reveal that phosphorylation of CPK28 at Ser318 is essential for ATL31 binding and subsequent ubiquitination. Further analyses confirm that CPK28 dimerization drives intermolecular autophosphorylation, creating a self-regulating loop that controls kinase stability through the ATL31/6 degradation machinery [54]. *VpCDPK9* and *VpCDPK13* activate salicylic acid and ethylene signaling pathways to confer powdery mildew resistance in grapevine [55]. In apple, multiple *CDPK* genes were significantly up-regulated after *Alternaria alternata* inoculation [56]. The potential biological function of apple *CDPK* genes in plant biotic stress responses remains unknown. Our study demonstrates that *MdCDPK24* transcription is induced by *C. gloeosporioides* infection in apple and also positively regulates plant disease resistance (Figures 4 and 6). However, the potential role of MdCDPK24 in regulating apple defense against *C. gloeosporioides* via protein–protein interaction and phosphorylation requires further experimental validation. Taken together, although an apple bitter rot resistance gene has been identified, substantial efforts remain to bridge the translational gap between theoretical discovery and breeding applications. Further investigation is required to determine whether this gene confers broad-spectrum resistance against major apple pathogens.

5. Conclusions

In this study, genome-wide identification in apple revealed 38 *CDPK* genes that were unevenly distributed across 14 chromosomes. Phylogenetic analysis with Arabidopsis orthologs classified these into four subgroups (I-IV). Following *C. gloeosporioides* infection, the expression level of *MdCDPK24* was induced, which indicates that it might participate in apple resistance to the bitter rot pathogen. Subcellular localization analysis confirmed cytoplasmic-specific localization of MdCDPK24. Functional analysis demonstrated that overexpression of *MdCDPK24* enhances resistance to bitter rot in apple calli and fruit. These findings provide valuable knowledge of the *MdCDPK* gene family and establish *MdCDPK24* as a candidate gene for the molecular breeding of bitter-rot-resistant apple cultivars.

Supplementary Materials: The following supporting information can be downloaded at https://www.mdpi.com/article/10.3390/horticulturae11080942/s1: Table S1: Characterization of CDPKs in apple. Table S2: List of the primers used in this study. Table S3: Analysis of the *cis*-elements in the promoter of *MdCDPK24*.

Author Contributions: Conceptualization, F.W. and D.W.; validation, Y.M.; investigation, Y.M.; resources, F.W.; writing—original draft preparation, J.S.; writing—review and editing, J.S. and F.W.; visualization, Y.M. and J.S.; supervision, F.W. and D.W.; project administration, F.W. and D.W.; funding acquisition, F.W. All authors have read and agreed to the published version of the manuscript.

Funding: This research was funded by the Nature Science Project of the Technology Department of Liaoning Province, grant number 2023-MSLH-277.

Data Availability Statement: Data is contained within the article or Supplementary Material.

Conflicts of Interest: The authors declare no conflicts of interest.

References

1. Wang, N.; Wolf, J.; Zhang, F. Towards Sustainable Intensification of Apple Production in China—Yield Gaps and Nutrient Use Efficiency in Apple Farming Systems. *J. Integr. Agric.* **2016**, *15*, 716–725. [CrossRef]
2. Liang, X.; Zhang, R.; Gleason, M.L.; Sun, G. Sustainable Apple Disease Management in China: Challenges and Future Directions for a Transforming Industry. *Plant Dis.* **2022**, *106*, 786–799. [CrossRef]
3. Singh, A.; Dhiman, N.; Kar, A.K.; Singh, D.; Purohit, M.P.; Ghosh, D.; Patnaik, S. Advances in Controlled Release Pesticide Formulations: Prospects to Safer Integrated Pest Management and Sustainable Agriculture. *J. Hazard. Mater.* **2020**, *385*, 121525. [CrossRef]
4. Romeis, T.; Ludwig, A.A.; Martin, R.; Jones, J.D.G. Calcium-dependent Protein Kinases Play an Essential Role in a Plant Defence Response. *EMBO J.* **2001**, *20*, 5556–5567. [CrossRef] [PubMed]
5. Su, Z.; Han, C.; Qiao, Q.; Li, C.; Dong, H.; Wang, X.; Qi, K.; Xie, Z.; Huang, X.; Zhang, S. Genome-Wide Analysis of the Family 10 Plant Pathogenesis-Related Proteins in Pyrus Bretschneideri and Functional Analysis of PbrMLP for Colletotrichum Fructicola Resistance. *Hortic. Adv.* **2024**, *2*, 21. [CrossRef]
6. Margaritopoulou, T.; Baira, E.; Anagnostopoulos, C.; Vichou, K.-E.; Markellou, E. Phospholipid Production and Signaling by a Plant Defense Inducer against Podosphaera Xanthii Is Genotype-Dependent. *Hortic. Res.* **2024**, *11*, uhae190. [CrossRef]
7. Kudla, J.; Becker, D.; Grill, E.; Hedrich, R.; Hippler, M.; Kummer, U.; Parniske, M.; Romeis, T.; Schumacher, K. Advances and Current Challenges in Calcium Signaling. *New Phytol.* **2018**, *218*, 414–431. [CrossRef] [PubMed]
8. Romeis, T.; Herde, M. From Local to Global: CDPKs in Systemic Defense Signaling upon Microbial and Herbivore Attack. *Curr. Opin. Plant Biol.* **2014**, *20*, 1–10. [CrossRef]
9. Bhar, A.; Chakraborty, A.; Roy, A. The Captivating Role of Calcium in Plant-Microbe Interaction. *Front. Plant Sci.* **2023**, *14*, 1138252. [CrossRef]
10. Lee, J.; Rudd, J.J. Calcium-Dependent Protein Kinases: Versatile Plant Signalling Components Necessary for Pathogen Defence. *Trends Plant Sci.* **2002**, *7*, 97–98. [CrossRef]

11. Bredow, M.; Monaghan, J. Cross-Kingdom Regulation of Calcium- and/or Calmodulin-Dependent Protein Kinases by Phospho-Switches That Relieve Autoinhibition. *Curr. Opin. Plant Biol.* **2022**, *68*, 102251. [CrossRef]
12. Cheng, S.-H.; Willmann, M.R.; Chen, H.-C.; Sheen, J. Calcium Signaling through Protein Kinases. The Arabidopsis Calcium-Dependent Protein Kinase Gene Family. *Plant Physiol.* **2002**, *129*, 469–485. [CrossRef]
13. Asano, T.; Tanaka, N.; Yang, G.; Hayashi, N.; Komatsu, S. Genome-Wide Identification of the Rice Calcium-Dependent Protein Kinase and Its Closely Related Kinase Gene Families: Comprehensive Analysis of the CDPKs Gene Family in Rice. *Plant Cell Physiol.* **2005**, *46*, 356–366. [CrossRef]
14. Liu, S.; Li, J.; Li, N.; Zhou, P.; Li, L. Genome-Wide Identification of Calcium-Dependent Protein Kinases (CDPKs) in Pear (Pyrus Bretschneideri Rehd) and Characterization of Their Responses to Venturia Nashicola Infection. *Hortic. Environ. Biotechnol.* **2022**, *63*, 903–915. [CrossRef]
15. Dekomah, S.D.; Wang, Y.; Qin, T.; Xu, D.; Sun, C.; Yao, P.; Liu, Y.; Bi, Z.; Bai, J. Identification and Expression Analysis of Calcium-Dependent Protein Kinases Gene Family in Potato Under Drought Stress. *Front. Genet.* **2022**, *13*, 874397. [CrossRef]
16. Zuo, R.; Hu, R.; Chai, G.; Xu, M.; Qi, G.; Kong, Y.; Zhou, G. Genome-Wide Identification, Classification, and Expression Analysis of CDPK and Its Closely Related Gene Families in Poplar (Populus Trichocarpa). *Mol. Biol. Rep.* **2013**, *40*, 2645–2662. [CrossRef] [PubMed]
17. Yip Delormel, T.; Boudsocq, M. Properties and Functions of Calcium-Dependent Protein Kinases and Their Relatives in Arabidopsis Thaliana. *New Phytol.* **2019**, *224*, 585–604. [CrossRef] [PubMed]
18. Ito, T.; Nakata, M.; Fukazawa, J.; Ishida, S.; Takahashi, Y. Alteration of Substrate Specificity: The Variable N-Terminal Domain of Tobacco Ca2+-Dependent Protein Kinase Is Important for Substrate Recognition. *Plant Cell* **2010**, *22*, 1592–1604. [CrossRef] [PubMed]
19. Schulz, P.; Herde, M.; Romeis, T. Calcium-Dependent Protein Kinases: Hubs in Plant Stress Signaling and Development. *Plant Physiol.* **2013**, *163*, 523–530. [CrossRef]
20. Chandran, V.; Stollar, E.J.; Lindorff-Larsen, K.; Harper, J.F.; Chazin, W.J.; Dobson, C.M.; Luisi, B.F.; Christodoulou, J. Structure of the Regulatory Apparatus of a Calcium-Dependent Protein Kinase (CDPK): A Novel Mode of Calmodulin-Target Recognition. *J. Mol. Biol.* **2006**, *357*, 400–410. [CrossRef]
21. Wernimont, A.K.; Artz, J.D.; Finerty, P.; Lin, Y.-H.; Amani, M.; Allali-Hassani, A.; Senisterra, G.; Vedadi, M.; Tempel, W.; Mackenzie, F.; et al. Structures of Apicomplexan Calcium-Dependent Protein Kinases Reveal Mechanism of Activation by Calcium. *Nat. Struct. Mol. Biol.* **2010**, *17*, 596–601. [CrossRef] [PubMed]
22. Meng, L.; Zhang, Q.; Yang, J.; Xie, G.; Liu, J.-H. PtrCDPK10 of *Poncirus Trifoliata* Functions in Dehydration and Drought Tolerance by Reducing ROS Accumulation via Phosphorylating PtrAPX. *Plant Sci.* **2020**, *291*, 110320. [CrossRef] [PubMed]
23. Wu, M.; Liu, H.; Wang, L.; Zhang, X.; He, W.; Xiang, Y. Comparative Genomic Analysis of the CPK Gene Family in Moso Bamboo (Phyllostachys Edulis) and the Functions of PheCPK1 in Drought Stress. *Protoplasma* **2023**, *260*, 171–187. [CrossRef] [PubMed]
24. Zhang, H.; Zhang, Y.; Deng, C.; Deng, S.; Li, N.; Zhao, C.; Zhao, R.; Liang, S.; Chen, S. The Arabidopsis Ca2+-Dependent Protein Kinase CPK12 Is Involved in Plant Response to Salt Stress. *Int. J. Mol. Sci.* **2018**, *19*, 4062. [CrossRef]
25. Borkiewicz, L.; Polkowska-Kowalczyk, L.; Cieśla, J.; Sowiński, P.; Jończyk, M.; Rymaszewski, W.; Szymańska, K.P.; Jaźwiec, R.; Muszyńska, G.; Szczegielniak, J. Expression of Maize Calcium-Dependent Protein Kinase (ZmCPK11) Improves Salt Tolerance in Transgenic Arabidopsis Plants by Regulating Sodium and Potassium Homeostasis and Stabilizing Photosystem II. *Physiol. Plant.* **2020**, *168*, 38–57. [CrossRef]
26. Qin, L.; Zhang, X.; Ren, S.; Zhao, J.; Zhu, Z.; Wang, Y. Overexpression of *AcCDPK1* and *AcCDPK5* from *Allium Cepa.* L Enhances *Phytophthora nicotianae* Resistance in *Nicotiana benthamiana*. *Sci. Hortic.* **2024**, *326*, 112743. [CrossRef]
27. Mou, B.; Zhao, G.; Wang, J.; Wang, S.; He, F.; Ning, Y.; Li, D.; Zheng, X.; Cui, F.; Xue, F.; et al. The OsCPK17-OsPUB12-OsRLCK176 Module Regulates Immune Homeostasis in Rice. *Plant Cell* **2024**, *36*, 987–1006. [CrossRef]
28. Yang, S.; Cai, W.; Shen, L.; Cao, J.; Liu, C.; Hu, J.; Guan, D.; He, S. A CaCDPK29–CaWRKY27b Module Promotes CaWRKY40-Mediated Thermotolerance and Immunity to Ralstonia Solanacearum in Pepper. *New Phytol.* **2022**, *233*, 1843–1863. [CrossRef]
29. Jung, S.; Lee, T.; Cheng, C.-H.; Buble, K.; Zheng, P.; Yu, J.; Humann, J.; Ficklin, S.P.; Gasic, K.; Scott, K.; et al. 15 Years of GDR: New Data and Functionality in the Genome Database for Rosaceae. *Nucleic Acids Res* **2019**, *47*, D1137–D1145. [CrossRef]
30. Wang, J.; Chitsaz, F.; Derbyshire, M.K.; Gonzales, N.R.; Gwadz, M.; Lu, S.; Marchler, G.H.; Song, J.S.; Thanki, N.; Yamashita, R.A.; et al. The Conserved Domain Database in 2023. *Nucleic Acids Res.* **2023**, *51*, D384–D388. [CrossRef]
31. Gasteiger, E.; Gattiker, A.; Hoogland, C.; Ivanyi, I.; Appel, R.D.; Bairoch, A. ExPASy: The Proteomics Server for in-Depth Protein Knowledge and Analysis. *Nucleic Acids Res.* **2003**, *31*, 3784–3788. [CrossRef] [PubMed]
32. Chen, C.; Wu, Y.; Li, J.; Wang, X.; Zeng, Z.; Xu, J.; Liu, Y.; Feng, J.; Chen, H.; He, Y.; et al. TBtools-II: A "One for All, All for One" Bioinformatics Platform for Biological Big-Data Mining. *Mol. Plant* **2023**, *16*, 1733–1742. [CrossRef] [PubMed]

33. Kumar, S.; Stecher, G.; Li, M.; Knyaz, C.; Tamura, K. MEGA X: Molecular Evolutionary Genetics Analysis across Computing Platforms. *Mol. Biol. Evol.* **2018**, *35*, 1547–1549. [CrossRef] [PubMed]
34. Bailey, T.L.; Johnson, J.; Grant, C.E.; Noble, W.S. The MEME Suite. *Nucleic Acids Res.* **2015**, *43*, W39–W49. [CrossRef]
35. Zhang, F.; Wang, F.; Yang, S.; Zhang, Y.; Xue, H.; Wang, Y.; Yan, S.; Wang, Y.; Zhang, Z.; Ma, Y. MdWRKY100 Encodes a Group I WRKY Transcription Factor in Malus Domestica That Positively Regulates Resistance to Colletotrichum Gloeosporioides Infection. *Plant Sci.* **2019**, *286*, 68–77. [CrossRef]
36. Ma, Y.; Xue, H.; Zhang, F.; Jiang, Q.; Yang, S.; Yue, P.; Wang, F.; Zhang, Y.; Li, L.; He, P.; et al. The miR156/SPL Module Regulates Apple Salt Stress Tolerance by Activating MdWRKY100 Expression. *Plant Biotechnol. J.* **2021**, *19*, 311–323. [CrossRef]
37. Liu, X.; Dai, H.; Zhang, F.; Wang, J.; Shi, J.; Chen, J.; He, P.; Wang, F.; Ma, Y. The miR7125-MdARF1 Module Enhances the Resistance of Apple to Colletotrichum Gloeosporioides by Promoting Lignin Synthesis in Response to Salicylic Acid Signalling. *Plant Biotechnol. J.* **2024**, *22*, 2741–2755. [CrossRef]
38. Liese, A.; Romeis, T. Biochemical Regulation of *in Vivo* Function of Plant Calcium-Dependent Protein Kinases (CDPK). *Biochim. Et Biophys. Acta (BBA)-Mol. Cell Res.* **2013**, *1833*, 1582–1589. [CrossRef]
39. Zhao, R.; Sun, H.-L.; Mei, C.; Wang, X.-J.; Yan, L.; Liu, R.; Zhang, X.-F.; Wang, X.-F.; Zhang, D.-P. The Arabidopsis Ca2+-Dependent Protein Kinase CPK12 Negatively Regulates Abscisic Acid Signaling in Seed Germination and Post-Germination Growth. *New Phytol.* **2011**, *192*, 61–73. [CrossRef]
40. Almadanim, M.C.; Alexandre, B.M.; Rosa, M.T.G.; Sapeta, H.; Leitão, A.E.; Ramalho, J.C.; Lam, T.T.; Negrão, S.; Abreu, I.A.; Oliveira, M.M. Rice Calcium-Dependent Protein Kinase OsCPK17 Targets Plasma Membrane Intrinsic Protein and Sucrose-Phosphate Synthase and Is Required for a Proper Cold Stress Response. *Plant Cell Environ.* **2017**, *40*, 1197–1213. [CrossRef]
41. Fantino, E.; Segretin, M.E.; Santin, F.; Mirkin, F.G.; Ulloa, R.M. Analysis of the Potato Calcium-Dependent Protein Kinase Family and Characterization of StCDPK7, a Member Induced upon Infection with Phytophthora Infestans. *Plant Cell Rep.* **2017**, *36*, 1137–1157. [CrossRef]
42. Demir, F.; Horntrich, C.; Blachutzik, J.O.; Scherzer, S.; Reinders, Y.; Kierszniowska, S.; Schulze, W.X.; Harms, G.S.; Hedrich, R.; Geiger, D.; et al. Arabidopsis Nanodomain-Delimited ABA Signaling Pathway Regulates the Anion Channel SLAH3. *Proc. Natl. Acad. Sci. USA* **2013**, *110*, 8296–8301. [CrossRef]
43. Zhu, S.-Y.; Yu, X.-C.; Wang, X.-J.; Zhao, R.; Li, Y.; Fan, R.-C.; Shang, Y.; Du, S.-Y.; Wang, X.-F.; Wu, F.-Q.; et al. Two Calcium-Dependent Protein Kinases, CPK4 and CPK11, Regulate Abscisic Acid Signal Transduction in Arabidopsis. *Plant Cell* **2007**, *19*, 3019–3036. [CrossRef]
44. Yang, Z.; Rannala, B. Molecular Phylogenetics: Principles and Practice. *Nat. Rev. Genet.* **2012**, *13*, 303–314. [CrossRef] [PubMed]
45. Lu, Y.-J.; Li, P.; Shimono, M.; Corrion, A.; Higaki, T.; He, S.Y.; Day, B. Arabidopsis Calcium-Dependent Protein Kinase 3 Regulates Actin Cytoskeleton Organization and Immunity. *Nat. Commun.* **2020**, *11*, 1–12. [CrossRef] [PubMed]
46. Savojardo, C.; Martelli, P.L.; Fariselli, P.; Profiti, G.; Casadio, R. BUSCA: An Integrative Web Server to Predict Subcellular Localization of Proteins. *Nucleic Acids Res.* **2018**, *46*, W459–W466. [CrossRef] [PubMed]
47. Dammann, C.; Ichida, A.; Hong, B.; Romanowsky, S.M.; Hrabak, E.M.; Harmon, A.C.; Pickard, B.G.; Harper, J.F. Subcellular Targeting of Nine Calcium-Dependent Protein Kinase Isoforms from Arabidopsis. *Plant Physiol.* **2003**, *132*, 1840–1848. [CrossRef]
48. Li, Y.; Zhang, H.; Liang, S.; Chen, X.; Liu, J.; Zhang, Y.; Wang, A. Identification of CDPK Gene Family in Solanum Habrochaites and Its Function Analysis under Stress. *Int. J. Mol. Sci.* **2022**, *23*, 4227. [CrossRef]
49. Zhou, G.; Liu, C.; Cheng, Y.; Ruan, M.; Ye, Q.; Wang, R.; Yao, Z.; Wan, H. Molecular Evolution and Functional Divergence of Stress-Responsive Cu/Zn Superoxide Dismutases in Plants. *Int. J. Mol. Sci.* **2022**, *23*, 7082. [CrossRef]
50. Chmelová, Ľ.; Kraeva, N.; Saura, A.; Krayzel, A.; Vieira, C.S.; Ferreira, T.N.; Soares, R.P.; Bučková, B.; Galan, A.; Horáková, E.; et al. Intricate Balance of Dually-Localized Catalase Modulates Infectivity of *Leptomonas Seymouri* (Kinetoplastea: Trypanosomatidae). *Int. J. Parasitol.* **2024**, *54*, 391–400. [CrossRef]
51. Yan, Y.; Wang, P.; Lu, Y.; Bai, Y.; Wei, Y.; Liu, G.; Shi, H. MeRAV5 Promotes Drought Stress Resistance in Cassava by Modulating Hydrogen Peroxide and Lignin Accumulation. *Plant J.* **2021**, *107*, 847–860. [CrossRef]
52. Dubiella, U.; Seybold, H.; Durian, G.; Komander, E.; Lassig, R.; Witte, C.-P.; Schulze, W.X.; Romeis, T. Calcium-Dependent Protein Kinase/NADPH Oxidase Activation Circuit Is Required for Rapid Defense Signal Propagation. *Proc. Natl. Acad. Sci. USA* **2013**, *110*, 8744–8749. [CrossRef]
53. Liu, X.; Zhou, Y.; Du, M.; Liang, X.; Fan, F.; Huang, G.; Zou, Y.; Bai, J.; Lu, D. The Calcium-Dependent Protein Kinase CPK28 Is Targeted by the Ubiquitin Ligases ATL31 and ATL6 for Proteasome-Mediated Degradation to Fine-Tune Immune Signaling in Arabidopsis. *Plant Cell* **2022**, *34*, 679–697. [CrossRef]
54. Liu, X.; Zhou, Y.; Chen, K.; Xiao, Z.; Liang, X.; Lu, D. Phosphorylation Status of CPK28 Affects Its Ubiquitination and Protein Stability. *New Phytol.* **2023**, *237*, 1270–1284. [CrossRef]

55. Hu, Y.; Cheng, Y.; Yu, X.; Liu, J.; Yang, L.; Gao, Y.; Ke, G.; Zhou, M.; Mu, B.; Xiao, S.; et al. Overexpression of Two CDPKs from Wild Chinese Grapevine Enhances Powdery Mildew Resistance in Vitis Vinifera and Arabidopsis. *New Phytol.* **2021**, *230*, 2029–2046. [CrossRef]
56. Wei, M.; Wang, S.; Dong, H.; Cai, B.; Tao, J. Characterization and Comparison of the CPK Gene Family in the Apple (Malus × Domestica) and Other Rosaceae Species and Its Response to Alternaria Alternata Infection. *PLoS ONE* **2016**, *11*, e0155590. [CrossRef]

horticulturae

MDPI

Article

Integrated Transcriptomics and Metabolomics Analyses Provide Insights into Heat Resistance in Passion Fruit (*P. edulis* f. *flavicarpa*)

Lin-Hua Chen [1], Jiong Dong [1], Bing-Liang Fan [1], Yongcai Huang [2], Liu Yang [2], Wenguo Cai [1] and Ling-Ling Chen [1,*]

1 State Key Laboratory for Conservation and Utilization of Subtropical Agro-Bioresources, College of Life Science and Technology, Guangxi University, Nanning 530004, China; linhuachenchn@outlook.com (L.-H.C.); Dongwith@outlook.com (J.D.); blfan@st.gxu.edu.cn (B.-L.F.); wcai@gxu.edu.cn (W.C.)

2 Guangxi Crop Genetic Improvement and Biotechnology Laboratory, Guangxi Academy of Agricultural Science, Nanning 530007, China; huangyongcai1981@126.com (Y.H.); yangliutibs@126.com (L.Y.)

* Correspondence: llchen@gxu.edu.cn

Abstract

Passion fruit (*Passiflora edulis*) is an economically important fruit worldwide. However, heat stress severely threatens its production, particularly in tropical and subtropical regions. To elucidate the molecular and metabolic mechanisms underlying heat tolerance, comparative physiological, transcriptomic, and metabolomic analyses were conducted between two yellow passion fruit cultivars: heat-tolerant 'Summer Queen' (F2) and heat-sensitive 'Qinmi 9' (QM9). Physiological evaluations demonstrated that QM9 exhibited significantly lower heat tolerance than F2, manifesting as severe leaf wilting, impaired photosynthetic efficiency, and elevated reactive oxygen species (ROS) accumulation. F2 exhibited distinct metabolic and transcriptional adaptations under heat stress, particularly in purine metabolism and flavonoid biosynthesis. Metabolites such as glutamine, xanthine, luteoloside, and trifolin were enriched in F2, alongside the upregulation of genes like adenosine kinase (*AK*), xanthine dehydrogenase (*XDH*), guanine deaminase (*GDA*), and flavonoid 3′-hydroxylase (F3′H). Weighted gene co-expression network analysis (WGCNA) highlighted strong associations between these pathways and transcription factors (e.g., MYB, HSF, WRKY), suggesting their pivotal roles in heat adaptation. Exogenous application of xanthine and trifolin markedly enhanced heat tolerance in passion fruit. Furthermore, knockdown of *PeGDA* and *PeXDH* markedly altered the heat tolerance of F2. These findings reveal that elevated metabolites in purine metabolism and flavonoid biosynthesis enhance heat tolerance in passion fruit, offering new insights into the molecular mechanisms of heat tolerance and potential targets for breeding climate-resilient passion fruit varieties.

Keywords: heat tolerance; purine metabolism; flavonoid biosynthesis; xanthine; trifolin; xanthine dehydrogenase (XDH); guanine deaminase (GDA)

1. Introduction

Global warming and the intensification of the greenhouse effect pose significant challenges to modern agricultural production. Elevated temperatures directly disrupt critical physiological processes in plants, including photosynthesis, respiration, and nutrient assimilation, ultimately leading to reduced crop yields [1]. Among various abiotic stressors, heat stress has become a predominant threat to global food security. As the frequency and intensity of extreme heat events continue to rise, deciphering the molecular and physiological mechanisms of plant heat tolerance is imperative for developing climate-resilient crops. This concern is particularly acute for tropical and subtropical species, which routinely endure intense heat stress that severely constrains their productivity and economic value [2]. Thus, research on plant heat resistance is not only scientifically significant but also urgently needed to address the increasing impacts of climate change on agriculture [3].

Yellow passion fruit (*P. edulis* f. *flavicarpa*), a climacteric fruit widely cultivated in tropical and subtropical regions, is highly valuable for its nutritional, medicinal, and economic value [4]. Renowned for its rich content of vitamins, carotenoids, and polyphenols, it has become a staple in various industries [5]. However, its commercial potential, particularly in major producing regions such as Brazil and China, is increasingly threatened by heat stress, which induces flower abscission and fruit set failure under elevated temperatures [1,6]. While research on passion fruit has predominantly focused on purple varieties, yellow passion fruit has recently gained attention because of its distinct flavor profile, high sweetness ratio, and unique growth characteristics, positioning it as a promising candidate for breeding heat-resistant cultivars [5,7]. At the molecular level, transcription factors (TFs) are key regulators of plant responses to abiotic stress, including heat stress. TFs such as heat stress transcription factors (HSFs), MYB, WRKY, and NAC are well-documented regulators that orchestrate the defense mechanisms of plants against environmental stressors [8]. HSFs, in particular, play a central role in the heat response by modulating the expression of heat shock proteins (HSPs), which safeguard cellular integrity under high-temperature conditions [9,10]. Other TF families, including DREB and bZIP, are involved in mediating stress signaling pathways that contribute to heat tolerance by activating stress-responsive genes [11]. Given the pivotal roles of TFs in stress adaptation, elucidating their regulatory networks in yellow passion fruit is essential for advancing the understanding of heat stress tolerance and facilitating the development of climate-resilient varieties.

Although extensive research has been conducted on heat resistance mechanisms in various plant species, studies on passion fruit remain limited. The integration of transcriptomics and metabolomics provides a comprehensive approach to unravel molecular and metabolic responses to heat stress, enabling the identification of differentially expressed genes and metabolic shifts for a deeper understanding of heat resistance mechanisms [12]. Transcriptomics and metabolomics analyses of two yellow passion fruit varieties with contrasting heat tolerances, Qinmi9 (QM9, conventional varieties) and Summer Queen (F2, heat-tolerant), revealed significant differences in gene expression and metabolic accumulation. This study identified key genes and metabolites linked to heat resistance, highlighting the critical roles of purine metabolism and flavone and flavonol biosynthesis pathways. These findings advance the understanding of heat tolerance in passion fruit and provide valuable insights for breeding climate-resilient varieties.

2. Materials and Methods

2.1. Seedling Responses to Controlled Heat Stress

Seedling heat stress assay under controlled conditions: To determine differences in heat tolerance among passion fruit varieties, seedlings of two yellow passion fruit cultivars (*P. edulis* f. *flavicarpa*: QM9 and F2) were collected from experimental fields at the Guangxi Academy of Agricultural Sciences and transplanted into soil-filled pots. These plants were acclimated for 3 days in a growth chamber under controlled conditions (25 °C, 16 h light/8 h dark photoperiod) before heat treatment. To detect differences in heat stress between two varieties, seedlings were transferred to a growth chamber at 40 °C for 48 h. Each treatment group included three biological replicates.

2.2. Field-Grown Plant Responses to Natural High-Temperature Conditions

Field sample collection under natural high-temperature conditions for transcriptomics and metabolomics analyses: To investigate differences in gene expression and metabolite accumulation among passion fruit varieties under naturally high-temperature stress, samples from mature QM9 and F2 plants were collected under naturally high temperature conditions (32–36 °C) in the same open-air experimental field at the Guangxi Academy of Agricultural Sciences. Leaves, stems, and fruits were collected during the ripening stage. All fruits were immediately dissected into pericarp and pulp components for subsequent analyses. This analysis focused on comparing transcriptomic and metabolomic differences between the conventional variety QM9 and the heat-tolerant variety F2 under natural high-temperature conditions, rather than examining responses of the same variety to both normal and heat-stress environments. Each sample type was collected with three biological replicates.

2.3. Physiological Measurements

Electrolyte leakage (EL) was measured as described by Dahro et al. [13], with minor modifications. Leaf discs (0.5 cm diameter) were incubated in 20 mL deionized water at 25 °C for 24 h, and initial conductivity (EC1) was recorded. Samples were then heated at 95 °C for 20 min to obtain the final conductivity (EC2). EL (%) was calculated as EC1/(EC1 + EC2) × 100%. The *Fv/Fm* ratio was assessed using Imaging WinGege software V2.56zn.

Malondialdehyde (MDA) content was measured using the thiobarbituric acid (TBA) colorimetric method. Fresh passion fruit leaves (0.1 g) were placed in grinding tubes containing stainless steel beads, flash-frozen in liquid nitrogen, and homogenized at 60 Hz for 120 s. The samples were mixed with 1 mL of 5% (*w*/*v*) trichloroacetic acid (TCA) and incubated on ice for 1 h. After centrifugation at 8000× *g* for 10 min at 4 °C, 200 μL of the supernatant was combined with 400 μL of 0.6% TBA solution. The mixture was incubated at 95 °C for 60 min, cooled on ice, and centrifuged at 10,000× *g* for 10 min at room temperature. Absorbance was measured at 532 nm and 600 nm using a microplate reader, with distilled water as a blank. Absorbance differences were calculated as $\Delta A = (A_{532} - A_{532}\text{ blank}) - (A_{600} - A_{600}\text{ blank})$, and MDA content was calculated using the instructions provided with the MDA assay kit (BC0025, Beijing Solarbio Science & Technology Co., Ltd., Beijing, China).

To detect reactive oxygen species (ROS), histochemical staining was performed using 3,3′-diaminobenzidine (DAB, CAS#: 91-95-2) and nitroblue tetrazolium (NBT, CAS#: 298-83-9). Heat-stressed leaves were incubated in DAB or NBT staining solutions at 37 °C in the dark on a shaker at 100 rpm for 24 h. When visible staining appeared, the staining solutions were discarded, and the leaves were decolorized in 65 °C anhydrous ethanol until chlorophyll was completely removed. The decolorized leaves were then flattened and photographed.

2.4. Metabolite Extraction and Profiling

To assess metabolite accumulation differences between QM9 and F2, stem, leaf, pericarp, and pulp samples were obtained from field-grown plants under natural high-temperature stress. Three biological replicates were analyzed for widely targeted metabolites and volatile organic compounds. Metabolite extraction, detection, and quantification were performed by Novogene Co., Ltd., using UPLC-MS/MS and GC-MS platforms. Data were processed with Analyst 1.6.3 software. An Orthogonal Partial Least Squares Discriminant Analysis (OPLS-DA) model was used to distinguish metabolic profiles. Metabolites with VIP ≥ 1 and $|\log_2 \text{fold change}| \geq 1$ were defined as differentially abundant metabolites (DAMs). The identified metabolites were annotated using the KEGG compound database and mapped to metabolic pathways via the KEGG pathway database.

2.5. Transcriptome Sequencing and Profiling

To assess gene expression differences between QM9 and F2, stem, leaf, pericarp, and pulp samples were obtained from field-grown plants under natural high-temperature stress. Total RNA was extracted using the RNAprep Pure Plant Kit, and genomic DNA was removed with RNase-Free DNase I (Takara Bio Inc., Kusatsu, Japan) following the manufacturer's protocol. RNA integrity was confirmed via a 1.0% EB-stained agarose gel, while quality and concentration were measured by Novogene Co., Ltd. (Beijing, China) using the NanoDrop 2000C and Agilent 2100. For cDNA library construction, 3 μg of RNA per sample was processed with the NEBNext Ultra RNA Library Prep Kit (New England Biolabs, Inc., Ipswich, MA, USA). PCR products were purified using AMPure XP beads (Beckman Coulter, Inc., Brea, CA, USA), and library quality was assessed on the Agilent Bioanalyzer 2100 (Agilent Technologies, Inc., Santa Clara, CA, USA). Sequencing was carried out on Illumina HiSeq 2500 or HiSeq X Ten platforms (Illumina, Inc., San Diego, CA, USA), depending on the experimental setup.

Raw RNA-seq reads were quality-checked using FastQC [14]. Adapters and low-quality reads were removed with Trimmomatic [15]. Clean reads were then mapped to the genome using HISAT2 [16] with default parameters. Transcripts were assembled with StringTie [17], and gene expression was quantified in fragments per kilobase of transcript per million fragments mapped (FPKM). DEGs were identified using DESeq2 [18], with thresholds of $|\log_2 \text{fold change}| \geq 1$ and adjusted $p < 0.01$. DEGs were subsequently annotated with Gene Ontology (GO) and KEGG pathway terms.

2.6. Mapping Correlation Networks Between DEGs and DAMs

Principal component analysis (PCA) and hierarchical clustering were employed to provide a comprehensive comparison of protein-coding gene and metabolite profiles in the sampled tissues (stem, leaf, pericarp, and pulp) of QM9 and F2. The pheatmap package in R was utilized for this analysis. To ensure comparability across samples, gene and metabolite accumulation data underwent transformation and Z-score normalization, facilitating clearer visualization of dataset variance. Enrichment analysis was conducted using the hypergeometric test to identify statistically significant over-representations of specific genes or metabolites. This was further enhanced by pathway enrichment analysis based on the KEGG database, which elucidated the biological pathways influenced over the time course. Additionally, co-expression networks were constructed to investigate co-expression relationships among protein-coding genes, employing the WGCNA package [19]. Visualization of the gene interaction network was accomplished using Cytoscape software V3.10.1 [20].

2.7. Exogenous Spraying Experiment

In the exogenous spraying experiment, the optimal concentrations of xanthine (10 mM) and trifolin (2 mM) were determined through preliminary experiments. 10 mM xanthine solution was prepared by dissolving 15.2 mg of xanthine in 10 mL of 0.025 M NaOH. 2 mM trifolin solution was prepared by dissolving 10 mg of trifolin in a total volume of 11.15 mL, using a solvent mixture containing 5% (v/v) anhydrous ethanol and 0.1% (v/v) Tween-80.

At room temperature, QM9 and gene-silenced F2 plants were exogenously sprayed with water, a 10 mM xanthine solution, and a 2 mM trifolin solution, respectively, using a spray bottle. Spraying was performed twice: a second spray was conducted 24 h after the first spray, and the subsequent heat stress experiment (42 °C for 24 h) was conducted 24 h later.

2.8. Construction of the Gene Silencing System

The VIGS (Virus-Induced Gene Silencing) system was established to target two key genes involved in purine metabolism: guanine deaminase (GDA) and xanthine dehydrogenase (XDH). Target gene fragments were amplified using the following primers to construct the pTRV2 vectors: GDA-F: 5′-aaggttaccgaattctctagaATCGTGACCATGGACCCCG-3′, GDA-R: 5′-cgtgagctcggtaccggatccTTCGTAAGGCCAGATGCGG-3′; XDH-F: 5′-aaggttaccgaattctctaga GAGTATCTTAGAGACTTGGGTCTGACTG-3′, XDH-R: 5′-cgtgagctcggtaccggatccTCCAG GAGTACAAAAACCACACTG-3′.

Recombinants were screened on LB medium containing kanamycin (50 mg/L), and positive clones were verified by colony PCR and sequencing. Correct recombinant plasmids were introduced into *Agrobacterium tumefaciens* strain GV3101 (carrying the pSoup-p19 helper plasmid). Transformants were selected on LB medium supplemented with kanamycin (50 mg/L) and rifampicin (25 mg/L).

Agrobacterium cultures carrying pTRV1, pTRV2, pTRV2-GDA, and pTRV2-XDH plasmids were grown in YEP medium. Cells were harvested, resuspended in infiltration buffer (10 mM MES, 10 mM $MgCl_2$, 200 μM acetosyringone), and adjusted to an OD_{600} of ~1.0, then incubated in the dark for 4 h. Equal volumes of Agrobacterium with pTRV1 were mixed with those containing pTRV2-GDA or pTRV2-XDH and infiltrated into one-month-old F2 passion fruit seedlings as experimental groups. A mixture of pTRV1 and empty pTRV2 was used as a negative control, and uninjected plants were used as blanks, with three biological replicates each. After infiltration, plants were kept in the dark at 25 °C for 1 day, then transferred to 25 °C, 50% humidity, under a 16 h light/8 h dark cycle for 2 days before sampling to evaluate gene silencing.

2.9. RT-qPCR Analysis

To validate the successful silencing of the target genes via VIGS, total RNA was extracted from leaves of F2 representing three experimental groups: (1) VIGS-treated plants (pTRV2-GDA and pTRV2-XDH), (2) vector control plants (empty pTRV2), and (3) untreated wild-type plants (mock control). Total RNA was isolated from these samples using TRIzol (Labled, Beijing, China) reagent according to the manufacturer's instructions. First-strand cDNA was synthesized using the Lablead First-Strand cDNA Synthesis Mix kit following the provided protocol. The reference gene primers [21] were as follows: Forward 5′-AGCTCTTCTACATCTGCGCT-3′ and Reverse 5′-TTCTTGTGCATCTTCCCCCG-3′. Gene-specific primers were as follows: qGDA-F: 5′-AATAACACGGCAGACGGGAGG-3′, qGDA-R: 5′-TTGACCCAAACTGTATGAGCTGCC-3′; qXDH-F: 5′-GGAGGTATTTGCTTCAGAG TTTGTG-3′, qXDH-R: 5′-CCCACCTGAACTTCCCCTTCT-3′. Quantitative PCR reactions were performed using the Lablead 2× Realab Green PCR Fast Mixture according to the manufacturer's instructions. Relative gene expression levels were calculated using the $2^{-\Delta\Delta Ct}$ method.

2.10. Statistical Analysis

All experiments were conducted with a minimum of two independent repetitions, each containing three biological replicates. Data are presented as mean ± standard deviation (SD). Statistical significance was determined using Student's *t*-test, with significance thresholds set at * $p < 0.05$, ** $p < 0.01$, and *** $p < 0.001$.

3. Results

3.1. F2 Passion Fruit Exhibits Greater Heat Tolerance Compared with QM9

Two yellow-fruited passion fruit (*P. edulis* f. *flavicarpa*) cultivars were utilized: 'Qinmi 9′ (QM9), identified as the normal (heat-sensitive) control, and 'Summer Queen' (F2), characterized as heat-tolerant. Both cultivars were authenticated and provided by the Guangxi Academy of Agricultural Sciences. To evaluate heat tolerance, one-month F2 and QM9 passion fruit seedlings were subjected to heat stress (40 °C for 48 h). While both varieties showed comparable performance under normal conditions, QM9 exhibited severe water-soaking and leaf wilting after stress, whereas F2 maintained better morphological integrity (Figure 1a). Contrary to the phenotypic observations, electrolyte leakage (EL) measurements revealed higher values in F2 than in QM9 after heat stress (Figure 1b). However, malondialdehyde (MDA) content—a more specific indicator of membrane lipid peroxidation—showed significantly greater accumulation in QM9 (Figure 1c). This apparent discrepancy suggests that while F2 experienced enhanced ion efflux under heat stress, QM9 suffered more severe oxidative damage to cellular membranes. Chlorophyll fluorescence measurements further corroborated these findings. QM9 exhibited severely impaired fluorescence parameters and significantly reduced *Fv/Fm* ratios following heat stress, whereas F2 plants maintained relatively stable photosynthetic efficiency with only minor perturbations (Figure 1d). Histochemical staining for reactive oxygen species (ROS) provided additional evidence of differential stress responses. Nitro blue tetrazolium (NBT) and 3′3-diaminobenzidine (DAB) staining revealed substantially lower accumulation of $O_2^{\bullet-}$ and H_2O_2 in F2 compared to QM9 after heat treatment (Figure 1e). Collectively, these results demonstrate that F2 possesses superior heat tolerance relative to QM9, as evidenced by (1) better maintenance of cellular membrane integrity, (2) preserved photosynthetic function, and (3) more effective ROS scavenging under heat stress conditions.

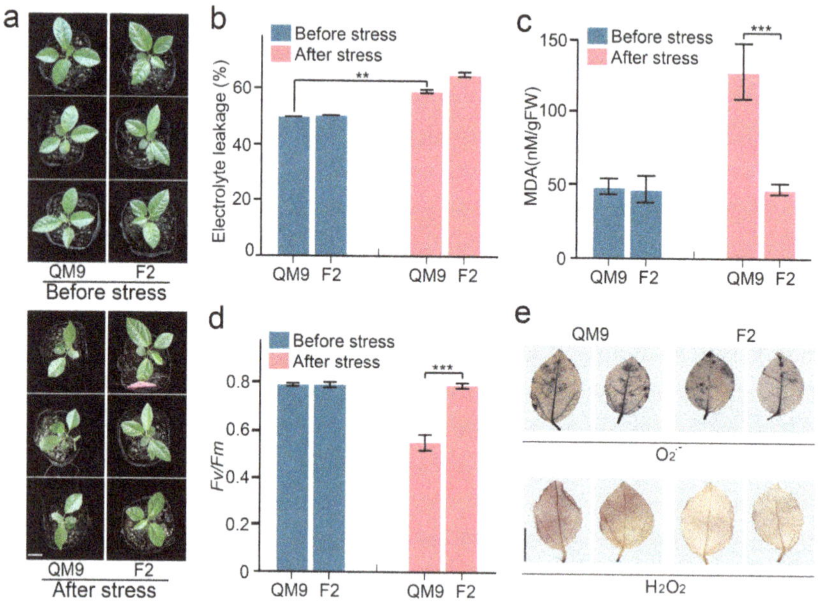

Figure 1. Comparative heat tolerance analysis of F2 and QM9 passion fruit cultivars: (**a**) Phenotypic comparison before and after 40 °C heat treatment. (**b**–**d**) Physiological indices measured pre- and post-treatment: electrolyte leakage (**b**), malondialdehyde (MDA) content (**c**), and maximum quantum yield of PSII (*Fv/Fm*) (**d**). Data represent mean ± SD (n = 3 biological replicates). Asterisks indicate statistically significant differences between cultivars (** $p < 0.01$, *** $p < 0.001$; Student's t-test) (**e**). Histochemical detection of reactive oxygen species using nitroblue tetrazolium (NBT, upper panels) for superoxide and 3,3′-diaminobenzidine (DAB, lower panels) for hydrogen peroxide. Scale bars = 2 cm (applies to a and e images within respective panels).

3.2. Metabolic Profiling of Heat-Tolerant and Conventional Passion Fruit

Metabolite detection mainly investigates the variation in heat tolerance between the two cultivars under these natural high-temperature conditions. To investigate heat-associated metabolic differences between QM9 and F2, samples were collected under natural high-temperature conditions (32–36 °C) with three biological replicates per group (Figure 2a). For comprehensive metabolic profiling, widely targeted metabolites and volatile organic compounds (VOCs) were studied in pericarp (P), pulp (J), stem (S) and leaf (L) tissues using complementary analytical platforms: liquid chromatography–mass spectrometry (LC-MS) for metabolite characterization and gas chromatography–mass spectrometry (GC-MS) for VOC detection [22,23].

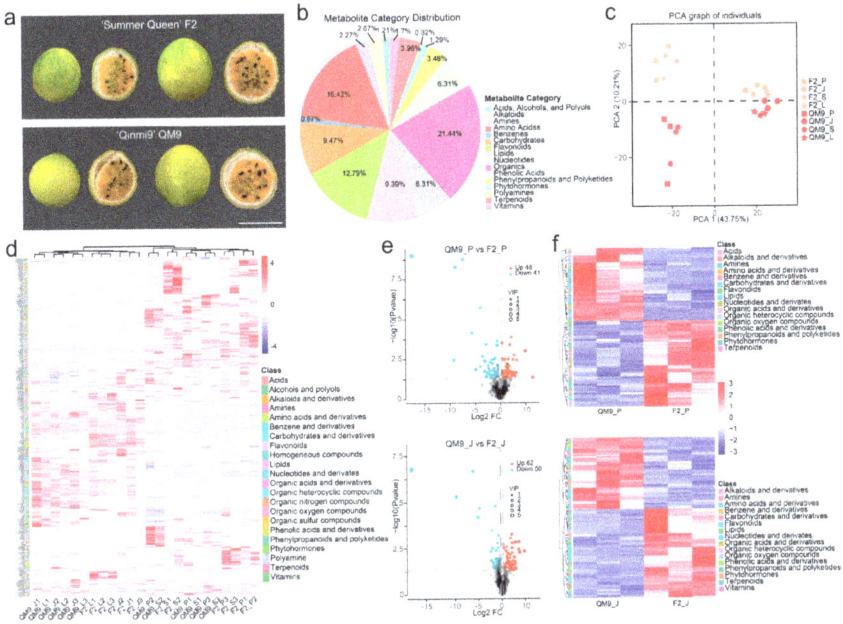

Figure 2. Metabolic profiling of F2 and QM9 passion fruit: (**a**) Mature fruits of F2 and QM9 showing size differences. Scale bars = 5 cm. (**b**) Proportional distribution of detected metabolites. (**c**) Principal component analysis (PCA) of metabolites in the pericarp and pulp of QM9 and F2. P, J, S, and L represent pericarp, pulp, stem, and leaf, respectively. (**d**) Hierarchical clustering heatmap of 1236 metabolites identified in QM9 and F2. (**e**) Volcano plot of differentially accumulated metabolites (DAMs) in the pericarp (top) and pulp (bottom) between QM9 and F2. (**f**) Heatmaps of DAMs in the pericarp (top) and pulp (bottom) between QM9 and F2.

A total of 1236 metabolites were identified and categorized into 16 classes, including amino acids and derivatives (16.42%), organic derivatives (21.44%), flavonoids (12.79%), carbohydrates (9.47%), lipids (9.39%), phenolic acids and derivatives (6.31%), nucleotides and derivatives (6.31%), terpenoids (3.96%), phenylpropanoids and polyketides (3.48%), alkaloids (2.67%), amines (2.27%), vitamins (1.70%), phytohormones (1.29%), acids, alcohols, and polyols (1.21%), benzene and derivatives (0.97%), and polyamines (0.32%) (Figure 2b, Table S1). Principal component analysis (PCA) revealed distinct clustering of biological replicates within each group, confirming the separation between groups and validating the reliability and robustness of the metabolomic data for downstream analysis (Figure 2c). Cluster analysis revealed that there were significant differences in metabolite accumulation between the two varieties and their respective tissues

(Figure 2d). To identify heat-resistant metabolic changes, differentially accumulated metabolites (DAMs) were compared in each tissue of QM9 and F2 (Table S2). In the pericarp, a total of 89 differentially accumulated metabolites (DAMs) were identified in F2 compared to QM9, including 48 upregulated and 41 downregulated metabolites (Figure 2e). Similarly, in the pulp, 112 DAMs were detected, of which 62 were upregulated and 50 were downregulated in F2 (Figure 2e). Additionally, stem and leaf tissues demonstrated the most DAMs between QM9 and F2 in the stem and leaf (Figure S1a). Cluster heatmap showed that there were significant differences in metabolite accumulation between the two varieties and their respective tissues (Figures 2f and S1b). These findings highlight the distinct metabolic profiles between the two cultivars, providing preliminary insights into the metabolic mechanisms underlying their differential heat tolerances.

3.3. *Specific Differences in Metabolites Between Two Passion Fruit Varieties*

In the comparison between QM9_P vs. F2_P and QM9_J vs. F2_J, A total of 22 shared DAMs were identified, with 179 DAMs detected across both comparison groups (Figure 3a). To systematically characterize these metabolic changes in response to heat stress, K-means clustering was applied to the 179 DAMs, which were categorized into nine distinct clusters (Figure 3a–c, Table S3). Notably, metabolites in clusters 2, 3, 4, and 8 exhibited significantly higher accumulation in the pericarp and/or pulp of the heat-tolerant F2 compared to QM9. Similarly, analysis of 371 DAMs in stem and leaf tissues revealed nine distinct clusters, with clusters 1, 3, 6, 7, and 8 showing increased accumulation in F2 (Figure S2, Table S3). These heat-induced metabolites were predominantly composed of organic acids and derivatives, phenolic acids and derivatives, flavonoids—compounds known to play critical roles in plant responses to heat stress. Their upregulation in F2 underscores their role as conserved stress-responsive metabolites, potentially contributing to enhanced thermotolerance.

Further metabolomic analysis revealed significant differences in the accumulation of key metabolite classes between QM9 and the heat-tolerant F2 variety. Strikingly, F2 displayed significantly elevated levels of terpenoids, organic acids, alkaloids, flavonoids, heterocyclic compounds, and nitrogen-containing compounds in both the pericarp and pulp compared to QM9 (Figure 3d,e). This enhanced accumulation suggests an active metabolic reprogramming in F2, potentially contributing to its stress-responsive phenotype. A similar trend was observed in stem and leaf tissues, where F2 accumulated higher amounts of these key metabolite classes (Figures S3 and S4). The consistent upregulation of these compounds across different tissues implies a systemic adaptation mechanism in F2, possibly linked to improved stress resilience. Conversely, the levels of alcohol and polyols, organic oxygen compounds, carbohydrates, and lipids in F2 were lower than those in QM9 (Figures 3d,e, S3 and S4). This differential metabolic profile may reflect a trade-off in resource allocation, where F2 prioritizes the biosynthesis of stress-protective secondary metabolites.

3.4. *Transcriptomic Analysis of Heat Stress Responses of Heat-Tolerant and Conventional Passion Fruit*

To explore the molecular mechanisms underlying metabolic adaptation to heat stress in QM9 and F2 passion fruit, RNA-Seq analysis was conducted alongside metabolome profiling. High-throughput RNA sequencing via the Illumina HiSeq platform generated an average of 7.07 Gb of high-quality reads for QM9 and 6.56 Gb for F2, with average mapping rates of 93.42% and 89.31%, respectively, to the previously assembled yellow passion fruit genome [24] (Table S4). Consequently, a comprehensive dataset was generated for gene expression analysis (Table S5). The heatmap (Figure 4a) depicts the gene expression profiles of QM9 and F2, demonstrating consistent expression patterns within biological replicates and significant differential expression between the two cultivars. PCA revealed clear separation between the groups, with distinct clustering of biological replicates within each group, further validating the reliability of the transcriptomic data (Figure 4b). Additionally, correlation analysis indicated high reproducibility among biological replicates for each treatment, underscoring the robustness of the dataset (Figure 4c). Collectively, these results indicate that gene expression changes under heat stress are both species- and tissue-specific, reflecting corresponding patterns of metabolite accumulation.

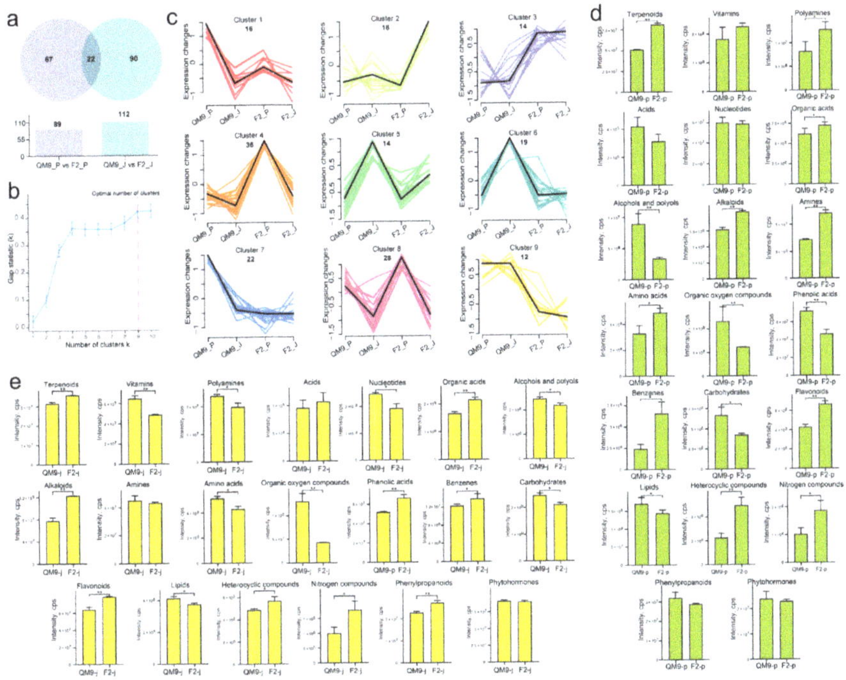

Figure 3. Specific differences in metabolites compared between QM9 and F2: (**a**) Venn diagram analysis of DAMs between QM9_P vs. F2_P and QM9_J vs. F2_J. (**b**) The gap statistic determined that the optimal number of clusters (k) for K-means clustering of DAMs between QM9 and F2 is 9, as indicated by the highest gap statistic at k = 9. (**c**) K-means clustering of the 9 clusters of DAMs between QM9 and F2. The X-axis depicts every group, and the Y-axis depicts the centralized and normalized per DAMs. (**d**) Histograms of metabolite contents in several classes of pericarp between QM9 and F2. (**e**) Histograms of metabolite contents in several classes of pulp between QM9 and F2. Data represent mean ± SD (n = 3 biological replicates). Asterisks indicate statistically significant differences between cultivars (* $p < 0.05$, ** $p < 0.01$; Student's t-test).

To further investigate the transcriptional responses to heat stress in QM9 and F2 passion fruit varieties, differentially expressed genes (DEGs) were identified (Figure 4d). Compared with those in QM9, 3894 genes in F2 were upregulated and 3465 were downregulated in the pericarp. Similarly, in the pulp, 1809 genes were upregulated, whereas 1113 were downregulated. In the stem, 5535 genes presented increased expression, and 4991 genes presented decreased expression. In the leaf tissue, 3901 genes were upregulated, and 3165 were downregulated. Both upregulated and downregulated DEGs were observed in heatmaps across all tissues—pericarp, pulp, stem, and leaf—indicating substantial and tissue-specific shifts in gene expression profiles in response to heat stress (Figure 4e). These findings collectively demonstrate that heat stress induces widespread and tissue-specific transcriptional reprogramming in passion fruit, with the heat-tolerant variety F2 exhibiting distinct gene expression patterns compared with those of QM9.

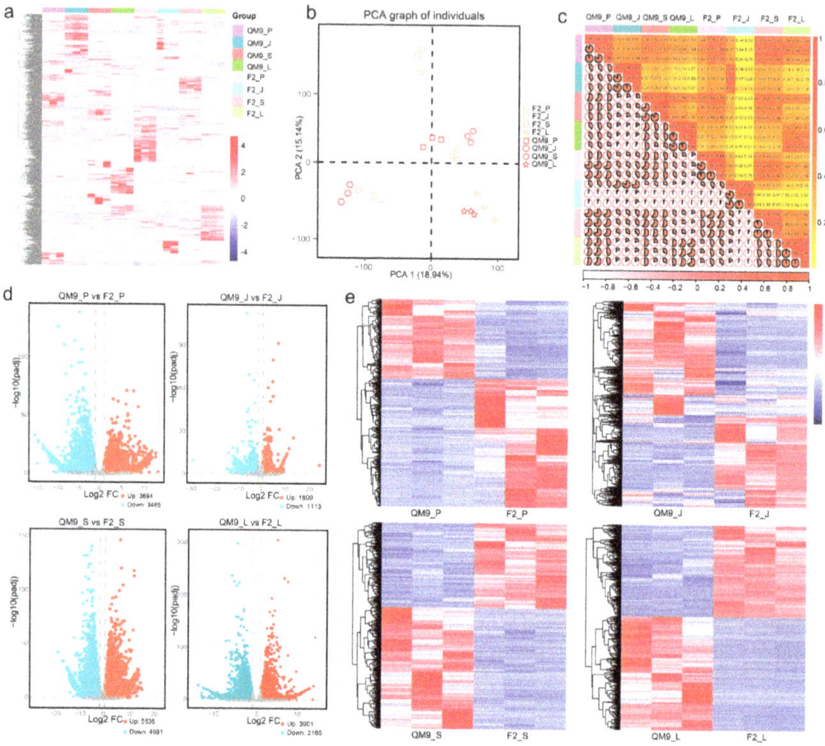

Figure 4. Transcriptomic analysis of QM9 and F2: (**a**) Hierarchically clustered heatmap of gene transcripts from QM9 and F2 passion fruits. (**b**) PCA of gene transcripts of QM9 and F2. (**c**) Hierarchically clustered heatmap of the gene transcripts. The completeness of the pie chart corresponds to the magnitude of the correlation value. (**d**) Volcano plot of up- and downregulated differentially expressed genes (DEGs) in pericarp, pulp, stem, and leaf of QM9 vs. F2. (**e**) Heatmap of up- and downregulated DEGs in QM9 vs. F2.

Gene Ontology (GO) analysis of DEGs showed significant enrichment in lipid metabolism (GO:0006629), cell wall biogenesis (GO:0042546), and transmembrane transporter activity (GO:0022857) across all tissues. Tissue-specific GO terms were also identified: in the pulp, purine nucleoside transmembrane transporter activity (GO:0015211) and the superoxide metabolic process (GO:0006801) were significantly enriched, whereas ribosome biogenesis (GO:0042254) and fatty acid metabolic processes (GO:0006631) were enriched in the pericarp (Figure S5a, Table S6). Kyoto Encyclopedia of Genes and Genomes (KEGG) pathway analysis highlighted amino acid biosynthesis, carbon metabolism, and glycolysis/gluconeogenesis as common heat-responsive pathways. Tissue-specific pathways included calcium and MAPK signaling in the pulp; fatty acid degradation and unsaturated fatty acid biosynthesis in the pericarp; photosynthesis and TCA cycle in the leaf; and carotenoid biosynthesis and flavonoid degradation in the stem (Figure S5b, Table S7). These functional enrichment analyses demonstrated that heat stress triggers extensive reprogramming of metabolic and signaling pathways in passion fruit, with tissue-specific adaptations in the pulp, pericarp, leaf, and stem. The enrichment of pathways related to lipid metabolism, amino acid biosynthesis, and stress signaling underscores the complex regulatory networks involved in heat tolerance, providing valuable insights into the molecular mechanisms underlying the adaptation of passion fruit to high-temperature conditions.

3.5. Investigating the Heat Tolerance Mechanisms in Passion Fruit

To further elucidate the regulatory mechanisms of heat stress-induced metabolic changes and explore the co-expression network of DEGs, a gene co-expression network was constructed using Weighted Gene Co-expression Network Analysis (WGCNA). Genes with similar expression patterns were grouped into modules, and 15 co-expression modules were identified based on gene expression similarity (Figure 5a, Table S8). Compared to QM9, the highly expressed genes in F2 were predominantly enriched in the cyan, turquoise, and green modules (Figure 5a). The module–trait correlation heatmap revealed that genes in the cyan and turquoise modules were significantly correlated with key metabolites in the purine metabolic pathway (including glutamine and xanthine) and the flavonoid and flavonol biosynthesis pathway (including luteoloside and trifoline) (Figure 5b). This strong correlation highlights the potential roles of these pathways in the heat stress response, prompting further investigation into their regulatory mechanisms.

Integrated analysis of DEGs, DAMs, and WGCNA highlighted purine metabolism and flavonoid biosynthesis as key heat tolerance mechanisms in passion fruit. While these pathways are known in other species, their roles in passion fruit are newly identified, suggesting conserved stress responses. Through WGCNA, genes in the cyan and turquoise modules were linked to purine pathways, supporting their role in heat stress adaptation. Notably, three adenosine kinase (AK) genes (02G04075, 01G01990, and 09G01909), xanthine dehydrogenase (XDH) gene (08G01335) guanine deaminase (GDA) gene (05G01808) were upregulated in the F2 (Figure 5c). Additionally, higher levels of metabolites such as glutamine and xanthine were detected in the pericarp and/or pulp of F2 compared to QM9 (Figure 5c). These findings suggest that F2 exhibits adaptive modifications in purine metabolism in response to heat stress. MYB transcription factors are known to interact with HSFs to regulate the expression of HSPs, which function as molecular chaperones involved in protein assembly, stabilization, and maturation [25,26]. HSPs play critical roles not only in plant development but also in responses to abiotic and biotic stresses [27]. In our recently sequenced passion fruit genome, over 700 transcription factors were identified, including 170 belonging to the MYB family [24]. Regulatory network analysis revealed strong associations between key genes in the purine metabolism pathway and several transcription factor families, such as MYB, HSF, WRKY, bZIP, bHLH, and MADS (Figure 5d). This suggests that purine metabolism-related genes may play a pivotal role in the response to heat stress of passion fruit.

In addition to upregulated genes in the purine metabolism pathway, significant upregulation was observed for two flavonoid 3′-hydroxylase (F3′H) genes (08G01441 and 02G00366) in heat-tolerant F2 varieties, highlighting their potential role in flavonoid and flavonol biosynthesis. Downstream metabolites, including luteoloside and trifolin, were highly accumulated in F2, particularly in the pericarp (Figure 5e). Regulatory network analysis revealed strong associations between the F3′H gene and multiple transcription factor families, such as HSF, MYB, and WRKY (Figure 5f). These findings suggest that F3′H and its associated metabolites play a pivotal role in enhancing heat tolerance in passion fruit.

3.6. Functional Validation via Exogenous Application and Gene Silencing

To investigate the role of downstream metabolites from the purine and flavonoid pathways in heat tolerance of passion fruit, optimal concentrations of xanthine (10 mM) and trifoliate (2 mM) were established through preliminary experiments. Exogenous spraying experiments were subsequently performed on QM9 seedlings using these concentrations (10 mM xanthine and 2 mM trifolin, respectively). Under heat stress conditions (42 °C for 24 h), water-pretreated control seedlings exhibited more severe leaf wilting and lower chlorophyll *Fv/Fm* ratios compared to those pretreated with xanthine or trifolin (Figure 6a,b). Histochemical staining for ROS further corroborated the differential stress responses between treatments. Notably, NBT and DAB staining demonstrated significantly attenuated accumulation of $O_2^{\bullet-}$ and H_2O_2 in xanthine- and trifolin-pretreated QM9 seedlings compared to water-treated controls following heat stress (Figure 6c). These findings reveal that xanthine and trifolin, as key downstream metabolites in the purine and flavonoid pathways, may play regulatory roles in enhancing passion fruit thermotolerance.

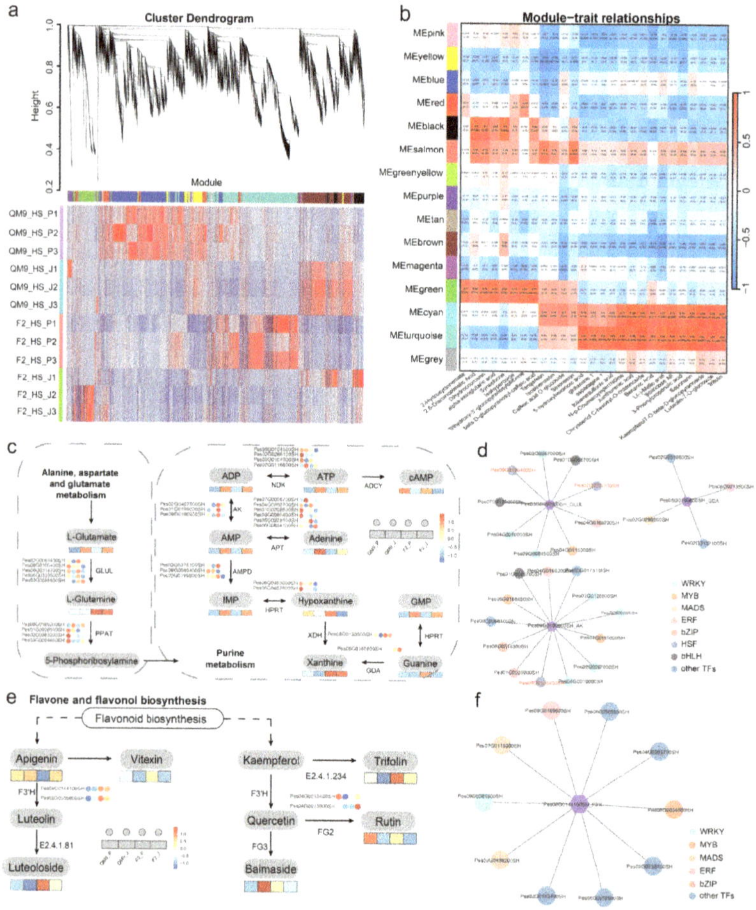

Figure 5. Integrated transcriptomic and metabolomic analysis of heat tolerance mechanisms in passion fruit varieties: (**a**) Co-expression modules (clusters) identified by weighted correlation network analysis (WGCNA). (**b**) Heatmap illustrating module–trait correlations. Each row corresponds to a module indicated by different colors. Each column corresponds to a heat-related compound. Red color indicates a positive correlation between the cluster and the compound. Blue color indicates a negative correlation. (**c**) Metabolic pathway of purine metabolism. Circles and squares represent gene expression and metabolite accumulation, respectively. (**d**) Transcriptional regulation network of AK and PPAT. The colored circles represent different transcription factors, while the lines indicate correlations greater than 0.5. (**e**) Metabolic pathway of flavone and flavonol biosynthesis. Circles and squares represent gene expression and metabolite accumulation, respectively. (**f**) Transcriptional regulation network of F3′H. The colored circles represent different transcription factors, while the lines indicate correlations greater than 0.3.

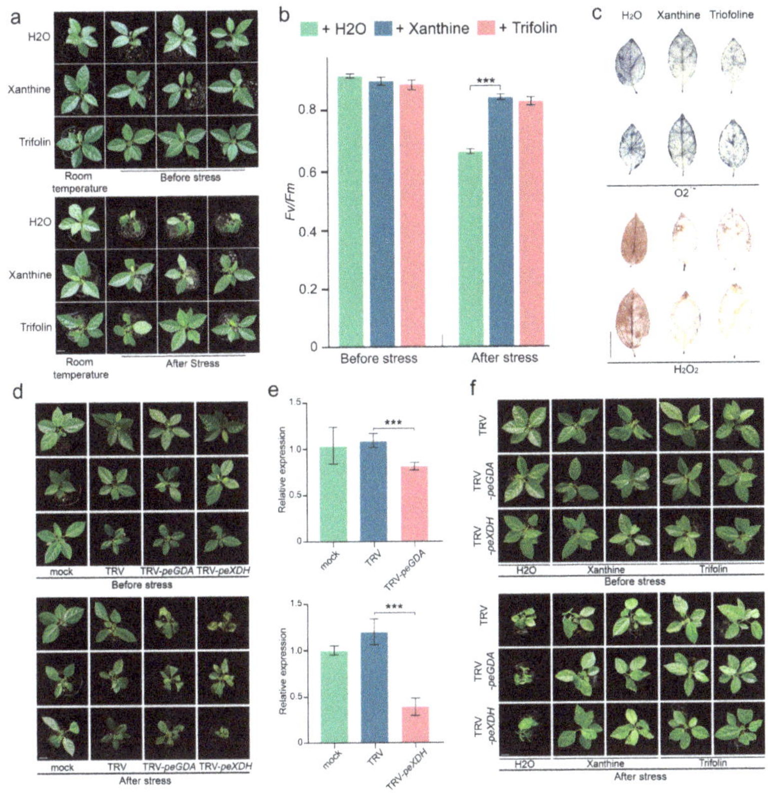

Figure 6. Exogenous xanthine and trifolin improved the heat tolerance of QM9, and silencing of *peGDA* and *peXDH* impaired the heat tolerance of F2: (**a**) Phenotypes of QM9 plant treated with water, xanthine (10 mM), and trifolin (2 mM) before and after heat treatment (42 °C for 24 h). (**b**) *Fv/Fm* ratios measured before and after heat treatment. Error bars indicate ± SD (n = 3). (**c**) Histochemical detection of reactive oxygen species using NBT (upper panels) for superoxide and DAB (lower panels) for hydrogen peroxide. Scale bars = 2 cm (applies to all images within respective panels). (**d**) Phenotypes of TRV-*peGDA*, TRV-*peXDH*, mock control, and TRV control of F2 before and after heat treatment. (**e**) RT-qPCR results of TRV-*peGDA*, TRV-*peXDH*, mock control, and TRV control. F2 of TRV-*peGDA* and TRV-*peXDH* showed significantly reduced expression levels of *peGDA* and *peXDH*, respectively, compared to both mock and TRV controls (Error bars indicate ± SD (n = 3), *** $p < 0.001$). (**f**) Phenotypes of TRV-*peGDA*, TRV-*peXDH*, and TRV control of F2, pretreated with water, xanthine (10 mM), and trifolin (2 mM) before and after heat treatment. Scale bars = 2 cm (applies to all images within respective panels).

To further investigate the key genes involved in xanthine biosynthesis, *peGDA* and *peXDH* were silenced in the F2 cultivar using virus-induced gene silencing (VIGS) to assess their functional roles in heat tolerance. Compared with the mock and TRV control plants, the TRV-*peGDA* and TRV-*peXDH* F2 plants exhibited pronounced water-soaking and leaf wilting following heat stress (Figure 6d). The transcript levels of *peGDA* and *peXDH* were significantly reduced in the respective silenced lines (Figure 6e). Additionally, exogenous application experiments were performed on F2 seedlings after VIGS treatment using 10 mM xanthine and 2 mM trifolin, respectively. Under heat stress conditions (42 °C for 24 h), water-treated control seedlings showed more severe leaf wilting compared to those pretreated with xanthine or trifolin (Figure 6f). These results suggest that knockdown of *peGDA* and *peXDH* may affect the accumulation of xanthine, leading to decreased heat tolerance, while the heat tolerance of gene-silenced F2 can be reversed by the application of clofoline, further emphasizing the role of xanthines and clofoline in the response of passion fruit to heat stress.

4. Discussion

High-temperature stress, a prevalent abiotic factor, significantly disrupts plant metabolic processes, primarily through protein degradation, consequently inhibiting growth and development [28]. With the persistent escalation of global temperatures, heat stress is anticipated to increasingly constrain plant growth [29]. Passion fruit, a tropical and subtropical crop renowned for its distinctive aroma, occupies a pivotal position in the tropical fruit market. However, throughout its growth cycle, the plant is subjected to various environmental stressors, including heat, cold, drought, and pathogen infections [11]. Although passion fruit is native to tropical regions, it exhibits marked sensitivity to high-temperature stress. With the acceleration of global warming, elevated temperatures have become a critical factor threatening passion fruit production. Studies demonstrate that heat stress significantly inhibits photosynthetic efficiency, reduces antioxidant enzyme activity, and modulates the expression of heat-resistant genes such as heat shock proteins (HSPs) [3]. Considerable interspecific variation exists in thermal responses among cultivars: for instance, the optimal photosynthetic temperature for 'Summer Queen' and 'Ruby Star' is 30 °C [30], whereas the hybrid cultivar 'Minami-jujisei' shows pollen germination optimum at 36 °C [31]. Such germplasm diversity provides valuable genetic resources for breeding thermotolerant varieties. However, current research on passion fruit thermotolerance remains predominantly focused on physiological responses, leaving the molecular mechanisms and key regulatory pathways largely unexplored. Further in-depth investigations are urgently needed to elucidate these fundamental aspects.

This study elucidates the molecular and metabolic mechanisms underlying heat tolerance in passion fruit by integrating comparison of physiological differences, transcriptomic, and metabolomic approaches. The results demonstrate that the heat-tolerant variety F2 exhibited enhanced heat tolerance compared with the heat-sensitive QM9, and F2 displayed unique metabolic and gene expression profiles, particularly in purine metabolism and flavonoid biosynthesis. This metabolic profile suggests that heat-tolerant passion fruit varieties may leverage these compounds as part of their adaptive response to heat stress. The increased accumulation of flavonoids and phenolic acids likely enhances antioxidant capacity, protecting cellular structures from heat-induced oxidative damage. F2 exhibits increased organic acids and nitrogen compounds alongside decreased alcohols, polyols, carbohydrates, and lipids, indicating a metabolic adaptation that enhances osmotic regulation and energy metabolism under thermal stress. This metabolic shift may reflect a strategic reallocation of resources in F2, where energy and carbon are preferentially diverted toward the synthesis of heat-protective metabolites (e.g., flavonoids and terpenoids) rather than storage compounds like carbohydrates and lipids. The reduced accumulation of osmoprotective alcohols and polyols in F2 suggests a shift in osmotic adjustment strategy, potentially indicating greater reliance on alternative osmoprotectants such as proline and other stress-responsive amino acids during heat stress. These metabolic adjustments demonstrate that thermotolerant varieties such as F2 selectively enhance biosynthesis of protective metabolites while suppressing metabolic pathways involved in maintaining non-stress-related functions. These insights not only advance the comprehension of passion fruit's response to elevated temperatures but also provide potential targets for enhancing heat tolerance in other crops.

Purine metabolism regulates essential cellular processes, including signaling, redox balance, and energy metabolism [32,33]. Purine metabolites play vital roles in cellular functions, including energy production, signal transduction, and redox regulation [34]. Previous studies have demonstrated the importance of purine metabolism in plant responses to heat stress. For instance, heat stress in quinoa enriched the purine metabolic pathway and upregulated related metabolites in both heat-sensitive and heat-tolerant varieties [28]. Similarly, tobacco shoots exhibited significant increases in purine metabolites such as adenine, xanthine, hypoxanthine, and guanine under heat stress [35]. Notably, 3′,5′-Cyclic AMP, a key second messenger, has been identified as a regulator of heat stress responses, influencing protein homeostasis, ion regulation, and the ubiquitin–proteasome system [36]. This study revealed purine metabolism as a critical factor in the heat tolerance of the F2 variety. Purine compounds like glutamine and xanthine were more abundant in F2, and exogenous application of xanthine could improve the heat tolerance of QM9. These metabolites play crucial roles in regulating energy metabolism and maintaining redox balance—key processes for mitigating heat-induced stress. The upregulation of purine metabolism-related genes, such as GDA and XDH, further underscores their role in heat adaptation. Gene silencing experiments further demonstrated that downstream genes in this metabolic pathway can influence heat tolerance in passion fruit. Flavonoid biosynthesis was another pathway significantly enriched in F2. Flavonoids are well known for their significant role in enhancing plant stress resistance, contributing to both individual growth and increased tolerance to abiotic stresses. These compounds provide a theoretical foundation for understanding plant resilience mechanisms [37]. Flavonoids, including luteolin and trifolin, accumulated at higher levels in F2, and exogenous application of trifolin could improve the heat resistance of QM9, although the effect was not as obvious as that of xanthine. As potent antioxidants, flavonoids scavenge reactive oxygen species and alleviate oxidative stress, a major consequence of heat exposure [38]. The elevated levels of these compounds in F2 suggest that flavonoid biosynthesis contributes to heat tolerance by reducing oxidative damage, consistent with findings in other plant species under abiotic stress. Integrated transcriptomic and metabolomic analyses established correlations between key flavonoid pathway genes (e.g., F3′H) and differential flavonoid accumulation in F2, further underscoring this pathway's role in thermal adaptation.

Numerous transcription factors (TFs) play critical roles in plant adaptation to elevated temperatures [39]. Among these, the WRKY gene family is well known for its involvement in stress responses [40,41], while HSFs regulate stress tolerance by controlling the expression of HSPs [42]. Under thermal stress, plants activate the heat shock response, a conserved mechanism that induces heat-stimulated proteins to enhance thermal tolerance [43]. HSFs, as plant-specific TFs, are central to this process, mediating the activation of HSPs and other protective genes during heat stress [44]. In this study, several TFs associated with plant adaptability were identified, many of which are linked to purine metabolism and flavonoid biosynthesis pathways, including MYB, WRKY, HSF, bZIP, bHLH, and MADS. These findings underscore the importance of TFs connected to key metabolic pathways in regulating heat stress responses, providing new insights into the molecular mechanisms of heat tolerance in passion fruit.

However, a limitation of this study is that the conclusions are drawn from comparisons between only two passion fruit varieties (F2 and QM9). While this approach identified key metabolic and transcriptional differences associated with heat tolerance, the generalizability of these findings may be constrained by the limited genetic diversity represented. Future studies incorporating a broader range of heat-tolerant and heat-sensitive varieties would strengthen the validity of the proposed mechanisms and help distinguish universal adaptive traits from variety-specific responses.

5. Conclusions

This study reveals that purine metabolism and flavonoid biosynthesis are key pathways contributing to heat tolerance in yellow passion fruit, with the heat-tolerant variety F2 exhibiting elevated levels of metabolites like xanthine and flavonoids (e.g., trifolin) and upregulation of associated genes. Exogenous application of these metabolites enhanced heat tolerance, while silencing downstream pathway genes compromised thermotolerance, further validating this mechanism. Transcription factors such as WRKY, HSF, and MYB may further regulate these adaptive responses. These findings provide critical molecular targets for breeding heat-resistant passion fruit varieties, offering a strategic foundation to mitigate the impacts of rising global temperatures on tropical crop production. Future research should validate these genes and metabolites to optimize heat stress resilience.

Supplementary Materials: The following supporting information can be downloaded at: https://www.mdpi.com/article/10.3390/horticulturae11091037/s1, Figure S1: Volcano and heatmap plot of differentially accumulated metabolites (DAMs) in stem and leaf between QM9 and F2. Figure S2: Specific differences in metabolites of stem and leaf between QM9 and F2. Figure S3: Histograms of metabolite contents in several classes of stems in QM9 and F2. Figure S4: Histograms of metabolite contents in several classes of leaf in QM9 and F2. Figure S5: The GO and KEGG enrichment analysis of differentially expressed genes (DEGs) in QM9 and F2. Table S1: Summary of metabolome results. Table S2: Differentially accumulated metabolites (DAMs) of QM9 and F2 passion fruit. Table S3: K-means clustering results. Table S4: Summary of RNA sequencing and alignment data for each sample. Table S5: The Fragments Per Kilobase of exon model per Million mapped fragment (FPKM) values of the DEGs. Table S6: Gene Ontology (GO) enrichment for DEGs comparing F2 vs. QM9. Table S7: Kyoto Encyclopedia of Genes and Genomes (KEGG) enrichment for DEGs comparing cold-tolerant vs. cold-sensitive varieties. Table S8. WGCNA analysis.

Author Contributions: Conceptualization, L.-L.C. and W.C.; methodology, L.-L.C. and W.C.; validation, L.-H.C. and J.D.; formal analysis, L.-H.C.; investigation, L.-H.C. and B.-L.F.; resources, Y.H. and L.Y.; data curation, L.-H.C., J.D. and B.-L.F.; writing—original draft preparation, L.-H.C.; writing—review and editing, L.-L.C.; visualization, L.-H.C. and J.D.; supervision, L.-L.C. and W.C.; project administration, L.-L.C.; funding acquisition, L.-L.C. All authors have read and agreed to the published version of the manuscript.

Funding: This research was funded by the National Natural Science Foundation of China (32270712), Guangxi Science and Technology Major Program (guikeAA23062085), the Guangxi Natural Science Foundation (2024GXNSFGA010003), and the State Key Laboratory for Conservation and Utilization of Subtropical Agro-Bioresources (SKLCUSA-a202306).

Data Availability Statement: The datasets generated and analyzed during this study are included in this published article and its Supplementary Materials. All relevant data supporting the findings are available from the corresponding author upon reasonable request.

Conflicts of Interest: The authors declare no conflicts of interest.

References

1. Chen, G.; Xu, Y.; Gui, J.; Huang, Y.; Ma, F.; Wu, W.; Han, T.; Qiu, W.; Yang, L.; Song, S. Characterization of dof transcription factors and the heat-tolerant function of pedof-11 in passion fruit (*Passiflora edulis*). *Int. J. Mol. Sci.* **2023**, *24*, 12091. [CrossRef]
2. Santos, E.A.; Souza, M.M.; Abreu, P.P.; da Conceição, L.D.H.C.S.; Araújo, I.S.; Viana, A.P.; de Almeida, A.-A.F.; Freitas, J.C.d.O. Confirmation and characterization of interspecific hybrids of *Passiflora* L. (Passifloraceae) for ornamental use. *Euphytica* **2012**, *184*, 389–399. [CrossRef]
3. Wang, H.; Zhao, J.; Lai, M.; Zhang, Y.; Qiu, W.; Li, Y.; Tu, H.; Ling, Q.; Fu, X. Differential gene expression analysis and physiological response characteristics of passion fruit (*Passiflora edulis*) buds under high-temperature stress. *PeerJ* **2023**, *11*, e14839. [CrossRef] [PubMed]

4. Abreu, P.P.; Souza, M.M.; Santos, E.A.; Pires, M.V.; Pires, M.M.; de Almeida, A.-A.F. Passion flower hybrids and their use in the ornamental plant market: Perspectives for sustainable development with emphasis on Brazil. *Euphytica* **2009**, *166*, 307–315. [CrossRef]
5. Dos Reis, L.C.R.; Facco, E.M.P.; Salvador, M.; Flôres, S.H.; de Oliveira Rios, A. Antioxidant potential and physicochemical characterization of yellow, purple and orange passion fruit. *J. Food Sci. Technol.* **2018**, *55*, 2679–2691. [CrossRef]
6. Abreu, E.F.M.; Aragão, F.J.L. Isolation and characterization of a myo -inositol-1-phosphate synthase gene from yellow passion fruit (*Passiflora edulis* f. *flavicarpa*) expressed during seed development and environmental stress. *Ann. Bot.* **2006**, *99*, 285–292. [CrossRef]
7. Liu, Y.; Yuan, J.; Zhang, D.; Deng, K.; Chai, G.; Huang, Y.; Ma, S.; Qin, Y.; Wang, L. Genome-wide identification and characterization of the sbp gene family in passion fruit (*Passiflora edulis Sims*). *Int. J. Mol. Sci.* **2022**, *23*, 14153. [CrossRef]
8. Ohama, N.; Sato, H.; Shinozaki, K.; Yamaguchi-Shinozaki, K. Transcriptional regulatory network of plant heat stress response. *Trends Plant Sci.* **2017**, *22*, 53–65. [CrossRef]
9. Guo, H.; Li, Z.; Zhou, M.; Cheng, H. cDNA-AFLP analysis reveals heat shock proteins play important roles in mediating cold, heat, and drought tolerance in *Ammopiptanthus mongolicus*. *Funct. Integr. Genom.* **2014**, *14*, 127–133. [CrossRef]
10. Wang, X.; Yan, B.; Shi, M.; Zhou, W.; Zekria, D.; Wang, H.; Kai, G. Overexpression of a brassica campestris hsp70 in tobacco confers enhanced tolerance to heat stress. *Protoplasma* **2016**, *253*, 637–645. [CrossRef]
11. Wang, L.; Liu, Y.; Chai, G.; Zhang, D.; Fang, Y.; Deng, K.; Aslam, M.; Niu, X.; Zhang, W.; Qin, Y.; et al. Identification of passion fruit HSF gene family and the functional analysis of PeHSF-C1a in response to heat and osmotic stress. *Plant Physiol. Biochem.* **2023**, *200*, 107800. [CrossRef]
12. Garcia, E.; Koh, J.; Wu, X.; Sarkhosh, A.; Liu, T. Tissue-specific proteome profile analysis reveals regulatory and stress responsive networks in passion fruit during storage. *Sci. Rep.* **2024**, *14*, 3564. [CrossRef] [PubMed]
13. Dahro, B.; Wang, F.; Peng, T.; Liu, J.H. PtrA/NINV, an alkaline/neutral invertase gene of *Poncirus trifoliata*, confers enhanced tolerance to multiple abiotic stresses by modulating ROS levels and maintaining photosynthetic efficiency. *BMC Plant Biol.* **2016**, *16*, 76. [CrossRef] [PubMed]
14. Wingett, S.W.; Andrews, S. FastQ Screen: A tool for multi-genome mapping and quality control. *F1000Research* **2018**, *7*, 1338. [CrossRef] [PubMed]
15. Bolger, A.M.; Lohse, M.; Usadel, B. Trimmomatic: A flexible trimmer for Illumina sequence data. *Bioinformatics* **2014**, *30*, 2114–2120. [CrossRef]
16. Kim, D.; Langmead, B.; Salzberg, S.L. HISAT: A fast spliced aligner with low memory requirements. *Nat. Methods* **2015**, *12*, 357–360. [CrossRef]
17. Pertea, M.; Pertea, G.M.; Antonescu, C.M.; Chang, T.C.; Mendell, J.T.; Salzberg, S.L. StringTie enables improved reconstruction of a transcriptome from RNA-seq reads. *Nat. Biotechnol.* **2015**, *33*, 290–295. [CrossRef]
18. Love, M.I.; Huber, W.; Anders, S. Moderated estimation of fold change and dispersion for RNA-seq data with DESeq2. *Genome Biol.* **2014**, *15*, 550. [CrossRef]
19. Langfelder, P.; Horvath, S. WGCNA: An R package for weighted correlation network analysis. *BMC Bioinform.* **2008**, *9*, 559. [CrossRef]
20. Shannon, P.; Markiel, A.; Ozier, O.; Baliga, N.S.; Wang, J.T.; Ramage, D.; Amin, N.; Schwikowski, B.; Ideker, T. Cytoscape: A software environment for integrated models of biomolecular interaction networks. *Genome Res.* **2003**, *13*, 2498–2504. [CrossRef]
21. Wu, Y.; Tian, Q.; Huang, W.; Liu, J.; Xia, X.; Yang, X.; Mou, H. Identification and evaluation of reference genes for quantitative real-time PCR analysis in *Passiflora edulis* under stem rot condition. *Mol. Biol. Rep.* **2020**, *47*, 2951–2962. [CrossRef]
22. Honour, J.W. Gas chromatography-mass spectrometry. *Methods Mol. Biol.* **2006**, *324*, 53–74. [CrossRef]
23. Want, E.J.; Wilson, I.D.; Gika, H.; Theodoridis, G.; Plumb, R.S.; Shockcor, J.; Holmes, E.; Nicholson, J.K. Global metabolic profiling procedures for urine using UPLC-MS. *Nat. Protoc.* **2010**, *5*, 1005–1018. [CrossRef]
24. Zheng, Y.Y.; Chen, L.H.; Fan, B.L.; Xu, Z.; Wang, Q.; Zhao, B.Y.; Gao, M.; Yuan, M.H.; Tahir Ul Qamar, M.; Jiang, Y.; et al. Integrative multiomics profiling of passion fruit reveals the genetic basis for fruit color and aroma. *Plant Physiol.* **2024**, *194*, 2491–2510. [CrossRef] [PubMed]
25. Jacob, P.; Hirt, H.; Bendahmane, A. The heat-shock protein/chaperone network and multiple stress resistance. *Plant Biotechnol. J.* **2017**, *15*, 405–414. [CrossRef] [PubMed]
26. Priya, M.; Dhanker, O.P.; Siddique, K.H.M.; HanumanthaRao, B.; Nair, R.M.; Pandey, S.; Singh, S.; Varshney, R.K.; Prasad, P.V.V.; Nayyar, H. Drought and heat stress-related proteins: An update about their functional relevance in imparting stress tolerance in agricultural crops. *Theor. Appl. Genet.* **2019**, *132*, 1607–1638. [CrossRef]
27. Ul Haq, S.; Khan, A.; Ali, M.; Khattak, A.M.; Gai, W.X.; Zhang, H.X.; Wei, A.M.; Gong, Z.H. Heat shock proteins: Dynamic biomolecules to counter plant biotic and abiotic stresses. *Int. J. Mol. Sci.* **2019**, *20*, 5321. [CrossRef]
28. Xie, H.; Zhang, P.; Jiang, C.; Wang, Q.; Guo, Y.; Zhang, X.; Huang, T.; Liu, J.; Li, L.; Li, H.; et al. Combined transcriptomic and metabolomic analyses of high temperature stress response of quinoa seedlings. *BMC Plant Biol.* **2023**, *23*, 292. [CrossRef]

29. Sadok, W.; Lopez, J.R.; Smith, K.P. Transpiration increases under high-temperature stress: Potential mechanisms, trade-offs and prospects for crop resilience in a warming world. *Plant Cell Environ.* **2021**, *44*, 2102–2116. [CrossRef]
30. Shimada, A.; Kubo, T.; Tominaga, S.; Yamamoto, M. Effect of temperature on photosynthesis characteristics in the passion fruits 'summer queen' and 'ruby star'. *Hortic. J.* **2017**, *86*, 194–199. [CrossRef]
31. Matsuda, H.; Ogata, T. Varietal differences in thermal response of passion fruit pollen germination. *Trop. Agric. Dev.* **2020**, *64*, 90–96. [CrossRef]
32. Dewulf, J.P.; Marie, S.; Nassogne, M.-C. Disorders of purine biosynthesis metabolism. *Mol. Genet. Metab.* **2022**, *136*, 190–198. [CrossRef]
33. Kokina, A.; Ozolina, Z.; Liepins, J. Purine auxotrophy: Possible applications beyond genetic marker. *Yeast* **2019**, *36*, 649–656. [CrossRef] [PubMed]
34. Linden, J.; Koch-Nolte, F.; Dahl, G. Purine release, metabolism, and signaling in the inflammatory response. *Annu. Rev. Immunol.* **2019**, *37*, 325–347. [CrossRef] [PubMed]
35. Chen, H.; Qiu, S.; Chen, Y.; Li, J.; Xu, T.; Zhong, P.; Shao, X.; Xu, S.; Ma, Z.; Huang, Z.; et al. Integrated transcriptomics and metabolomics provides insights into the *Nicotiana tabacum* response to heat stress. *Front. Plant Sci.* **2024**, *15*, 1425944. [CrossRef]
36. Liang, S.; Sun, J.; Luo, Y.; Lv, S.; Chen, J.; Liu, Y.; Hu, X. cAMP Is a Promising regulatory molecule for plant adaptation to heat stress. *Life* **2022**, *12*, 885. [CrossRef] [PubMed]
37. Wang, M.; Zhang, Y.; Zhu, C.; Yao, X.; Zheng, Z.; Tian, Z.; Cai, X. EkFLS overexpression promotes flavonoid accumulation and abiotic stress tolerance in plant. *Physiol. Plant.* **2021**, *172*, 1966–1982. [CrossRef]
38. Dias, M.C.; Pinto, D.; Silva, A.M.S. Plant flavonoids: Chemical characteristics and biological activity. *Molecules* **2021**, *26*, 5377. [CrossRef]
39. Wei, S.; Zhang, L.; Huo, G.; Ge, G.; Luo, L.; Yang, Q.; Yang, X.; Long, P. Comparative transcriptomics and metabolomics analyses provide insights into thermal resistance in lettuce (*Lactuca sativa* L.). *Sci. Hortic.* **2021**, *289*, 110423. [CrossRef]
40. Gao, Y.F.; Liu, J.K.; Yang, F.M.; Zhang, G.Y.; Wang, D.; Zhang, L.; Ou, Y.B.; Yao, Y.A. The WRKY transcription factor WRKY8 promotes resistance to pathogen infection and mediates drought and salt stress tolerance in *Solanum lycopersicum*. *Physiol. Plant.* **2020**, *168*, 98–117. [CrossRef]
41. Jiang, J.; Ma, S.; Ye, N.; Jiang, M.; Cao, J.; Zhang, J. WRKY transcription factors in plant responses to stresses. *J. Integr. Plant Biol.* **2017**, *59*, 86–101. [CrossRef]
42. Hu, W.; Hu, G.; Han, B. Genome-wide survey and expression profiling of heat shock proteins and heat shock factors revealed overlapped and stress specific response under abiotic stresses in rice. *Plant Sci.* **2009**, *176*, 583–590. [CrossRef]
43. Haider, S.; Iqbal, J.; Naseer, S.; Yaseen, T.; Shaukat, M.; Bibi, H.; Ahmad, Y.; Daud, H.; Abbasi, N.L.; Mahmood, T. Molecular mechanisms of plant tolerance to heat stress: Current landscape and future perspectives. *Plant Cell Rep.* **2021**, *40*, 2247–2271. [CrossRef]
44. Li, Z.; Howell, S.H. Heat stress responses and thermotolerance in maize. *Int. J. Mol. Sci.* **2021**, *22*, 948. [CrossRef]

horticulturae

MDPI

Article

Manganese Deficiency Exacerbates Boron Deficiency-Induced Corky Split Vein in Citrus by Disrupting Photosynthetic Physiology and Enhancing Lignin Metabolism

Yanhong Li [1], Yiping Fu [1], Zhili Gan [1], Qingjing Wei [2], Mei Yang [1,3], Fengxian Yao [1] and Gaofeng Zhou [1,*]

[1] National Navel Orange Engineering Research Center, College of Navel Orange, Gannan Normal University, Ganzhou 341000, China; yanhongli208@163.com (Y.L.); fuyiping2024@163.com (Y.F.); zlgan_208@163.com (Z.G.); yangmei_208@163.com (M.Y.); yfengxian_919@163.com (F.Y.)

[2] College of Agronomy, Jiangxi Agricultural University, Nanchang 330045, China; qjwell@126.com

[3] College of Tropical Agriculture and Forestry, Hainan University, Danzhou 570228, China

* Correspondence: zhougaofeng428@163.com

Abstract

Corky split vein (CSV) is a common physiological disease in citrus that can result from multiple types of stresses. Preliminary field investigation found that more severe CSV in citrus cultivated in orchards lacking both boron (B) and other photosynthesis-related nutrients, including manganese (Mn). In this study, two-year-old 'Newhall' navel orange seedlings were treated with control (CK), B deficiency (BD), Mn deficiency (MnD), and combined B and Mn deficiency (BD + MnD). After 31 weeks, typical CSV symptoms appeared on old leaves (OLs) and secondary new leaves (SLs) in BD, while BD + MnD symptoms were more severe. BD and BD + MnD significantly reduced B concentrations in all leaf types, but there were no significant differences between them. Except for OLs in MnD, the net photosynthetic rate (P_n) of all leaf types significantly decreased in all treatments, with BD + MnD showing significantly lower P_n values than BD. Compared with BD, BD + MnD significantly increased minimal fluorescence (F_o) of all leaves at the later stage and significantly decreased Y(II) of new leaves. BD significantly increased sucrose and starch contents in all type leaves, while the OL starch content was significantly higher in BD + MnD than that in BD. BD + MnD significantly decreased the enzyme activities of Rubisco, TK, and FBA in OLs, FBPase and NI in PLs, and Rubisco in SLs compared with BD, while the activities of NI and AI in OLs and SS in SLs were significantly increased. BD + MnD significantly enhanced lignin concentrations and the expression of key lignin synthesis genes in leaves compared with BD. In conclusion, Mn deficiency exacerbates B-deficiency-induced CSV not only by intensifying photosynthetic dysfunction and carbohydrate accumulation but also by promoting lignin biosynthesis. These findings highlight the synergistic nature of B and Mn deficiencies in impairing leaf function and structure, providing new insights into the physiological and molecular mechanisms underlying CSV development.

Keywords: citrus; corky split vein; boron deficiency; manganese deficiency; photosynthetic physiologic characteristics; lignin

1. Introduction

Boron (B) is an essential micronutrient for the growth and development of higher plants [1,2]. It regulates plant cell elongation and division, as well as the formation and development of reproductive organs, through contributing to biological functions including the structure and function of plant cell walls, sugar transport, auxin synthesis and transport, carbohydrate and phenolic metabolism, and lignin synthesis [3–8]. On a global scale, B deficiency is the most widespread plant micronutrient deficiency and represents one of the main limiting factors for the production of 132 crops in many regions [9,10]. Studies have shown that 60–98% of B in plants is bound to the cell wall. When plants are deficient in B, their cell walls thicken and the arrangement of the cell wall becomes disordered [4,11]. Therefore, B deficiency symptoms typically first appear in growing regions rather than mature tissues, leading to stem tip necrosis, inhibited root elongation, suppressed leaf expansion, and decreased fertility [12].

As one of the most important fruit tree species in southern China, citrus often experiences growth and development issues due to B deficiency, resulting in reduced yield and quality [13–15]. The major navel orange production areas of China are concentrated in the region encompassing southern Jiangxi, southern Hunan, northern Guangxi, and Three Gorges, where soil conditions are relatively poor. In particular, acidic soil areas in southern Jiangxi are deficient in a variety of micronutrients [16–18]. Due to frequent high temperatures and rainy weather in southern Jiangxi, B is easily leached from the soil, leading to widespread B deficiency in the soils of navel orange orchards [17]. B deficiency symptoms include stem tip necrosis, thickened new leaves, malformed leaf veins, inhibited root development, and root tip swelling, among which the most typical symptom is corky split vein (CSV). CSV has been reported in many citrus species, including navel orange [8,19,20], sweet orange [14], pummelo [15,21], trifoliate orange, and fragrant citrus [22,23]. The causes of CSV are diverse. For example, both magnesium (Mg) deficiency and Huanglongbing disease can induce CSV [24–26]. Recent studies have shown that lignin metabolism plays an important role in the development of CSV symptoms caused by B deficiency or Mg deficiency [8,25,26].

Photosynthesis serves as the foundation of tree growth and fruiting. More than 90% of the dry matter content of roots, stems, leaves, flowers, and fruits is contributed by photosynthetic products produced by leaves. Both the economical yield and fruit quality of fruit trees are closely related to photosynthesis [27]. B plays a fundamental role in governing the efficiency of photosynthetic processes and thereby facilitates the subsequent translocation of photoassimilates [28,29]. Current research indicates that deficient or excessive B affects the photosynthetic parameters of plants, especially the synthesis and transport of photosynthetic products [10,30–32]. Therefore, regulating the application of B fertilizer can improve the growth and development of crops. For instance, applying B can enhance the photosynthetic efficiency of soybeans during their vegetative growth period [33]. In citrus, B deficiency reduces the photosynthetic pigment content of leaves. For example, research has demonstrated that B deficiency significantly reduces the contents of chlorophyll a (Chl a), chlorophyll b (Chl b), and carotenoid (Car) photosynthetic pigments in sweet orange leaves [14]. B deficiency also induces a decrease in photosynthetic pigment content in the leaves of a variety of citrus rootstocks, including trifoliate orange, Carrizo citrange, Chongyi tangerine, red tangerine, Cleopatra mandarin, and sour orange [22,30]. Furthermore, deficient or excessive B affects photosynthetic gas exchange parameters such as the leaf net photosynthetic rate (P_n). In citrus plants, the net assimilation of CO_2 and the stomatal conductance (G_s) leaf gas exchange parameters were found to be reduced under excess B in the leaves of Verna lemon trees grafted onto four different rootstocks (Carrizo citrange, Cleopatra mandarin, *Citrus macrophylla*, and sour orange), and this reduction was less pronounced for trees grafted onto *Citrus macrophylla* and Cleopatra mandarin rootstock. However, analysis of the intercellular CO_2 concentration (C_i) and chlorophyll fluorescence showed that the reduction in net CO_2 assimilation under excess B is mainly induced by non-stomatal factors [34]. B deficiency also affects the photosynthetic characteristics of citrus. For example, research has shown that B deficiency significantly decreases the P_n value of sweet orange leaves [14], significantly reduces CO_2 assimilation efficiency and photosystem II (PSII) photochemical efficiency in the seedling leaves of *Citrus grandis* [14,35], significantly inhibits leaf gas exchange parameters in 'HB' pummelo [21], and significantly affects the seedling leaf gas exchange parameters and diurnal photosynthetic dynamics of 'Newhall' navel orange [36,37]. Moreover, B deficiency may decrease the P_n value of the leaves of trifoliate orange and Carrizo citrange, which are widely utilized as citrus rootstocks [32]. In addition, B deficiency also affects the contents of photosynthetic products. In previous work, anatomical observations showed that B deficiency resulted in the obvious accumulation of starch granules in leaf cells [23]. Further investigation demonstrated that the contents of photosynthetic products such as soluble sugars, sucrose, fructose, and starch significantly increased under B deficiency stress [10,14,32,35]. The accumulation of large quantities of photosynthetic products in leaf cells may result in the feedback inhibition of photosynthesis, thereby leading to a decline in photosynthetic performance. In summary, deficient or excessive B impacts citrus photosynthesis through three main pathways: (i) reducing the contents of Chl a, Chl b, and Car photosynthetic pigments in leaves to weaken the capability of citrus leaves to capture light energy; (ii) altering photosynthesis-related leaf gas exchange parameters such as the net CO_2 assimilation, G_s, and C_i to reduce the photosynthetic performance of citrus leaves; and (iii) influencing the synthesis and transport of photosynthetic products to induce the feedback inhibition of photosynthesis in citrus leaves.

In addition to B, other micronutrients also play crucial roles in citrus photosynthesis. These micronutrients act as enzyme cofactors or components of the electron transport chain, participating directly or indirectly in the photosynthetic process [37–39]. Because changes in citrus photosynthetic characteristics induced by micronutrient deficiency usually occur before the appearance of visible leaf symptoms, such changes may provide as an important basis for the early diagnosis of micronutrient deficiency [37]. Therefore, investigating the relationship between micronutrients and photosynthetic characteristics in citrus is of great practical significance. Manganese (Mn), a key micronutrient, is directly involved in photosynthesis, serving as a structural component of PSII. The oxygen-evolving complex composed of Mn_4CaO_5 forms in the thylakoid lumen, playing a key role in the water-splitting reaction of PSII and the photosynthetic electron transport process. Therefore, Mn is a core element in carbon assimilation [40–42], as well as an essential element for plant growth and development. In addition to its direct involvement in photosynthesis, Mn acts as a cofactor for a variety of enzymes, participating in the regulation of secondary metabolite synthesis, including proteins, lipids, carbohydrates, and lignin [42–44]. Furthermore, Mn is involved in the formation of superoxide dismutase in mitochondria and peroxisomes, playing a key role in the scavenging of reactive oxygen species [41,42,45]. Soil pH is the most important factor affecting Mn availability. Alkaline soils usually cause Mn deficiency, whereas acidic soils may lead to Mn toxicity. Mn availability is also influenced by other factors, such as soil microbial activity, soil moisture content, and soil organic matter content [42,46–50]. Therefore, even citrus grown in acidic soils may encounter Mn deficiency. There are also the citrus orchards deficient in both Mn and B [16,51,52]. In citrus, Mn deficiency symptoms mainly appear in newly emerged leaves. The typical symptom in the early stage of Mn deficiency is interveinal chlorosis with green veins forming a net-like pattern, followed by the appearance of brown spots in the later stage, while CSV does not appear [39].

Previous field investigations of citrus leaf nutrient status showed that mineral nutrient deficiencies were usually not limited to a single element, and the lack of two or more elements was common [16,53]. Deficiency in multiple mineral nutrients is caused by poor soil nutrient status and antagonistic or synergistic interactions between mineral elements. Plant growth and development rely on a balance of essential mineral nutrients. A deficiency in one element can alter the uptake and transport of other minerals and may trigger multiple regulatory pathways to redistribute minerals to support normal plant growth and development [54]. Therefore, examining the interactions between mineral elements and other types of stresses is of great importance for understanding their physiological and molecular functions in crops. For example, research on nitrogen (N)–calcium (Ca) interactions in nectarine has shown that N–Ca imbalance hinders the development of photosynthetic organs and reduces photosynthetic efficiency [55]. There are also clear interactions between phosphorus (P) and zinc (Zn) in plant roots. Zn deficiency may result in P toxicity to barley, while P deficiency can cause an increase in the Zn content of barley [56]. Previous studies have shown that B can interact with a variety of mineral nutrients and stress factors [33]. B deficiency stress acts simultaneously with other stresses (extreme temperatures, excessive light, high CO_2 concentrations, drought, salinity, and heavy metal contamination), which can increase the sensitivity of plants to B toxicity or deficiency. B deficiency can alter the content of multiple mineral elements in plants. For example, many researchers have shown that the nitrate (N) contents of plant roots and leaves are markedly decreased under B deficiency [57,58]. B deficiency was reported to promote the uptake of Ca^{2+} by tobacco BY-2 cells, increase cytoplasmic Ca^{2+} concentrations in *Arabidopsis* roots, and enhance the Ca content in red orange leaves [22,59,60]. It is worth noting that B deficiency can change the content of multiple elements in citrus leaves and roots, notably causing a significant reduction in Mn content [22].

In recent years, the application of B fertilizer has been adopted by citrus growers and researchers to alleviate CSV symptoms. While such treatments have improved the nutritional status of trees and partially reduced symptom severity, CSV remains an unresolved issue under field conditions. Field observations indicated that CSV symptoms were more severe in orchards co-deficient in B and Mn (an essential nutrients involved in photosynthesis). Building on these findings, the present study employed 'Newhall' navel orange, a cultivar highly susceptible to CSV, to elucidate how Mn deficiency exacerbates B-deficiency-induced CSV. Through integrated physiological, biochemical, and molecular analyses—including assessments of symptom severity, photosynthetic performance, accumulation of photosynthetic products, activities of key metabolic enzymes, lignin content, and expression of lignin biosynthesis genes—we reveal the synergistic detrimental effects of combined B and Mn deficiency. Our results provide novel insights into the mechanisms underlying CSV development and offer a theoretical foundation for sustainable management strategies based on balanced nutrient regulation rather than B supplementation alone.

2. Materials and Methods

2.1. Plant Materials and Treatments

Two-year-old 'Newhall' navel orange [*Citrus sinensis* (L.) Osb. CV. Newhall] grafted onto trifoliate orange (*Citrus trifoliata* L.) rootstock was used in this experiment. Uniform size seedlings with relatively consistent growth were selected. Each seedling retains one scion stem; this scion stem was defined as old scion stem (OS) and the leaves on this scion stem were defined as old leaves (OLs). The residual substrate on the roots was washed with deionized water, and approximately one-third of the root system was pruned before transplanting into black plastic pots (8 L volume) filled with quartz sand: perlite (1:1, *v*/*v*). Each pot contained one plant, with four replicates per treatment. The seedlings were cultivated in a greenhouse with temperatures maintained at 22–28 °C and relative humidity at 50–75%. According to published method [8,61], one-week pretreatment with deionized water was conducted to acclimatize the plants to the sand culture system. The modified full-strength complete Hoagland's No.2 nutrient solution contained: 6 mM KNO_3, 4 mM $Ca(NO_3)_2$, 1 mM $NH_4H_2PO_4$, 2 mM $MgSO_4$, 50 μM Fe-EDTA, 25 μM H_3BO_3, 9 μM $MnCl_2$, 0.8 μM $ZnSO_4$, 0.3 μM $CuSO_4$, and 0.01 μM H_2MoO_4.

The experiment set up three treatments and one control. Control (CK): Plants were cultivated with a 1/2-strength complete nutrient solution [8]; Boron deficiency (BD): 1/2-strength complete nutrient solution without H_3BO_3; Manganese deficiency (MnD): 1/2-strength complete nutrient solution without $MnCl_2$, and KCl with the same molar concentration was added; Combined boron and manganese deficiency (BD + MnD): 1/2-strength complete nutrient solution without H_3BO_3 and $MnCl_2$, supplemented with the same concentration of KCl. Based on the previous method, each pot was irrigated with nutrient solution twice a week, with 500 mL each time [8]. To prevent salt accumulation in the substrate, plants were irrigated with 10 L of deionized water every four weeks, ensuring that an excessive amount of solution drained out from the bottom of the pot [39]. The experiment started in December 2023 and ended in July 2024. At 10th weeks after treatment, two primary new scion stems (PNSs) were retained per plant, and their leaves were defined as primary new leaves (PLs). Similarly, at 21 weeks, two secondary new scion stems (SNSs) were retained on the primary new scion stems, and their leaves were designated as secondary new leaves (SLs).

2.2. Determination of Gas Exchange and Chlorophyll Fluorescence Parameters

The gas exchange parameters of 'Newhall' navel oranges leaves were determined at the 10th, 21st and 31st weeks of the experimental treatment, following the method described by Zhou et al. [37]. Measurements were conducted between 9:30 and 11:30 AM on a clear and cloudless day. The gas exchange parameter indicators net photosynthetic rate (P_n), transpiration rate (T_r), intercellular CO_2 concentration (C_i), and stomatal conductance (G_s) were measured by portable photosynthesis measuring instrument (LI-6400XT, LI-COR, Lincoln, NE, USA) with the photosynthetic active radiation (*PAR*) set at 1000 $\mu mol \cdot m^{-2} \cdot s^{-1}$. OL was measured at the 10th, 21st, and 31st weeks, PL at the 21st and 31st weeks, and SL at the 31st week. For each plant, five leaves of each grade were selected for repeated measurement. After the data stabilized, the instantaneous gas exchange parameter values of each leaf were recorded five times.

After measuring the gas exchange parameters, the photochemical efficiency of the same leaves was determined using a Portable Pulse-Amplitude-Modulation Chlorophyll Fluorometer PAM-2500 (Heinz Walz GmbH, Effeltrich, BY, Germany), following the methods of Ling et al. [62] and Han et al. [35]. The leaves were dark-adapted for 30 min using a dark-adapted leaf clip. The optical fiber was then placed on the clip, and the 'F_v/F_m' button in the 'Field Screen' window was clicked. After one second, the maximal photochemical efficiency of PSII was measured. After the fluorescence value stabilized (approximately 2 min), 'Y(II)' button was clicked to measure the actual photochemical efficiency and other chlorophyll fluorescence parameters under the corresponding light intensity.

2.3. Determination of Chlorophyll Content

The determination of photosynthetic pigment (chlorophyll a and chlorophyll b) contents in leaves (OL, PL, and SL) was conducted following the method described by Zhou et al. [22]. The leaves were washed clean with deionized water, and the veins were removed. Fresh leaf samples (0.10 g) were taken and ground into a homogenate in a mortar with a small amount of quartz sand and 95% ethanol until the tissue turned white. The homogenate was then transferred and filtered into a 25 mL brown volumetric flask. A rubber-tipped dropper was used to rinse the residual debris in the mortar several times with a small amount of 85% acetone. The solution was brought to volume and thoroughly mixed. Samples were measured with a UV-1700 spectrophotometer (UV-1700, Shimadzu, Kyoto, Japan) at a wavelength of 663 and 644 nm, respectively. The concentration of chlorophyll a and b, in milligrams per gram of FW tissue, was calculated based on previous research [22].

2.4. Measurement of Plant Growth Parameters, Mineral Nutrient Concentrations, and Lignin Contents

After 31 weeks of treatment, all 'Newhall' navel orange seedlings were harvested and blotted with tissue paper. The plant height (cm) was measured using a scaled ruler. Then, the collected materials were divided into leaf (OL, PL, and SL), stem and root tissues. The fresh tissue materials were placed into a forced air oven at 105 °C for 15 min and then maintained at 75 °C until constant dry weights (g) were reached. All dried samples were then ground into fine powder for the determination of the mineral nutrient concentrations. The macro- and micro-nutrient concentrations of P, potassium (K), Ca, Mg, iron (Fe), Mn, B, Zn, copper (Cu) and molybdenum (Mo) in the leaves, stems, and roots of 'Newhall' navel orange plants were determined following the method described by Zhou et al. [61]. Briefly, 0.20 g of each sample was dry-ashed in a muffle furnace at 550 °C for 6 h, followed by dissolution in 5% HNO_3. Suitable dilutions were subsequently made for the determination of various mineral concentrations. Each sample was measured three times repeatedly. The mineral nutrient concentrations were then determined using an Inductively Coupled Plasma Mass Spectrometry ICP-MS 7900 (Agilent Technologies Inc., Santa Clara, CA, USA).

For N analysis, the H_2SO_4-H_2O_2 digestion method was employed. Briefly, 0.30 g of each dried sample (in triplicate) was weighed into a 50 mL digestion tube. Then, 5 mL of concentrated sulfuric acid (H_2SO_4) was added, and the mixture was shaken and left to stand overnight. The digestion process was carried out on an electric heating plate with gradual temperature increase to 300 °C. After brief cooling, 10–20 drops of H_2O_2 solution were added intermittently, and the tube was returned to the heating plate. This step was repeated several times until the solution turned clear. After cooling, the digest was diluted to 50 mL with ultrapure water and filtered. Finally, the N content was determined using a fully automated discrete chemical analyzer Smartchem 200 (Alliance Instruments Inc., Villebon-sur-Yvette, Essonne, France).

The lignin contents in 'Newhall' navel oranges leaves (OL, PL, and SL) were determined following the method described by Xiong et al. [63], using a 72% concentrated sulfuric acid (H_2SO_4) hydrolysis approach. For each treatment group, four biological replicates were prepared per leaf type, with three technical replicates performed for each biological replicate.

2.5. Determination of Fructose, Sucrose and Starch Contents

After 31 weeks of experimental treatment, samples of leaves (OL, PL, and SL) of 'Newhall' navel oranges of different grades were collected during sunny weather between 9:30 and 11:30 AM. The determination of fructose, sucrose, and starch content in the leaves was performed following the instructions of their respective assay kits. In brief, 0.10 g of ground freeze-dried leaf sample was weighed, and the extraction of fructose, sucrose, and starch was conducted according to the protocols of the respective assay kits (Enzyme-linked Biotechnology, Shanghai, China). Samples were measured with a UV-1700 spectrophotometer (Shimadzu Corp., Nakagyo-ku, Kyoto, Japan) at a wavelength of 480 nm, 480 nm, and 620 nm, respectively. The contents of fructose, sucrose, and starch ($mg{\cdot}kg^{-1}$, dry weight) were then calculated based on standard curves.

2.6. Measurement of Enzyme Activity Related to Photosynthesis

At the 31st week of treatment, leaf samples (OL, PL, and SL) of 'Newhall' navel oranges were collected between 9:30 and 11:30 AM on clear days. The samples were immediately frozen in liquid nitrogen and stored at −80 °C. Prior to analysis, the samples were ground into powder on ice. The activities of ribulose-1,5-bisphosphate carboxylase/oxygenase (Rubisco), Rubisco activase (RCA), fructose-1,6-bisphosphatase (FBPase), sedoheptulose-1,7-bisphosphatase (SBPase), fructose-1,6-bisphosphate aldolase (FBA), and transketolase (TK) were determined using ELISA kits (Enzyme-linked Biotechnology, Shanghai, China) following the manufacturer's instructions with slight modifications. Briefly, 0.10 g of sample was homogenized in 1 mL of 0.1 mol/L phosphate buffer, and the homogenate was centrifuged at 4500 $r \cdot min^{-1}$ for 20 min at 4 °C. The supernatant was collected for subsequent analysis. Enzyme activities were measured using a Spark Multimode Microplate Reader (Tecan Trading AG, Männedorf, Switzerland). Key enzyme activities in photosynthetic product synthesis were also carried out. The neutral invertase (NI), acid invertase (AI), sucrose synthetase (SS), and sucrose phosphate synthase (SPS) were measured using an Enzyme-linked Immunosorbent Assay (ELISA) kit following the manufacturer's instructions (Enzyme-linked Biotechnology, Shanghai, China) and a Spark Multimode Microplate Reader.

2.7. Expression Analysis of Key Enzyme Genes Involved in Lignin Biosynthesis

According to the method of Zhou et al. [8], the expression patterns of key lignin biosynthesis enzyme genes in leaves of different grades from 'Newhall' navel oranges subjected to different treatments was conducted. Quantitative real-time PCR (qRT-PCR) primers for the lignin pathway genes and the internal reference gene (*β-actin*) are shown in Supporting Information Table S1. Differences in gene expression, expressed as fold change relative to control, were calculated using the $2^{-\Delta\Delta CT}$ method. Each measurement was performed four times by using independently prepared RNA/cDNA templates, and the error bars represent the SE of the mean fold-change in the four replicates.

2.8. Experimental Design and Statistical Analysis

The experiment was set up in a completely randomized 2 × 2 factorial design with two B concentrations (0 and 20 μmol L^{-1} B) and two Mn concentrations (0 and 9 μmol L^{-1} Mn). Unless otherwise specified, the data in the chart represents the average ± standard error of four individual plants (repetitions). The data underwent analysis of variance (ANOVA) in SAS software SAS 8.1 (SAS Institute Inc., Cary, NC, USA), and the differences were compared using Duncan's test with a significance level of $p < 0.05$. The data distribution of each experimental group approximates a normal distribution, and the homogeneity of variance meets the requirements for further ANOVA. Microsoft Office Excel 2010 (Microsoft Corp., Redmond, WA, USA) was employed for basic statistics. Graphs were prepared using SigmaPlot 12.5 (Systat Softwares, San Jose, CA, USA) and OriginPro 8.0 (Origin Lab, Northampton, MA, USA).

3. Results

3.1. Visible Symptoms and Plant Growth

After 31 weeks of treatment, symptoms were visible on 'Newhall' navel orange plants in the BD, MnD, and BD + MnD treatment groups. Compared with the CK group, the SLs of the plants in the BD treatment group exhibited typical CSV symptoms in parallel with slight leaf yellowing; the SLs of the plants in the MnD treatment group showed interveinal chlorosis, but no CSV symptoms were observed. The SLs of the plants in the BD + MnD treatment group presented more severe CSV compared with the BD treatment group, as well as significantly curled leaf tips. BD-treated plants exhibited slight CSV on the tips of the OLs, and these leaves curled up. BD + MnD treated plants displayed CSV on OLs. The OLs of plants in the MnD treatment group did not display CSV. Compared with the CK group, the BD and BD + MnD treatments significantly inhibited the growth and development of root systems, while the MnD treatment had a lesser inhibitory effect on the root system (Figure 1). Compared with the CK, all three nutrient deficiency treatments significantly decreased the plant height, stem dry matter weight, and total dry matter weight. The BD and BD + MnD treatments exerted a significantly stronger effect on plant height than the MnD treatment; the BD and BD + MnD treatments significantly decreased leaf and root dry weight, while the MnD treatment did not show a significant effect; and the BD and BD + MnD treatments significantly increased the root: shoot ratio, while the MnD treatment did not have a significant impact (Table 1).

Figure 1. Symptoms of 'Newhall' navel orange [*Citrus sinensis* (L.) Osb. CV. Newhall] under boron and manganese co-deficiency treatment for 31 weeks. Two-year-old plants were grown in sand culture and treated under control (**A**), boron deficiency (**B**), manganese deficiency (**C**), and boron + manganese co-deficiency (**D**) for 31 weeks. I, whole plant; II, secondary new leaf; III, old leaf; IV, root.

Table 1. Effects of boron and manganese co-deficiency on growth of 'Newhall' navel orange [*Citrus sinensis* (L.) Osb. CV. Newhall] seedlings.

Treatments	Plant Height (cm)	Dry Weight (g plant^{-1})				Root-Shoot Ratio
		Leaf	Stem	Root	Total	
CK	74.8 ± 5.6 a	14.5 ± 1.9 a	20.3 ± 1.3 a	25.9 ± 1.6 a	60.7 ± 2.3 a	0.77 ± 0.11 b
BD	56.6 ± 1.3 c	8.6 ± 1.7 b	13.8 ± 1.0 c	20.4 ± 0.9 b	42.8 ± 2.3 c	0.92 ± 0.06 a
MnD	65.4 ± 2.3 b	12.6 ± 0.7 a	17.7 ± 1.4 b	22.6 ± 1.1 ab	52.9 ± 2.7 b	0.75 ± 0.06 b
BD + MnD	54.3 ± 2.8 c	6.6 ± 2.1 b	14.5 ± 2.3 c	18.0 ± 1.3 b	39.1 ± 3.6 c	0.90 ± 0.13 a

Note: Two-year-old 'Newhall' navel orange seedlings were grown under different micronutrient deficiency conditions for 31 weeks. CK: control, BD: boron deficiency, MnD: manganese deficiency, and BD + MnD: Boron + manganese co-deficiency. Data are presented as the mean ± standard error of four biological replicates. Different lowercase letters following the mean values indicate significant differences ($p < 0.05$) between different treatment conditions.

3.2. Mineral Nutrient Concentrations

Compared with the CK group, the BD and BD + MnD treatments significantly decreased B concentrations in OLs, PLs, and SLs of 'Newhall' navel orange seedlings, and in all types of leaves the differences in B concentrations between the BD and BD + MnD treatment groups were non-significant. The MnD treatment significantly decreased B concentrations in SLs but did not affect the B concentrations in OLs and PLs. Except for the OLs in the MnD treatment group, the Mn concentrations in the OLs, PLs, and SLs in the MnD and BD + MnD treatment groups significantly decreased compared with the CK group, and the Mn concentrations in the OLs and PLs in the BD + MnD treatment group were significantly lower than that in the MnD treatment group. In addition, the BD treatment significantly decreased Mn concentrations in OLs and SLs. For other micronutrients, Fe significantly increased only in the OLs in the MnD treatment group. The concentration of Cu significantly decreased in the OLs in the MnD treatment group and significantly increased in PLs and SLs. Zn was significantly elevated in the OLs in the MnD treatment group compared with the CK group and significantly decreased in the PLs in the BD + MnD treatment group. The Mo concentration significantly increased only in the SLs in the MnD treatment group compared with the CK group. It is worth noting that the differences in Fe, Cu, and Mo concentrations in all types of leaves between the BD and BD + MnD treatment groups were non-significant. The Zn concentrations in PLs and SLs were significantly lower in the BD + MnD treatment group than those in the BD treatment group. For macronutrients, none of the three treatments significantly affected the N concentrations in all types of leaves compared with the CK group. Except for the OLs in the BD treatment group, the P concentrations in all types of leaves in all three treatment groups were significantly increased compared with the CK group. The K concentration significantly increased in the OLs in the BD + MnD treatment group and significantly decreased in the PLs in the MnD and BD + MnD treatment groups compared with that in the CK group. The Ca concentrations in the PLs and SLs of the BD treatment group and in the SLs of the BD + MnD treatment group significantly decreased compared with the CK group. The Mg concentrations in the OLs and PLs in all treatment groups and in the SLs in the BD treatment group significantly decreased compared with the CK group (Figure 2).

The mineral nutrient concentrations in the stems and roots of 'Newhall' navel orange seedlings were also determined (Figures S1 and S2). The results indicated that the BD and BD + MnD treatments significantly decreased B concentrations in all types of stems, and the MnD treatment significantly increased B concentrations in OSs and SNSs compared with the CK treatment. Compared with the BD treatment, the BD + MnD treatment did not decrease B concentration farther, and the B concentration was even increased in OSs and PNSs. Except for the OSs in the BD treatment, the Mn concentrations in all types of shoots in all treatment groups significantly decreased compared with the CK group. The B concentrations in the primary and lateral roots in the BD and BD + MnD treatment groups were significantly lower than those in the CK and MnD treatment group, and B was also significantly decreased in the lateral roots in the MnD treatment group. The MnD and BD + MnD treatments significantly decreased the Mn concentration of lateral roots but had no significant impact on primary roots.

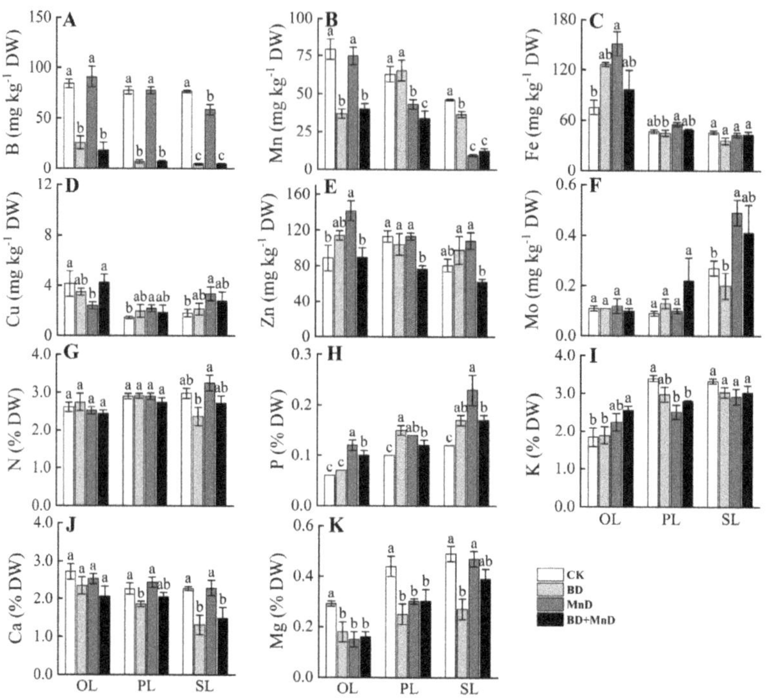

Figure 2. Effect of boron and manganese co-deficiency on mineral nutrient concentrations in the leaves of 'Newhall' navel orange [*Citrus sinensis* (L.) Osb. CV. Newhall] seedlings. (**A**) boron concentration; (**B**) manganese concentration; (**C**) iron concentration; (**D**) copper concentration; (**E**) zinc concentration; (**F**) molybdenum concentration; (**G**) nitrogen concentration; (**H**) phosphorus concentration; (**I**) potassium concentration; (**J**) calcium concentration; (**K**) magnesium concentration. Two-year-old 'Newhall' navel orange seedlings were grown under boron, manganese, or boron and manganese co-deficiency conditions for 31 weeks. CK: Control; BD: Boron deficiency; MnD: Manganese deficiency; BD + MnD: Boron + manganese co-deficiency. Data are presented as the mean ± standard error of four biological replicates. DW, dry weight. Different lowercase letters above the bars indicate significant differences ($p < 0.05$) between the different treatments. OL, old leaf; PL, primary new leaf; SL, secondary new leaf.

3.3. Gas Exchange Parameters and Photosynthetic Pigment Content

As shown in Figure 3, the P_n value of the OLs of 'Newhall' navel orange in the BD and BD + MnD treatment groups declined as the treatment duration increased compared with the CK group, and this increase became significant in the 21st and 31st weeks of treatment. In contrast, the MnD treatment had no significant effect on the P_n value of OLs. The P_n values of PLs in the BD and BD + MnD treatment groups significantly decreased in the 21st and 31st weeks of treatment, while a significant decrease in the P_n value of PLs in the MnD treatment group presented only in the 31st week of treatment. The P_n values of SLs in all treatment groups significantly decreased in the 31st week of treatment. It is worth noting that the P_n values of OLs, PLs, and SLs in the BD + MnD treatment group was significantly lower than that in the BD treatment group in the 31st week of treatment. Compared with the CK group, the MnD and BD + MnD treatments did not significantly affect the G_s value of OLs in the 10th week of treatment, but for all other types of leaves, treatment groups, and time points the G_s value significantly decreased. The G_s value of OLs in the 21st week and the G_s values of PLs and SLs in the 31st week of treatment were significantly lower in the BD + MnD treatment group than those in the BD treatment group. The C_i values of OLs in the 31st week of BD treatment and in the 21st and 31st week of BD + MnD treatment significantly increased compared with the CK group, and the differences in the C_i value of OLs between the BD and BD + MnD treatment groups were non-significant. Except for the 21st week of BD treatment, the C_i values of PLs and SLs in all treatment groups significantly increased, and the C_i value of PLs in the 31st week of treatment was significantly lower in the BD treatment group than that in the BD + MnD treatment group. The T_r values of OLs significantly decreased in the 31st week of BD treatment, in the 10th and 31st week of MnD treatment, and in the 21st week of BD + MnD treatment. The T_r value of PLs in the 21st week of BD + MnD treatment significantly decreased, whereas it significantly increased in the 31st week. The T_r value of SLs in the 31st week of all treatments significantly decreased.

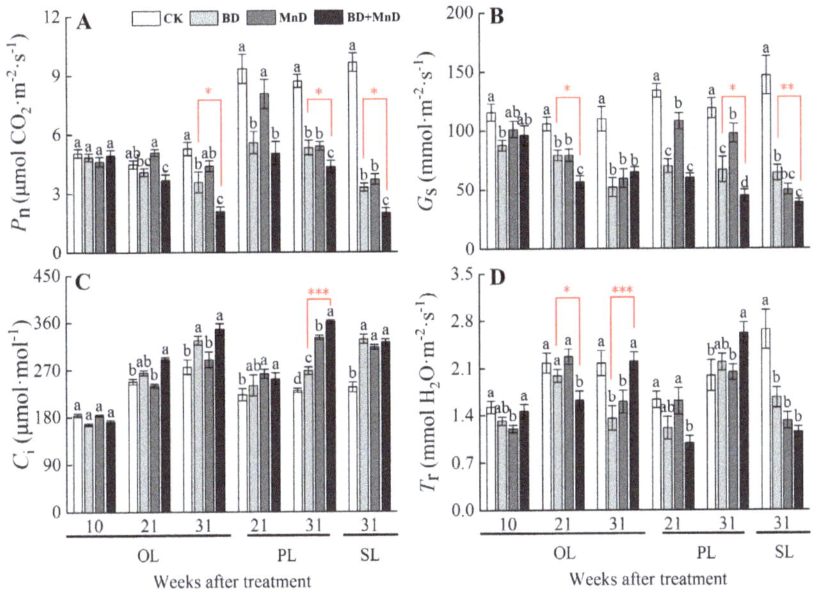

Figure 3. Effect of boron and manganese co-deficiency on gas exchange parameters of 'Newhall' navel orange [*Citrus sinensis* (L.) Osb. CV. Newhall] seedlings. (**A**) P_n, net photosynthetic rate; (**B**) G_s, stomatal conductance; (**C**) C_i, intercellular CO_2 concentration; (**D**) T_r, transpiration rate. CK: Control; BD: Boron deficiency; MnD: Manganese deficiency; BD + MnD: Boron + manganese co-deficiency. The data in the figure is the mean ± standard error. Lowercase letters indicate significant differences at the 5% level among different treatments at the same time. OL, old leaf; PL, primary new leaf; SL, secondary new leaf. The asterisk indicates the significance of the difference between BD and BD + MnD treatments, signiffcant level * $p < 0.05$, ** $p < 0.01$, *** $p < 0.001$.

After 31 weeks of treatment, the photosynthetic pigment contents of OLs, PLs, and SLs in 'Newhall' navel orange seedlings were measured (Table 2). The results revealed that Chl a, Chl b, and total chlorophyll contents displayed the same change trend. The Chl a, Chl b, and total chlorophyll contents of OLs were significantly reduced in the BD and BD + MnD treatment groups compared to the CK group, and were significantly lower in the BD + MnD treatment group than those in the BD treatment group. The Chl a, Chl b, and total chlorophyll contents of PLs were significantly lower in the MnD and BD + MnD treatment groups compared to those in the CK and BD groups, and the differences between the BD and CK groups were non-significant. The Chl a, Chl b, and total chlorophyll contents of SLs were significantly lower in all three treatment groups than those in the CK group, and the Chl b and total chlorophyll contents were significantly reduced in the BD + MnD treatment group compared to the BD group.

Table 2. Effects of boron and manganese co-deficiency on photosynthetic pigment contents in leaves of 'Newhall' navel orange [*Citrus sinensis* (L.) Osb. CV. Newhall] seedlings.

Photosynthetic Pigment		CK	BD	MnD	BD + MnD
Chlorophyll a ($mg\ g^{-1}$ FW)	OL	1.67 ± 0.08 a	1.32 ± 0.06 b	1.79 ± 0.07 a	1.10 ± 0.01 c
	PL	2.02 ± 0.11 a	2.07 ± 0.12 a	1.64 ± 0.06 b	1.78 ± 0.11 ab
	SL	1.98 ± 0.15 a	1.46 ± 0.21 b	1.16 ± 0.13 c	1.42 ± 0.13 b
Chlorophyll b ($mg\ g^{-1}$ FW)	OL	1.13 ± 0.07 a	0.83 ± 0.06 b	1.02 ± 0.11 ab	0.68 ± 0.07 c
	PL	1.08 ± 0.05 a	1.11 ± 0.10 a	0.65 ± 0.03 b	0.70 ± 0.04 b
	SL	1.19 ± 0.05 a	0.85 ± 0.09 b	0.74 ± 0.01 bc	0.62 ± 0.05 c
Total chlorophyll ($mg\ g^{-1}$ FW)	OL	2.76 ± 0.02 a	2.15 ± 0.09 b	2.81 ± 0.18 a	1.78 ± 0.07 c
	PL	3.10 ± 0.06 a	3.18 ± 0.18 a	2.30 ± 0.08 b	2.48 ± 0.15 b
	SL	3.17 ± 0.12 a	2.31 ± 0.12 b	1.89 ± 0.12 c	2.04 ± 0.18 c

Note: Two-year-old 'Newhall' navel orange seedlings were grown under boron, manganese, or boron and manganese co-deficiency conditions for 31 weeks. CK: Control; BD: Boron deficiency; MnD: Manganese deficiency; BD + MnD: Boron + manganese co-deficiency. OL, old leaf; PL, primary new leaf; SL, secondary new leaf. Data are presented as the mean ± standard error of four biological replicates. FW, fresh weight. Different lowercase letters following the mean values in same line indicate significant differences ($p < 0.05$) between different treatment conditions.

3.4. Chlorophyll Fluorescence Parameters

As shown in Figure 4 and Table S2, the chlorophyll fluorescence parameter minimal fluorescence (F_o) value of 'Newhall' navel orange began to increase in the 21st week of BD + MnD treatment and rose significantly in the 25th, 27th, 29th, and 31st weeks of treatment, while for the BD treatment the increase began in the 25th week and continued until the 31st week of treatment. The MnD treatment had no significant impact on the F_o value of OLs. For new leaves, except for the 21st week of the BD + MnD treatment, the F_o values of PLs and SLs at all treatment time points under the MnD and BD + MnD treatments were significantly elevated, while the BD treatment only significantly increased the F_o value of SLs in the 31st week. Compared with the BD treatment, the BD + MnD treatment significantly increased the F_o value of OLs in the 27th and 29th weeks, the F_o value of PLs in weeks 25–31, and the F_o value of SLs in the 27th and 31st weeks of treatment. In addition, the F_o values of OLs, PLs, and SLs began to decline 4 weeks earlier in the BD + MnD treatment group than those in the BD treatment group. The maximal fluorescence (F_m) value of OLs displayed a significant decreasing trend from the 21st week of BD treatment onward, apart from the 25th week, while the BD + MnD treatment showed a significant decrease starting in the 23rd week. The F_m value of PLs began to significantly decrease from the 25th week of the BD treatment onward, the 31st week of the MnD treatment onward, and in the 25th, 29th, and 31st weeks of the BD + MnD treatment. The F_m value of SLs significantly decreased in the 27th, 29th, and 31st weeks for all treatments. The variable fluorescence (F_v) and F_m values of 'Newhall' navel orange leaves, except for the PLs in the MnD treatment group, showed the same change trend in all treatment groups compared with that in the CK group. The F_v value of PLs significantly declined at all the time points under the MnD treatment. Compared with the CK group, the F_v/F_m value of OLs was significantly decreased in the 29th and 31st week under the BD treatment and in weeks 23–31 under the BD + MnD treatment. The F_v/F_m value of PLs significantly decreased in the 25th, 27th, 29th, and 31st weeks of the MnD treatment and in the 25th and 31st weeks of the BD + MnD treatment. The F_v/F_m values of SLs declined significantly in the 29th and 31st weeks of all treatments. Compared with the CK, the electron transport efficiency (ETR) values of OLs declined significantly in the 25th, 29th, and 31st weeks under the BD treatment and in weeks 21–31 under the BD + MnD treatment; the ETR values of PLs significantly decreased in the 29th and 31st weeks under the BD and MnD treatments and in weeks 27–31 under the BD + MnD treatment; and the ETR values of SL dropped significantly decreased in the 31st week of the BD treatment and in weeks 27–31 of the BD + MnD treatment. It is worth noting that the ETR values of OLs, PLs, and SLs decreased earlier in the BD + MnD treatment group than in the BD treatment group.

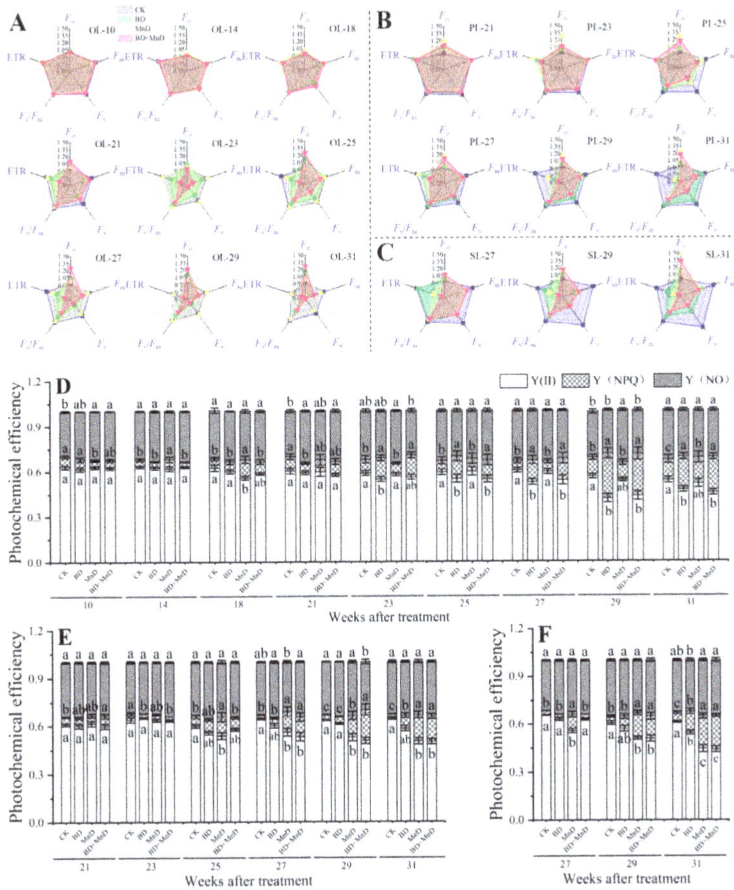

Figure 4. Effect of boron and manganese co-deficiency on chlorophyll fluorescence parameters in the leaves of 'Newhall' navel orange [*Citrus sinensis* (L.) Osb. CV. Newhall] seedlings. (**A**,**D**) Chlorophyll fluorescence parameters of old leaves (OL) after 10, 14, 18, 21, 23, 25, 27, 29, and 31 weeks treatment; (**B**,**E**) Chlorophyll fluorescence parameters of primary new leaves (PL) after 21, 23, 25, 27, 29, and 31 weeks treatment; (**C**,**F**) Chlorophyll fluorescence parameters of secondary new leaves (SL) after 27, 29, and 31 weeks treatment. CK: Control; BD: Boron deficiency; MnD: Manganese deficiency; BD + MnD: Boron + manganese co-deficiency. F_o minimal fluorescence; F_m, maximal fluorescence; F_v, variable fluorescence; F_v/F_m, maximal photochemical efficiency of PSII; ETR, electron transport efficiency; Y(II), actual photochemical efficiency; Y(NPQ), Quantum yield of regulated energy dissipation; Y(NO), Quantum yield of non-regulated energy dissipation. Data are presented as the mean ± standard error of 12 biological replicates. All the data in figure (**A**–**C**) are relative values, which were calculated based on the chlorophyll fluorescence value of the old leaves being 1.00 at the 10th week after treatment. Different lowercase letters above the bars in (**D**–**F**) indicate significant differences ($p < 0.05$) between the different treatments.

Compared with the CK group, the actual photochemical efficiency [Y(II)] values of OLs significantly decreased in weeks 23–31 under the BD treatment and in weeks 25–31 under the BD + MnD treatment; PLs displayed significantly reduced Y(II) values in weeks 25–31 of the MnD treatment and in weeks 27–31 of the BD + MnD treatment; and the Y(II) values of SLs dropped significantly in week 31 of the BD treatment, weeks 27–31 of the MnD treatment, and weeks 29–31 of the BD + MnD treatment. Interestingly, the Y(II) of PLs and SLs significantly decreased in the 29th and 31st weeks of the BD + MnD treatment, respectively. Compared with the CK group, the quantum yield of regulated energy dissipation [Y(NPQ)] values of OLs significantly declined in week 21 of the BD treatment and week 10 of the MnD treatment, whereas they significantly increased in weeks 23–31 of the BD and BD + MnD treatments and in weeks 14, 18, and 31 of the MnD treatment. In PLs, Y(NPQ) values significantly decreased in week 23 of the BD and BD + MnD treatments, while they significantly increased in week 31 of the BD treatment, weeks 25–31 of the MnD treatment, and weeks 21, 27, 29, and 31 of the BD + MnD treatment compared with the CK group. The Y(NPQ) values of SLs significantly increased in the 31st week under the BD treatment, in the 27th, 29th, and 31st weeks under the MnD treatment, and in the 29th and 31st weeks under the BD + MnD treatment compared with the CK group. The quantum yield of non-regulated energy dissipation [Y(NO)] values of OLs rose significantly in the 21st week of the BD treatment, the 10th and 29th weeks of the MnD treatment, and the 10th and 21st weeks of the BD + MnD treatment compared with the CK group. The Y(NO) values of PLs significantly declined in the 27th week of the MnD treatment and in the 29th week of the BD + MnD treatment compared with the CK (Figure 4D–F).

3.5. Leaf Photosynthetic Product Content and Related Enzyme Activity

The BD treatment group displayed significantly enhanced fructose content in 'Newhall' navel orange SLs, while the BD + MnD treatment significantly increased the fructose contents of OLs and SLs. In contrast, the fructose contents of other treatment groups and leaf types did not exhibit significant differences compared with the CK group. Apart from the finding that the fructose content of OLs was higher in the BD + MnD group than that in the BD group, the difference in fructose content between the BD and BD + MnD treatment groups was not significant. Compared with the CK, the BD treatment significantly promoted the sucrose contents of OLs, PLs, and SLs, while the MnD and BD + MnD treatments did not exhibit significantly increased sucrose contents, and the sucrose content of OLs was decreased under the BD + MnD treatment. Notably, the sucrose contents of OLs, PLs, and SLs were significantly higher in the BD treatment group compared to the BD + MnD treatment group. Except for PLs in the BD + MnD treatment group, OLs, PLs, and SLs contained significantly elevated starch contents in the BD and BD + MnD treatment groups compared with then CK group, but in the MnD treatment group the differences in starch content between leaf types were non-significant. In addition, OLs showed higher starch content in the BD + MnD treatment than in the BD treatment (Figure 5).

As shown in Figure 6, compared with the CK group, the BD treatment significantly reduced Rubisco activity in OLs, thymidine kinase (TK) activity in PLs, and Rubisco and fructose 1,6-bisphosphatase (FBPase) activity in SLs, whereas it significantly increased the activity of fructose 1,6-bisphosphate aldolase (FBA) in OLs. The MnD treatment significantly decreased the activity of TK in OLs; the activities of rubisco activase (RCA), TK, and FBPase in PLs; and the activities of Rubisco, RCA, TK, and FBPase in SLs. Under the BD + MnD treatment, the activity of RCA was significantly increased in OLs, while the activities of Rubisco, TK, and sedoheptulose 1,7-bisphosphatase (SBPase) were significantly decreased in OLs; Rubisco, RCA, TK, and FBPase displayed significantly reduced activities in PLs; and Rubisco, RCA, and FBPase showed significantly decreased activities in SLs.

Interestingly, compared with the BD treatment, the BD + MnD treatment significantly increased RCA activity in OLs but significantly decreased the activities of Rubisco, TK, and FBA in OLs, that of FBPase in PLs, and that of Rubisco in SLs. Compared with the CK group, the BD treatment significantly promoted the activities of neutral invertase (NI) and synthetase (SS) in PLs; the MnD treatment significantly increased the activities of acid invertase (AI), SS, and sucrose phosphate synthase (SPS) in PLs and that of AI in SLs. The BD + MnD treatment significantly increased the activities of NI and AI in OLs and those of SS and SPS in SLs, whereas the SS activity in OLs was significantly decreased. Moreover, compared with the BD treatment, the BD + MnD treatment significantly suppressed NI activity in PLs but significantly enhanced NI and AI activities in OLs and SS activity in SLs.

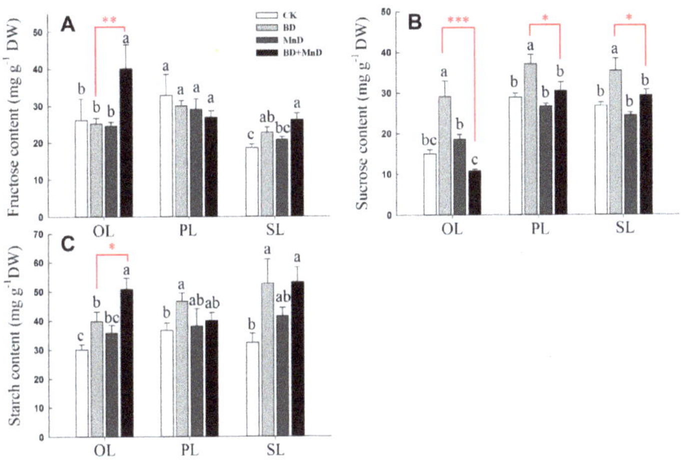

Figure 5. Effect of boron and manganese co-deficiency on the fructose (**A**), sucrose (**B**), and starch (**C**) contents in the leaves of 'Newhall' navel orange [*Citrus sinensis* (L.) Osb. CV. Newhall] seedlings. CK: Control; BD: Boron deficiency; MnD: Manganese deficiency; BD + MnD: Boron + manganese co-deficiency. DW, Dry weight. OL, old leaf; PL, primary new leaf; SL, secondary new leaf. Data are presented as the mean ± standard error of four biological replicates. Different lowercase letters above the bars indicate significant differences ($p < 0.05$) between the different treatments. The asterisk indicates the significance of the difference between BD and BD + MnD treatments, significance level * $p < 0.05$, ** $p < 0.01$, *** $p < 0.001$.

3.6. Lignin Content and Related Key Gene Expression Level

As shown in Figure 7A, compared with the CK group, the BD treatment significantly increased the lignin concentration in the veins of SLs; the BD + MnD treatment significantly enhanced the lignin concentration in the veins of OLs; and the MnD treatment had no significant effect on the lignin concentrations in the veins of all types of leaves. In addition, compared with the BD treatment, the BD + MnD treatment significantly promoted the lignin concentration in the veins of SL. The relative expression levels of key genes involved in lignin synthesis in OLs and SLs were analyzed. Compared with the CK group, in the BD treatment the *CsPAL1*, *Cs4CL1*, *CsCCR1*, *CsCCR2*, and *CsCAD3* genes were significantly upregulated in OLs, while the *CsCAD5* gene was significantly downregulated; *CsPAL2*, *CsCCR1*, and *CsCAD2* were significantly upregulated in OLs in the MnD treatment; and *CsPAL1*, *Cs4CL1*, *Cs4CL2*, *CsCCR1*, *CsCCR2*, *CsCAD1*, *CsCAD2*, and *CsCAD3* were significantly upregulated in OLs in the BD + MnD treatment. Compared with the BD treatment, in the BD + MnD treatment the expression levels of *CsC4H*, *Cs4CL2*, *CsCAD2*, *CsCAD4*, and *CsCAD5* in OLs were significantly upregulated, while the expression levels of *CsPAL3*, *CsPAL4*, *Cs4CL1*, and *CsCCR2* were downregulated. Compared with the CK group, in the BD treatment *CsC4H*, *Cs4CL1*, and *CsCAD3* were significantly upregulated in SLs and *CsCAD5* was significantly downregulated; in the MnD treatment, *CsC4H*, *CsCCR1*, *CsCCR2*, and *CsCAD2* were significantly upregulated in SLs; and in the BD + MnD treatment, the expression levels of all other genes except for *CsPAL2*, *CsC4H*, *Cs4CL3*, *CsCAD2*, and *CsCAD5* were significantly upregulated in SLs. Compared with the BD treatment, the BD + MnD treatment significantly promoted the expression levels of *CsPAL1*, *CsPAL3*, *CsPAL4*, *Cs4CL2*, *CsCCR1*, *CsCCR2*, *CsCAD1*, *CsCAD3*, *CsCAD4*, and *CsCAD5* in SLs, while *CsC4H* was significantly downregulated (Figure 7B,C).

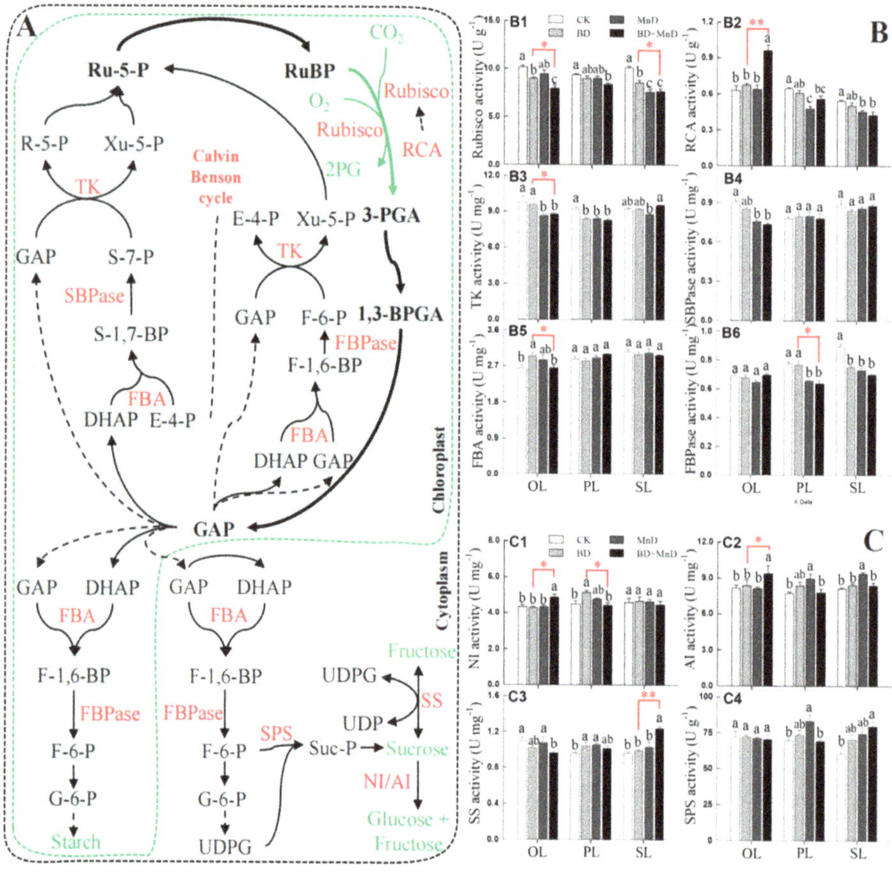

Figure 6. Effect of boron and manganese co-deficiency on the photosynthetic enzyme activity of 'Newhall' navel orange [*Citrus sinensis* (L.) Osb. CV. Newhall] seedlings. (**A**) Calvin Benson cycle and the synthetic pathways of starch and sucrose in chloroplasts and cytoplasmic, (**B**) photosynthetic enzyme activity and (**C**) the activity of key enzymes for the synthesis of photosynthetic products. Green arrow indicates carboxylation catalyzed by Rubisco resulting in the formation of 3-PGA and competing oxygenase reaction of Rubisco. (**B1**) Rubisco, ribulose bisphosphate carboxylase/oxygenase; (**B2**) RCA, rubisco activase; (**B3**) TK, thymidine kinase; (**B4**) SBPase, sedoheptulose 1,7-bisphosphatase; (**B5**) FBA, fructose 1,6-bisphosphate aldolase; (**B6**) FBPase, fructose 1,6-bisphosphatase; (**C1**) NI, neutral invertase; (**C2**) AI, acid invertase; (**C3**) SS, sucrose synthetase; (**C4**) SPS, sucrose phosphate synthase. Other abbreviations in (**A**): CO_2, carbon dioxide; 2GP, 2-phosphoglycerate; RuBP, ribulose 1,5-bisphosphate; 3-PGA, 3-phosphoglyceric acid; 1,3-PGA, glycerate 1,3-bisphosphate; GAP, glyceraldehyde 3-phosphate; Ru-5-P, ribulose 5-phosphate; R-5-P, ribose 5-phosphate; DHAP, dihydroxyacetone phosphate; F-1,6-BP, fructose 1,6-bisphosphate; F-6-P, fructose

6-phosphate; Xu-5-P, xylose 5-phosphate; E-4-P, erythritose 4-phosphate; S-1,7-BP, sedoheptulose 1,7-bisphosphate; S-7-P, sedoheptulose 7-phosphate; G-6-P, glucose 6-phosphate; Suc-P, sucrose phosphate; UDPG, uridine diphosphate glucose; UDP, uridine diphosphate. Data are presented as the mean ± standard error of four biological replicates. Different lowercase letters above the bars indicate significant differences ($p < 0.05$) between the different treatments. CK: Control; BD: Boron deficiency; MnD: Manganese deficiency; BD + MnD: Boron + manganese co-deficiency. The asterisk indicates the significance of the difference between BD and BD + MnD treatments, signiffcant level * $p < 0.05$, ** $p < 0.01$.

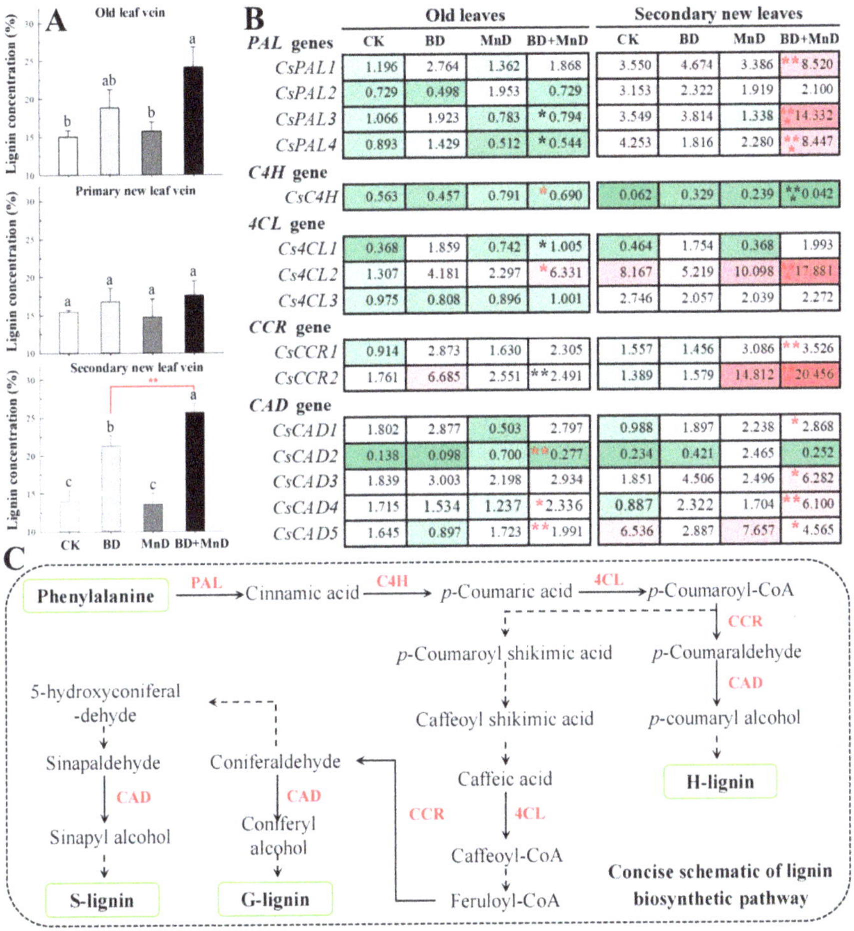

Figure 7. Effects of boron and manganese co-deficiency on lignin metabolism in the leaf of 'Newhall' navel orange [*Citrus sinensis* (L.) Osb. CV. Newhall] seedlings. (**A**) lignin concentrations in leaves, (**B**) relative expression levels of key enzyme genes involved in lignin synthesis in the old and secondary new leaves, (**C**) concise schematic of lignin synthetic pathway. Two-year-old 'Newhall' navel orange seedlings were grown under boron, manganese, or boron and manganese co-deficiency conditions for 31 weeks. Data are presented as the mean ± SE of four biological replicates. Different lowercase letters above the bars indicate significant differences ($p < 0.05$) between the different treatments. The asterisk indicates the significance of the difference between BD and BD + MnD treatments, significance level * $p < 0.05$, ** $p < 0.01$, *** $p < 0.001$; Red asterisk indicates significant increase, while black asterisk indicates significant decrease. CK: Control; BD: Boron deficiency; MnD: Manganese

deficiency; BD + MnD: Boron + manganese co-deficiency. PAL, phenylalanine ammonia-lyase; C4H, Cinnamic acid 4-hydroxylase; 4CL, 4-coumarate: coenzyme A ligase; CCR, Cinnamoyl-CoA reductase; CAD, cinnamyl alcohol dehydrogenase.

4. Discussion

CSV is a common physiological symptom of disease in citrus [8,10]. As early as 1950, it was reported that under B-deficient conditions, sour orange and 'Wendan' pomelo leaves developed CSV symptoms [64]. Apart from sour orange and 'Wendan' pomelo, CSV symptoms have been widely reported in various types of citrus, including trifoliate orange, fragrant citrus, navel orange, satsuma orange, *Citrus sinensis*, 'HB' pummelo, 'Guanximiyou' pummelo, and kumquat [8,14,15,21,22,37,65]. Research indicates that CSV occurs across all three genera (*Citrus, Poncirus*, and *Fortunella*) of cultivated citrus species. Among citrus species, *Citrus sinensis*, *Citrus trifoliata*, and *Citrus grandis* have been extensively studied. These citrus species are the most sensitive to nutritional deficiencies such as B deficiency, and their symptoms are the most pronounced [8,15,22]. However, CSV is not unique to citrus. For example, tip blight in *Eucalyptus* not only causes tip die-back and branch deformation but also results in raised and cracked leaf veins, resembling the CSV symptoms in citrus [66]; similarly, mulberry trees under B-deficient conditions exhibit swelling and cracking in leaf veins [67]. Our results showed that under the BD and BD + MnD treatments, 'Newhall' navel orange exhibited CSV symptoms in OLs and SLs, with greater severity under the BD + MnD treatment than under the BD treatment. In contrast, no CSV was observed in the MnD treatment group (Figure 1). Therefore, these findings suggest that B deficiency is the main cause of CSV symptoms. This conclusion is in accordance with previous studies of trifoliate orange, *Citrus sinensis*, and pummelo [8,14,21,22,68]. Although B deficiency results in CSV symptoms in multiple citrus species, these symptoms can be induced by a variety of factors [66]. For example, studies have shown that Mg deficiency also causes CSV symptoms in citrus [25,26,65,69]. An investigation in the citrus production area of Fujian Province showed that 86.2% of CSV symptoms were induced by Mg deficiency, with combined B and Mg deficiency being the second greatest contributor to CSV symptoms, while B deficiency only accounted for 2.3% of CSV symptoms. Although both Mg and B deficiency can induce CSV, these CSV symptoms display marked differences. CSV due to Mg deficiency is characterized by swelling and cracking veins on the leaf tip, which typically displays an inverse V-shaped chlorotic area, whereas B-deficiency-induced CSV exhibits cracking leaf veins, but the diseased leaves remain green [65]. Furthermore, some biological stresses, such as citrus Huanglongbing and citrus yellow vein clearing disease, can also lead to the occurrence of CSV symptoms when the diseases are severe [24,70].

Previous studies have shown that B deficiency not only affects the B concentrations in various parts of citrus plant, but also significantly influences the concentrations of other nutrients [22,39]. In this study, compared with the CK group, the BD and BD + MnD treatments significantly decreased B concentrations in leaves (OLs, PLs, and SLs), stems (OSs, PSs, and SSs) and roots (taproots and lateral roots) of 'Newhall' navel orange seedlings, while the MnD treatment alone significantly reduced B concentrations only in SLs and lateral roots (Figure 2, Figures S1 and S2). This result indicates that B deficiency treatment significantly reduces the B concentrations in various parts of citrus plants, which is consistent with the research results of previous studies [8,21,22,39]. In addition, in all types of leaves the differences in B concentrations between the BD and BD + MnD treatment groups were non-significant (Figure 2A). This indicates that while MnD alone can reduce B concentrations in SLs and lateral roots, the combination of B deficiency and Mn deficiency does not lead to a further decrease in B concentration in citrus leaves. Therefore, the exacerbation of CSV symptoms under BD + MnD treatment is not attributable to additional reduction in leaf B concentrations. In this study, the Mn concentrations in various parts of citrus were significantly decreased under MnD and BD + MnD conditions, except in OLs and taproots. In addition, the BD treatment significantly decreased Mn concentrations in OLs, SLs, PSs, and SSs (Figure 2, Figures S1 and S2). This result indicates that B deficiency treatment not only reduces the B concentrations in various parts of citrus plants, but also leads to significant reduction in the Mn concentration in the above-ground tissues. Similarly, previous studies have also shown that the Mn concentration was significantly reduced in the leaves of trifoliate orange, Carrizo citrange, Chongyi tangerine, and sour orange under B deficiency conditions [22]. The above research indicates that there is an interaction relationship between B and Mn in citrus, but the specific mechanism still needs further research. In addition to Mn, B deficiency also has a significant impact on the concentration of other mineral nutrients in citrus, particularly the Mg concentration in leaves. In this study, the Mg concentrations in the SLs, PLs, and OLs were significantly decreased under BD conditions compared with the CK (Figure 2). This is consistent with the previous research results on trifoliate orange, Carrizo citrange, Chongyi tangerine, Cleopatra mandarin, red tangerin, and 'Nanfeng' tangerine [22,39].

Both Mn and Mg directly participate in plant photosynthesis [40–42,71]. CSV symptoms induced by Mg deficiency disrupt the photosynthetic performance of citrus plants and significantly impact the synthesis and transport of photosynthetic products [26,69]. However, whether Mn deficiency aggravates CSV symptoms induced by B deficiency through influencing photosynthesis in citrus remains unclear. Our results showed that the BD, MnD, and BD + MnD treatments significantly affected the content of photosynthetic pigments, gas exchange parameters, chlorophyll fluorescence parameters, photosynthetic enzyme activities, photosynthetic products, and activity of key enzymes involved in the synthesis of photosynthetic products in the leaves of 'Newhall' navel orange (Table 2, Figures 3–6). Although B does not directly participate in the metabolism of photosynthetic pigments, studies have shown that B deficiency can lead to reduced photosynthetic pigment content in citrus leaves [14,22,32]. In our study, both the BD and BD + MnD treatments significantly reduced the Chl a, Chl b, and total chlorophyll content in the OLs and SLs of 'Newhall' navel orange (Table 2), which was consistent with previously reported findings in sweet orange and trifoliate orange [14,22,32]. However, our study also found that the contents of Chl a, Chl b, and total chlorophyll in OLs, as well as the Chl b and total chlorophyll contents in SLs, were significantly lower in the BD + MnD treatment group than those in the BD treatment group (Table 2). This result implies that under B deficiency, photosynthetic pigment content is closely associated with the severity of CSV symptoms. Further analysis of chlorophyll fluorescence parameters revealed that the BD, MnD, and BD + MnD treatments significantly affected the chlorophyll fluorescence values of citrus leaves (Figure 4). Specifically, the F_o value of the OLs of 'Newhall' navel orange began to rise significantly from the 25th week under the BD treatment and continued to increase until the end of the experiment. The F_o value of SLs also rose significantly in the 31st week under the BD treatment. These findings are consistent with previous results [14]. Notably, compared with the BD treatment, the BD + MnD treatment significantly increased the F_o value of OLs in the 27th and 29th weeks of treatment, the F_o value of PLs in the 25th through 31st weeks of treatment, and the F_o value of SLs in the 27th and 31st weeks of treatment, while the Y(II) values of PLs and SLs were significantly decreased in the 29th and 31st weeks of treatment. Furthermore, significant declines in the ETR values of OLs, PLs, and SLs occurred earlier in the BD + MnD treatment group than in the BD treatment group (Figure 4). Altogether, these results showed that the BD + MnD treatment had a significantly stronger impact on photosynthetic pigments and chlorophyll fluorescence parameters than the BD treatment.

B deficiency affects both the photosynthetic rate and photochemical efficiency of citrus leaves. In our study, both the BD and BD + MnD treatments displayed significantly decreased P_n values in the 31st week of treatment, and the P_n values of OLs, PLs, and SLs were significantly lower in the BD + MnD treatment group compared to the BD treatment group (Figure 3A). In accordance with our results, the finding that B deficiency significantly decreases P_n has already been confirmed in sweet orange, 'HB' pummelo, trifoliate orange, and Carrizo citrange [14,21,32,36]. Moreover, a previous study indicated that a significant decline in the P_n value of citrus leaves could be detected before CSV symptoms became obvious [37]. The above-described results indicate that B deficiency can reduce the P_n value of citrus leaves, which occurs prior to the appearance of CSV symptoms. Our study found that, at the same time point, the P_n values of 'Newhall' navel orange OLs, PLs, and SLs were significantly lower in the BD + MnD treatment group than those in the BD treatment group (Figure 3A), which suggested that Mn deficiency exacerbated the decline in the P_n value of citrus leaves caused by B deficiency. Further analysis of PSII revealed that B deficiency significantly decreased the photochemical efficiency of PSII in *Citrus grandis* seedling leaves [35]. We also found that the Y(II) value of 'Newhall' navel orange OLs was significantly reduced in weeks 23–31 of the BD treatment, and the Y(II) value of SLs was significantly reduced in the 31st week of the BD treatment compared with that in the CK group. The Y(II) values of PLs and SLs were significantly lower in the 29th and 31st weeks of the BD + MnD treatment compared to the BD treatment (Figure 4). Mn participates directly in photosynthesis, serving as a structural component of PSII, and is involved in the water-splitting reaction and photosynthetic electron transport [40–42]. Therefore, the MnD treatment significantly reduced the Y(II) values of PLs and SLs in 'Newhall' navel orange (Figure 4). The above findings suggest that compared with the BD treatment, the BD + MnD treatment had a stronger impact on the PSII photochemical efficiency of citrus leaves, and as a result, the Y(II) value of new leaves was significantly lower in the BD + MnD treatment group than that in the BD treatment group.

B deficiency affects photosynthetic products and the activities of key enzymes involved in their synthesis. In our study, the BD treatment significantly increased the sucrose and starch contents of OLs, PLs, and SLs in 'Newhall' navel orange, and the starch content of OLs was significantly higher in the BD + MnD treatment group than that in the BD treatment group (Figure 5C). Previous studies have demonstrated that B deficiency leads to the accumulation of photosynthetic products such as sucrose and starch in citrus leaves [10,14,23,32,35]. The excessive accumulation of photosynthetic products in mesophyll cells can inhibit the activity of related enzymes through feedback mechanisms, significantly reducing the efficiency of photosynthesis and carbon assimilation. The results of our study showed that, compared with the CK group, the BD treatment significantly suppressed the activity of Rubisco in OLs, TK in PLs, and Rubisco and FBPase in SLs, while the activities of FBA in OLs and NI and SS in PLs were significantly increased (Figure 6). Previous research demonstrated that Rubisco, NADP-glyceraldehyde-3-phosphate dehydrogenase, and FBPase activities were lower in B-deficient sweet orange leaves than in controls [14]. Our findings are consistent with results reported by other authors. However, we found that the BD + MnD treatment could intensify this phenomenon. Compared with the BD treatment, the BD + MnD treatment significantly increased the starch content of OLs; significantly decreased the activities of Rubisco, TK, and FBA in OLs, FBPase and NI in PLs, and Rubisco in SLs; and significantly enhanced NI and AI activities in OLs and SS activity in SLs (Figure 6). Mn is a key element in plant carbon assimilation and serves as a cofactor for multiple enzymes involved in the synthesis of secondary metabolites, including carbohydrates [42–44]. These results suggest that Mn plays an important role in the physiological changes described above, though the specific regulatory mechanisms require further investigation.

The accumulation of photosynthetic products in the mesophyll cells of citrus mainly occurs due to impaired transport. Photosynthetic products must be transported from inside the cell to the outside and then moved out of the leaf through the vascular tissues of the veins. B deficiency induces the thickening of citrus mesophyll cell walls and the abnormal development of vascular tissues in the veins, which leads to the accumulation of photosynthetic products in citrus mesophyll cells [8,15]. Lignin accumulation is a major cause of cell wall thickening and the abnormal development of vascular bundles in the veins of citrus leaves. Our study showed that, compared with the CK group, the BD treatment significantly increased leaf vein lignin concentrations in the SLs of 'Newhall' navel orange, while the BD + MnD treatment significantly enhanced lignin concentrations in the veins of both OLs and SLs (Figure 7). Lignin is a macromolecule aromatic heteropolymer second only to cellulose in terms of abundance in plants. Lignin closely binds with cellulose and acts as major filler in the middle lamella and secondary cell wall. In higher plants, lignin plays key roles in water transport, mechanical support, and defense against plant pathogens [72,73]. Lignin can also respond to various forms of biotic and abiotic stress [74]. Studies have shown that lignin metabolism plays an important role in the formation of CSV symptoms. In accordance with previous reports, our study demonstrated that B deficiency significantly increased the lignin content in the leaves of 'Newhall' navel orange, boosted the activity of lignin-related metabolic enzymes, and upregulated the expression of key lignin biosynthesis genes [8,19]. In addition, we found that both the lignin content and the relative expression levels of key lignin biosynthesis genes (*CsPAL1*, *CsPAL3*, *CsPAL4*, *Cs4CL2*, *CsCCR1*, *CsCCR2*, *CsCAD1*, *CsCAD3*, *CsCAD4*, and *CsCAD5*) were significantly higher in SLs under the BD + MnD treatment than under the BD treatment (Figure 7). Mn is essential for the biosynthesis of chlorophyll, aromatic amino acids, and secondary products such as lignin and flavonoids [43,75]. Accordingly, this study demonstrated that combined B and Mn deficiency promoted lignin accumulation in citrus leaves, consequently aggravating the CSV symptoms on the leaves. Moreover, previous studies have shown that lignin is involved in plant responses to a range of biotic and abiotic stresses [76–79]. For example, both the intensity and quality of light affect lignin accumulation in plants [80–82]. Therefore, Mn deficiency may promote lignin accumulation under combined B and Mn deficiency through altering the absorption and utilization of light, ultimately leading exacerbating CSV symptoms.

5. Conclusions

In summary, Mn deficiency aggravates B-deficiency-induced CSV in the OLs and SLs of 'Newhall' navel orange. In the present study, the B concentrations in different types of leaves were not significantly different between the BD and BD + MnD treatment groups, which indicated that Mn deficiency aggravated CSV symptoms without further reducing the B concentration in leaves. B deficiency disrupted the physiological functions of 'Newhall' navel orange leaves, including photosynthetic characteristics, the synthesis and transport of photosynthetic products, and lignin metabolism. Mn is a key element for photosynthesis, and combined Mn and B deficiency intensifies the decline in leaf photosynthetic pigments, photochemical efficiency, and the P_n and increases starch accumulation. Compared with B deficiency alone, the lack of both B and Mn significantly promoted lignin concentrations and the relative expression levels of key lignin synthesis genes in the leaves of 'Newhall' navel orange. Therefore, Mn deficiency may aggravate CSV symptoms induced by B deficiency through negatively impacting the leaf photosynthetic characteristics and increasing the accumulation of photosynthetic products and lignin, although further research is necessary to elucidate the specific mechanism. This study provides a theoretical basis for addressing CSV symptoms in citrus production through regulating nutrient balance.

Supplementary Materials: The following supporting information can be downloaded at: https://www.mdpi.com/article/10.3390/horticulturae11101172/s1, Table S1: Primers used for qRT-PCR analysis of lignin biosynthesis genes; Table S2: Effect of boron and manganese co-deficiency on chlorophyll fluorescence parameters in the leaves of 'Newhall' navel orange seedlings; Figure S1: Effect of boron and manganese co-deficiency on mineral nutrient concentrations in the stems of 'Newhall' navel orange seedlings; Figure S2: Effect of boron and manganese co-deficiency on mineral nutrient concentrations in the roots of 'Newhall' navel orange seedlings.

Author Contributions: Conceptualization, G.Z. and F.Y.; Writing—original draft preparation, Y.L. and Y.F.; Data curation, software, visualization, Y.L., Z.G. and M.Y.; Resources, Q.W.; Writing—review and editing, G.Z.; Supervision, F.Y. All authors have read and agreed to the published version of the manuscript.

Funding: This research was funded by the National Natural Science Foundation of China (Nos. 31960573 and 32160680).

Data Availability Statement: The original contributions presented in the study are included in the article/Supplementary Materials; further inquiries can be directed to the corresponding author.

Conflicts of Interest: The authors declare no conflicts of interest.

Abbreviations

The following abbreviations are used in this manuscript:

CSV	Corky split vein
BD	Boron deficiency
MnD	Manganese deficiency
BD + MnD	Combined boron and manganese deficiency
CK	Control
OL	Old leaf
PL	Primary new leaf
SL	Secondary new leaf
OS	Old scion stem
PNS	Primarynew scion stem
SNS	Secondary new scion stem
Chl a	Chlorophyll a
Chl b	Chlorophyll b
Car	Carotenoid
P_n	Net photosynthetic rate
T_r	Transpiration rate
C_i	Intercellular CO_2 concentration
G_s	Stomatal conductance
F_o	Minimal fluorescence
F_m	Maximal fluorescence
F_v	Variable fluorescence
F_v/F_m	Maximal photochemical efficiency of PSII
ETR	Electron transport efficiency
Y(II)	Actual photochemical efficiency
Y(NPQ)	Quantum yield of regulated energy dissipation
Y(NO)	Quantum yield of non-regulated energy dissipation
Rubisco	Ribulose bisphosphate carboxylase/oxygenase
RCA	Rubisco activase
TK	Rubisco, thymidine kinase

SBPase	Sedoheptulose 1,7-bisphosphatase
FBA	Fructose 1,6-bisphosphate aldolase
FBPase	Fructose 1,6-bisphosphatase
NI	Neutral invertas
AI	Acid invertase
SS	Sucrose synthetase

References

1. Matthes, M.S.; Robil, J.M.; Mcsteen, P. From element to development: The power of the essential micronutrient boron to shape morphological processes in plants. *J. Exp. Bot.* **2020**, *71*, 1681–1693. [CrossRef]
2. Bolaños, L.; Abreu, I.; Bonilla, I.; Camacho-Cristóbal, J.J.; Reguera, M. What can boron deficiency symptoms tell us about its function and regulation. *Plants* **2023**, *12*, 777. [CrossRef] [PubMed]
3. Brown, P.H.; Bellaloui, N.; Wimmer, M.A. Boron in plant biology. *Plant Biol.* **2002**, *4*, 205–223. [CrossRef]
4. O'Neill, M.A.; Ishii, T.; Albersheim, P.; Darvill, A.G. Rhamnogalacturonan II: Structure and function of a borate cross-linked cell wall pectic polysaccharide. *Annu. Rev. Plant Biol.* **2004**, *55*, 109–139. [CrossRef] [PubMed]
5. Lu, Y.B.; Yang, L.T.; Li, Y.; Xu, J.; Liao, T.T.; Chen, Y.B.; Chen, L.S. Effects of boron deficiency on major metabolites, key enzymes and gas exchange in leaves and roots of *Citrus sinensis* seedlings. *Tree Physiol.* **2014**, *34*, 608–618. [CrossRef]
6. Chatter, M.; Tabi, Z.; Galli, M.; Malcomber, S.; Buck, A.; Muszynski, M.; Gallavotti, A. The boron efflux transporter rotten ear is required for maize inflorescence development and fertility. *Plant Cell* **2014**, *26*, 2962–2977. [CrossRef]
7. Durbak, A.R.; Phillips, K.A.; Pike, S.; O'Neill, M.A.; Mares, J.; Gallavotti, A.; Malcomber, S.T.; Gassmann, W.; McSteen, P. Transport of boron by the *tassel-less1* aquaporin is critical for vegetative and reproductive development in maize. *Plant Cell* **2014**, *26*, 2978–2995. [CrossRef]
8. Zhou, G.F.; Sun, X.N.; Zhang, L.P.; Zeng, X.L.; Liu, G.D.; Sheng, O. Lignin metabolism plays an essential role in the formation of corky split vein caused by boron deficiency in 'Newhall' navel orange (*Citrus sinensis* Osb.). *Sci. Hortic.* **2022**, *294*, 110763. [CrossRef]
9. Shorrocks, V.M. The occurrence and correction of boron deficiency. *Plant Soil* **1997**, *193*, 121–148. [CrossRef]
10. Wang, N.N.; Yang, C.Q.; Pan, Z.Y.; Liu, Y.Z.; Peng, S.A. Boron deficiency in woody plants: Various responses and tolerance mechanisms. *Front. Plant Sci.* **2015**, *6*, 916. [CrossRef]
11. Cakmak, I.; Kurz, H.; Marschner, H. Short-term effects of boron, germanium and high light intensity on membrane permeability in boron deficient leaves of sunflower. *Physiol. Plant.* **1995**, *95*, 11–18. [CrossRef]
12. Dell, B.; Huang, L. Physiological response of plants to low boron. *Plant Soil* **1997**, *193*, 103–120. [CrossRef]
13. Xiao, J.X.; Yan, X.; Peng, S.A.; Deng, X.X.; Fang, Y.W. Relationship between boron deficiency occurrence and annual changes in contents of boron and sugar of 'Newhall' navel orange. *Acta Hortic. Sin.* **2006**, *33*, 356–359.
14. Han, S.; Chen, L.S.; Jiang, H.X.; Smith, B.R.; Yang, L.T.; Xie, C.Y. Boron deficiency decreases growth and photosynthesis, and increases starch and hexoses in leaves of citrus seedlings. *J. Plant Physiol.* **2008**, *165*, 1331–1341. [CrossRef] [PubMed]
15. Liu, Y.Z.; Li, S.; Yang, C.Q.; Peng, S.A. Effects of boron-deficiency on anatomical structures in the leaf main vein and fruit mesocarp of pummelo [*Citrus grandis* (L.) Osbeck]. *Korean J. Hortic. Sci.* **2013**, *88*, 693–700. [CrossRef]
16. Ling, L.L.; Peng, L.Z.; Chun, C.P.; Cao, L.; Jiang, C.L. Characteristics analysis of microelement contents in 'Newhall' navel orange leaves in Southern Jiangxi Province. *Acta Hortic. Sin.* **2010**, *37*, 1388–1394.
17. Wang, R.D.; Jiang, C.C.; Liu, G.D.; Wang, Y.H.; Peng, S.A.; Zeng, Q.L. Status and analysis on available boron content in soil of Gannan navel orange orchards. *S. China Fruits* **2011**, *40*, 1–3+7.
18. Liu, X.S.; Zhang, Z.Y.; Wei, J.P. Influencing factors and intensity change of soil pH value in South Jiangxi Province. *Bull. Soil Water Conserv.* **2021**, *41*, 100–105.
19. Yang, C.Q.; Liu, Y.Z.; An, J.C.; Li, S.; Jin, L.F.; Zhou, G.F.; Wei, Q.J.; Yan, H.Q.; Wang, N.N.; Fu, L.N.; et al. Digital gene expression analysis of corky split vein caused by boron deficiency in 'Newhall' navel orange (*Citrus sinensis* Osbeck) for selecting differentially expressed genes related to vascular hypertrophy. *PLoS ONE* **2013**, *8*, e65737. [CrossRef]
20. Liu, G.D.; Dong, X.C.; Liu, L.C.; Wu, L.S.; Peng, S.A.; Jiang, C.C. Metabolic profiling reveals altered pattern of central metabolism in navel orange plants as a result of boron deficiency. *Physiol. Plant.* **2015**, *153*, 513–524. [CrossRef]
21. Zhou, G.F.; Li, B.X.; Zhang, L.P.; Sheng, O.; Wei, Q.J.; Yao, F.X.; Guan, G.; Liu, G.D. Physiological and nutritional responses of 'HB' Pummelo [*Citrus grandis* (L.) Osbeck 'Hirado Buntan'] to the combined effects of low pH levels and boron deficiency. *HortScience* **2020**, *55*, 449–456. [CrossRef]
22. Zhou, G.F.; Peng, S.A.; Liu, Y.Z.; Wei, Q.J.; Han, J.; Islam, M.Z. The physiological and nutritional responses of seven different citrus rootstock seedlings to boron deficiency. *Trees-Struct. Funct.* **2014**, *28*, 295–307. [CrossRef]

23. Zhou, G.F.; Zhang, L.P.; Li, B.X.; Sheng, O.; Wei, Q.J.; Yao, F.X.; Guan, G.; Liu, G.D. Genome-wide identification of long non-coding RNA in trifoliate orange (*Poncirus trifoliata* L. Raf) leaves in response to boron deficiency. *Int. J. Mol. Sci.* **2019**, *20*, 5419. [CrossRef] [PubMed]
24. Folimonova, S.Y.; Achor, D.S. Early events of citrus greening (Huanglongbing) disease development at the ultrastructural level. *Phytopathology* **2010**, *100*, 949–958. [CrossRef]
25. Huang, J.H.; Xu, J.; Ye, X.; Luo, T.Y.; Ren, L.H.; Fan, G.C.; Qi, Y.P.; Li, Q.; Ferrarezi, R.S.; Chen, L.S. Magnesium deficiency affects secondary lignification of the vascular system in *Citrus sinensis* seedlings. *Trees-Struct. Funct.* **2019**, *33*, 171–182. [CrossRef]
26. Ye, X.; Chen, X.F.; Cai, L.Y.; Lai, N.W.; Deng, C.L.; Guo, J.X.; Yang, L.T.; Chen, L.S. Molecular and physiological mechanisms underlying magnesium-deficiency-induced enlargement, cracking and lignification of *Citrus sinensis* leaf veins. *Tree Physiol.* **2020**, *40*, 1277–1291. [CrossRef]
27. Liang, D.; Ma, F.W. Photosynthesis and assimilate transport in fruit trees. In *Modern Fruit Tree Biology*; Li, T.Z., Zhang, Z.H., Eds.; Science Press: Beijing, China, 2008; pp. 141–152. ISBN 978-7-030-20116-4.
28. Mishra, S.; Heckathorn, S.A.; Frantz, J.M.; Krause, C. The effect of boron availability, CO2, and irradiance on relative accumulation of the major boron transport proteins, bor1 and nip5;1. *Biol. Plant.* **2018**, *62*, 121–128. [CrossRef]
29. Sharma, P.N.; Ramchandra, T. Water relations and photosynthesis in mustard plants subjected to boron deficiency. *Indian J. Plant Physiol.* **1990**, *85*, 150–154.
30. Lu, X.P.; Jiang, C.C.; Dong, X.C.; Wu, X.W.; Yan, L. Leaf photosynthetic characteristics of citrus rootstocks with different boron efficiency. *Plant Nutr. Fert. Sci.* **2017**, *23*, 476–483.
31. García-sánchez, F.; Simón-grao, S.; Martínez-nicolás, J.J.; Alfosea-simón, M.; Liu, C.; Chatzissavvidis, C.; Pérez-pérez, J.G.; Cámara-zapata, J.M. Multiple stresses occurring with boron toxicity and deficiency in plants. *J. Hazard. Mater.* **2020**, *397*, 122713. [CrossRef]
32. Song, X.; Song, B.Q.; Huo, J.L.; Liu, H.J.; Adil, M.F.; Jia, Q.; Wu, W.Y.; Kuerban, A.; Wang, Y.; Huang, W.G. Effect of boron deficiency on the photosynthetic performance of sugar beet cultivars with contrasting boron efficiencies. *Front. Plant Sci.* **2023**, *13*, 1101171. [CrossRef]
33. Fujiyama, B.S.; Silva, A.R.B.E.; Júnior, M.L.D.S.; Cardoso, N.R.P.; Fonseca, A.B.D.; Viana, R.G.; Sampaio, L.S. Boron fertilization enhances photosynthesis and water use efficiency in soybean at vegetative growth stage. *J. Plant Nutr.* **2019**, *42*, 2498–2506. [CrossRef]
34. Gimeno, V.; Simón, I.; Nieves, M.; Martínez, V.; Cámara-zapata, J.M.; García, A.L.; García-Sánchez, F. The physiological and nutritional responses to an excess of boron by Verna lemon trees that were grafted on four contrasting rootstocks. *Trees-Struct. Funct.* **2012**, *26*, 1513–1526. [CrossRef]
35. Han, S.; Tang, N.; Jiang, H.X.; Yang, L.T.; Li, Y.; Chen, L.S. CO_2 assimilation, photosystem II photochemistry, carbohydrate metabolism and antioxidant system of citrus leaves in response to boron stress. *Plant Sci.* **2009**, *176*, 143–153. [CrossRef]
36. Sheng, O.; Song, S.W.; Peng, S.A.; Deng, X.X. The effects of low boron on growth, gas exchange, boron concentration and distribution of 'Newhall' navel orange (*Citrus sinensis* Osb.) plants grafted on two rootstocks. *Sci. Hortic.* **2009**, *121*, 278–283. [CrossRef]
37. Zhou, G.F.; Li, B.X.; Lin, H.Z.; Guan, G.; Liu, G.D.; Yao, F.X.; Zhong, B.L. Early and late symptoms of micronutrient deficiency and diurnal changes of photosynthesis of 'Newhall' navel orange. *Plant Nutr. Fert. Sci.* **2019**, *25*, 1957–1966.
38. Zhang, G.J.; Jiang, H.; Zheng, L.Q.; Chen, J.; Qiu, D.L.; Liu, X.H. Effect of copper stress on photosynthesis of navel orange seedlings. *Chin. J. Eco-Agric.* **2009**, *17*, 130–134. [CrossRef]
39. Zhou, G.F.; Li, B.X.; Fu, Y.L.; Guan, G.; Yao, F.X.; Liu, G.D. Effects of iron, manganese and zinc deficiency on the symptom, photosynthetic characteristics and nutrient status of 'Nanfeng' tangerine. *Acta Hortic. Sin.* **2019**, *46*, 691–700.
40. Schmidt, S.B.; Jensen, P.E.; Husted, S. Manganese deficiency in plants: The impact on photosystem II. *Trends Plant Sci.* **2016**, *21*, 622–632. [CrossRef]
41. Alejandro, S.; Höller, S.; Meier, B.; Peiter, E. Manganese in plants: From acquisition to subcellular allocation. *Front. Plant Sci.* **2020**, *11*, 300. [CrossRef]
42. Xie, D.X.; Ma, X.N.; Zhao, Y.Q.; Li, J.X.; Fu, D.L.; Zhang, Z.Q.; Ju, C.F.; Wang, C. Absorption, transport and regulation of manganese in plants. *Sci. Sin.* **2023**, *53*, 1199–1212. [CrossRef]
43. Millaleo, R.; Reyes-Diaz, M.; Ivanov, A.G.; Mora, M.L.; Alberdi, M. Manganese as essential and toxic element for plants: Transport, accumulation and resistance mechanisms. *J. Soil. Sci. Plant Nutr.* **2010**, *10*, 470–481. [CrossRef]
44. Tao, Y.Z.; Liu, C.Z.; Piao, L.; Yang, F.Q.; Liu, J.Q.; Faheem, J.M.; Li, M. Effect of Mn deficiency on carbon and nitrogen metabolism of different genotypes seedlings in maize (*Zea mays* L.). *Plants* **2023**, *12*, 1407. [CrossRef]
45. Pittman, J.K. Managing the manganese: Molecular mechanisms of manganese transport and homeostasis. *New Phytol.* **2005**, *167*, 733–742. [CrossRef]
46. Bradl, H.B. Adsorption of heavy metal ions on soils and soils constituents. *J. Colloid Interf. Sci.* **2004**, *277*, 1–18. [CrossRef]

47. Khabaz-Saberi, H.; Setter, T.L.; Waters, I. Waterlogging induces high to toxic concentrations of iron, aluminum, and manganese in wheat varieties on acidic soil. *J. Plant Nutr.* **2007**, *29*, 899–911. [CrossRef]
48. Hansel, C.M.; Zeiner, C.A.; Santelli, C.M.; Webb, S.M. Mn(II) oxidation by an ascomycete fungus is linked to superoxide production during asexual reproduction. *Proc. Natl. Acad. Sci. USA* **2012**, *109*, 12621–12625. [CrossRef] [PubMed]
49. Rengel, Z. Availability of Mn, Zn and Fe in the rhizosphere. *J. Soil. Sci. Plant Nutr.* **2015**, *15*, 397–409. [CrossRef]
50. Huang, X.; Shabala, S.; Shabala, L.; Rengel, Z.; Wu, X.; Zhang, G.; Zhou, M. Linking waterlogging tolerance with Mn^{2+} toxicity: A case study for barley. *Plant Biol.* **2015**, *17*, 26–33. [CrossRef]
51. Ying, J.G.; Liu, X.H.; Li, J.B.; Wu, Q.; Peng, S.A.; Jiang, C.C. Analysis of Cu and Mn contents in the soil and leaves of citrus orchards in Nanfeng and Quzhou. *S. China Fruits* **2016**, *45*, 15–18.
52. Cheng, J.J.; Ding, C.F.; Li, X.G.; Zhang, T.L.; Wang, X.X. Soil quality evaluation for navel orange production systems in central subtropical China. *Soil Till. Res.* **2016**, *155*, 225–232. [CrossRef]
53. Chun, C.P.; Peng, L.Z.; Ling, L.L.; Lai, J.J.; Cao, L.; Jiang, C.L. Study on the contents of macronutrients and medium nutrients in 'Newhall' navel orange leaves in Southern Jiangxi province of China. *J. Fruit Sci.* **2010**, *27*, 678–682.
54. Andresen, E.; Peiter, E.; Küpper, H. Trace metal metabolism in plants. *J. Exp. Bot.* **2018**, *69*, 909–954. [CrossRef]
55. Li, Z.Y.; Zhang, Y.; Han, L.H.; Xu, J.Z. The interactive effects of nitrogen and calcium on photosynthetic characteristics and chlorophyll fluorescence parameters of nectarine under protected culture. *Plant Nutr. Fert. Sci.* **2013**, *19*, 893–900.
56. Huang, C.; Barker, S.J.; Langridge, P.; Smith, F.W.; Graham, R.D. Zinc deficiency up-regulates expression of high-affinity phosphate transporter genes in both phosphate-sufficient and -deficient barley roots. *Plant Physiol.* **2000**, *124*, 415–422. [CrossRef]
57. Camacho-Cristóbal, J.J.; González-Fontes, A. Boron deficiency decreases plasmalemma H^+-ATPase expression and nitrate uptake, and promotes ammonium assimilation into asparagine in tobacco roots. *Planta* **2007**, *226*, 443–451. [CrossRef] [PubMed]
58. Matas, M.A.; González-Fontes, A.; Camacho-Cristóbal, J.J. Effect of boron supply on nitrate concentration and its reduction in roots and leaves of tobacco plants. *Biol. Plant.* **2009**, *53*, 120–124. [CrossRef]
59. Koshiba, T.; Kobayashi, M.; Ishihara, A.; Matoh, T. Boron nutrition of cultured tobacco BY-2 cells. VI. Calcium is involved in early responses to boron deprivation. *Plant Cell Physiol.* **2010**, *51*, 323–327. [CrossRef] [PubMed]
60. Quiles-Pando, C.; Rexach, J.; Navarro-Gochicoa, M.T.; Camacho-Cristóbal, J.J.; Herrera-Rodríguez, M.B.; González-Fontes, A. Boron deficiency increases the levels of cytosolic Ca^{2+} and expression of Ca^{2+}-related genes in *Arabidopsis thaliana* roots. *Plant Physiol. Biochem.* **2013**, *65*, 55–60. [CrossRef]
61. Zhou, G.F.; Wei, Q.J.; Li, B.X.; Zeng, X.L.; Liu, G.D. Establishment and optimization of hydroponic culture system for root morphologic and nutritional analysis of trifoliate orange [*Poncirus trifoliata* (L.) Raf]. *Sci. Agr.* **2020**, *77*, e20180261. [CrossRef]
62. Ling, L.L.; Huang, Y.; Peng, L.Z.; Wu, Y.T.; Jiang, C.L.; Cao, L.; Chun, C.P. Influence of magnesium deficiency and excess on chlorophyll fluorescence characteristics of 'Newhall' navel orange leaves. *Acta Ecol. Sin.* **2014**, *34*, 1672–1680. [CrossRef]
63. Xiong, S.M.; Zuo, X.F.; Zhu, Y.Y. Determination of cellulose, hemi-cellulose and lignin in rice hull. *Cereal Feed Ind.* **2005**, *8*, 40–41.
64. Chowdhury, S.; Dutt, S. Vein-splitting, a boron deficiency disease of citrus in Assam. *Sci. Cult.* **1950**, *15*, 358–359.
65. Li, J.; Xie, Z.C.; Xie, W.L.; Wu, X.M.; Shi, Q. Relationship between leaf vein splitting and mineral nutrition of citrus. *Acta Hortic. Sin.* **2011**, *38*, 425–433.
66. Xi, F.S.; Wu, Y.J.; Xiang, D.Y. The causes of severe physiological shoot blight of fast-growing eucalyptus and emergency control measures. *Guangxi For. Sci.* **2005**, *34*, 73–75.
67. Tewari, R.K.; Kumar, P.; Sharma, P.N. Morphology and oxidative physiology of boron-deficient mulberry plants. *Tree Physiol.* **2010**, *30*, 68–77. [CrossRef] [PubMed]
68. Liu, G.D.; Jiang, C.C.; Wang, Y.H. Distribution of boron and its forms in young 'Newhall' navel orange (*Citrus sinensis* Osb.) plants grafted on two rootstocks in response to deficient and excessive boron. *Soil Sci. Plant Nutr.* **2011**, *57*, 93–104. [CrossRef]
69. Ye, X.; Huang, H.Y.; Wu, F.L.; Cai, L.Y.; Chen, L.S. Molecular mechanisms for magnesium-deficiency-induced leaf vein lignification, enlargement and cracking in *Citrus sinensis* revealed by RNA-Seq. *Tree Physiol.* **2020**, *41*, 280–301. [CrossRef]
70. Bagherian, S.A.A.; Izadpanah, K. Two novel variants of hop stunt viroid associated with yellow corky vein disease of sweet orange and split bark disorder of sweet lime. *J. Kühn Arch.* **2010**, *427*, 105–113.
71. Ye, X.; Chen, X.F.; Deng, C.L.; Yang, L.T.; Guo, J.X.; Chen, L.S. Magnesium-deficiency effects on pigments, photosynthesis and photosynthetic electron transport of leaves, and nutrients of leaf blades and veins in Citrus sinensis seedlings. *Plants* **2019**, *8*, 389. [CrossRef]
72. Zhao, Q.; Dixon, R.A. Transcriptional networks for lignin biosynthesis: More complex than we thought? *Trends Plant Sci.* **2011**, *16*, 227–233. [CrossRef]
73. Liu, Q.Q.; Luo, L.; Zheng, L.Q. Lignins: Biosynthesis and biological functions in plants. *Int. J. Mol. Sci.* **2018**, *19*, 335. [CrossRef] [PubMed]
74. Moura, J.C.M.S.; Bonine, C.A.V.; Viana, J.D.O.F.; Dornelas, M.C.; Mazzafera, P. Abiotic and biotic stresses and changes in the lignin content and composition in plants. *J. Integr. Plant Biol.* **2020**, *52*, 360–376. [CrossRef] [PubMed]

75. Lidon, F.C.; Barreiro, M.; Ramalho, J. Manganese accumulation in rice: Implications for photosynthetic functioning. *J. Plant Physiol.* **2004**, *161*, 1235–1244. [CrossRef] [PubMed]
76. Fan, L.; Linker, R.; Gepstein, S.; Tanimoto, E.; Yamamoto, R.; Neumann, P.M. Progressive inhibition by water deficit of cell wall extensibility and growth along the elongation zone of maize roots is related to increased lignin metabolism and progressive stelar accumulation of wall phenolics. *Plant Physiol.* **2006**, *140*, 603–612. [CrossRef]
77. Wei, H.; Dhanaraj, A.L.; Arora, R.; Rowland, L.J.; Fu, Y.; Sun, L. Identification of cold acclimation-responsive Rhododendron genes for lipid metabolism, membrane transport and lignin biosynthesis: Importance of moderately abundant ESTs in genomic studies. *Plant Cell Environ.* **2006**, *29*, 558–570. [CrossRef]
78. Moura-Sobczak, J.; Souza, U.; Mazzafera, P. Drought stress and changes in the lignin content and composition in *Eucalyptus*. *BMC Proc.* **2011**, *5*, 103. [CrossRef]
79. Wang, Y.J.; Sheng, L.P.; Zhang, H.R.; Du, X.P.; An, C.; Xia, X.L.; Chen, F.D.; Jiang, J.F.; Chen, S.M. *CmMYB19* over-expression improves aphid tolerance in chrysanthemum by promoting lignin synthesis. *Int. J. Mol. Sci.* **2017**, *18*, 619. [CrossRef]
80. Liu, W.G.; Ren, M.L.; Liu, T.; Du, Y.L.; Zhou, T.; Liu, X.M.; Liu, J.; Hussain, S.; Yang, W.Y. Effect of shade stress on lignin biosynthesis in soybean stems. *J. Integr. Agr.* **2018**, *17*, 1594–1604. [CrossRef]
81. Wen, B.X.; Zhang, Y.; Hussain, S.; Wang, S.; Zhang, X.W.; Yang, J.Y.; Xu, M.; Qin, S.S.; Yang, W.Y.; Liu, W.G. Slight shading stress at seedling stage does not reduce lignin biosynthesis or affect lodging resistance of soybean stems. *Agronomy* **2020**, *10*, 544. [CrossRef]
82. Li, C.H.; Luo, Y.L.; Jin, M.; Sun, S.F.; Wang, Z.L.; Li, Y. Response of lignin metabolism to light quality in wheat population. *Front. Plant Sci.* **2021**, *12*, 729647. [CrossRef]

MDPI AG
Grosspeteranlage 5
4052 Basel
Switzerland
Tel.: +41 61 683 77 34

Horticulturae Editorial Office
E-mail: horticulturae@mdpi.com
www.mdpi.com/journal/horticulturae

www.ingramcontent.com/pod-product-compliance
Lightning Source LLC
Chambersburg PA
CBHW040810250926
51850CB00007BA/1130
9783725862801